Machine Learning for Disease Detection, Prediction, and Diagnosis

Tanupriya Choudhury • Avita Katal
Editors

Machine Learning for Disease Detection, Prediction, and Diagnosis

Challenges and Opportunities

Springer

Editors
Tanupriya Choudhury
School of Computer Science
University of Petroleum and Energy Studies (UPES)
Dehradun, Uttarakhand, India

Avita Katal
School of Computer Science
University of Petroleum and Energy Studies (UPES)
Dehradun, Uttarakhand, India

ISBN 978-981-96-4240-3 ISBN 978-981-96-4241-0 (eBook)
https://doi.org/10.1007/978-981-96-4241-0

This Springer imprint is published by the registered company Springer Nature Singapore Pte Ltd.
The registered company address is: 152 Beach Road, #21-01/04 Gateway East, Singapore 189721, Singapore

The editorial team dedicates this volume to the valiant Indian Army, whose unwavering commitment, sacrifices, and exemplary service to our nation stand as a beacon of honor. We also extend our heartfelt gratitude to our families—parents, spouses, and children—for their boundless support throughout the creation of this work.

Preface

Advancements in machine learning (ML) and image processing have tremendously revolutionized disease detection, prediction, or diagnosis. Using image processing to decipher medical images like X-rays and MRI provides early diagnosis of cancer, heart disease, and neurological disorders. Machine learning algorithms detect subtle changes within large datasets, allowing a correct diagnosis with higher precision and specificity.

ML algorithms help to diagnose conditions such as diabetes and kidney disease by analyzing patient data in the chronic disease management field. Early detection of a change in condition can lead to proactive care and ultimately save lives. Using predictive models to identify risk factors guides individualized treatment strategies. ML helps in predicting neurological and cardiac diseases and monitoring speech biomarkers for Parkinson's disease and risk factors of heart condition and chronic kidney diseases. Deep learning (DL, a subset of ML) made immense contributions in personalized medicine, especially for analyzing genomic-related disease susceptible genes. Data mining can assist swimmingly therein by locating these patterns within the medical records that otherwise nobody will see. Computer-aided diagnosis systems with ML support increase sensitivity in the detection of abnormalities that might be overlooked by human doctors.

This can be applied outside of the human health domain as well. ML in agriculture identifies pests and diseases in plants quickly by examining a leaf image to increase food security. AI continues to progress, and its development has the promise of revolutionizing healthcare in this modern world by enabling more accurate diagnoses and personalized treatments plans for each patient offering better health outcomes worldwide.

This book focuses on the progressive concepts, methods, and methodologies that are advancing the practicability of this already widespread trend in machine learning techniques. Artificial Intelligence is growing with the onset of new technologies for disease detection and diagnosis. The book talks about image classification techniques and machine learning techniques that are used in disease detection, prediction, or diagnosis.

- It starts with a discussion on the fundamentals of machine learning image classification methods for disease detection.
- It enlightens readers with various machine learning algorithms to detect diseases for a better understanding of the technology
- It evaluates multiple machine learning algorithms for detection of various diseases.
- It provides in-depth knowledge of different machine learning, deep learning, and image classification techniques along with the prediction, detection, and diagnosis of human and crop diseases.

Chapter 1: Introduction to Machine Learning and Image Processing for Disease Detection

Chapter 1 introduces the application of image processing and machine learning for disease detection. The importance of disease detection, the function of machine learning in the identification of diseases, the various machine learning methodologies for identifying illnesses, and the various data sources that are employed are covered in detail in addition to the numerous applications of machine learning and image processing.

Chapter 2: Comparative Study of Various Deep Learning-Based Methods for Prediction of Disease

Chapter 2 explains the disease prediction system for diagnosing medical problems. This chapter conducts a thorough analysis of the performance measures of several supervised machine learning models in the context of disease identification. Numerous algorithms have been taken into consideration, including K-Nearest Neighbor (KNN), Decision Trees, and Naive Bayes.

Chapter 3: Introduction to Deep Learning for Disease Prediction

Chapter 3 explores a number of practical uses, demonstrating deep learning's amazing ability to forecast a wide range of illnesses, like cancer, heart conditions, and neurological disorders. It also explores the obstacles and ethical issues surrounding the incorporation of deep learning models into healthcare facilities. Readers will have a thorough concept of the revolutionary role deep learning plays in disease prognosis after completing this chapter, enabling them to investigate this emerging topic and positively impact precision medicine and healthcare in the future.

Chapter 4: A Survey of Image Classification Techniques for the Prediction of Diseases

Chapter 4 introduces a novel methodology for categorizing pneumothorax in radiographic images of the chest, utilizing advanced deep convolutional neural networks. This research study presents a unique deep convolutional neural network-based approach for classifying pneumothorax in chest radiography pictures.

Chapter 5: Prediction of Disease Related to Heart by Using Different Techniques: Survey

Cardiologists may better manage their patients by guaranteeing proper categorization via the diagnosis and treatment of cardiovascular illness, which are vital medical activities. Chapter 5 provides the details about performance of various models based on machine learning algorithms and methodologies for detecting heart disease as much as earlier and lowering mortality rates.

Chapter 6: Automated Plant Disease Diagnosis with Machine Learning

In order to accurately classify or diagnose plant species, the scientists have successfully used a variety of machine learning techniques, including convolutional neural networks (CNNs), support vector machines (SVMs), decision trees, and ensemble methods. Chapter 6 examines and compares these methods for detection of plant diseases, highlighting their strengths and weaknesses and their application to real-world situations.

Chapter 7: Exploring Disease Prediction Techniques Through Data Mining: A Comprehensive Overview

Chapter 7 presents insights into the techniques in the domain of data mining that are transforming disease prediction. The complex and voluminous healthcare data facilitate disease prediction before clinical manifestation. The chapter presents the significance of data mining principles. It unravels the intricacies of data preprocessing, feature selection, and dimensionality reduction, presenting perceptions of these initial stages for precise disease prediction models.

Chapter 8: Detection of Parkinson's Disease Using Different Machine Learning Techniques: A comparative Analysis

Chapter 8 discusses the different tools and software that use ML and deep learning approaches for efficient development of Artificial Intelligence tool for the detection of Parkinson's disease.

Chapter 9: Kidney Disease Prediction by Machine Learning Techniques

A critical worldwide health concern, chronic kidney disease (CKD) must be detected early to prevent the disease from progressing. In the past, a lot of diagnostic techniques used were always more time-consuming, expensive, and invasive than those utilized now. SVM, Decision Trees, Random Forest, k-NN, and Neural Networks are the algorithms used in this chapter to model the likelihood of developing renal disease. To enhance the efficacy of the models, the study also applies feature selection and data preprocessing techniques on the CKD datasets that are made publicly available. To evaluate and contrast their performance, metrics including accuracy, precision, recall, F1 score, and area under the ROC curve are used. The results displayed here demonstrate that a variety of machine learning algorithms, with a focus on ensemble techniques, can be used to attain the high predicted accuracy models.

Chapter 10: Prediction of Diabetes Using Different Machine Learning Algorithms

Chapter 10 offers a thorough summary of the state of research in the field of Diabetes prediction and explores the use of different machine learning algorithms too. By examining algorithms such as SVM, DT, neural networks, and ensemble methods, this chapter critically evaluates their strengths, weaknesses, and suitability for diabetes prediction. Furthermore, it discusses the diverse features and data sources utilized in diabetes prediction, exploring their impact on model accuracy and generalization.

Chapter 11: Investigation of Machine Learning Algorithms in Detecting Chronic Kidney Disorder

Chapter 11 explores traditional machine learning algorithms for improving diagnosis and prognosis of chronic kidney disease (CKD). The comparative study invokes analysis of the proposed models in uncovering patterns, extracting vital components from the diverse patient information and medical imaging to generate accurate prognostic insights. The study not only assists in the early identification of CKD but also simplifies the process of treatment strategy.

Chapter 12: Skin Disease Prediction Using Machine Learning Techniques

Chapter 12 aims to provide a comprehensive review of the application of ML techniques in dermatology, with a specific focus on skin disease prediction and diagnosis. The chapter covers the basics of ML, the preparation and refinement of dermatological data, the exploration of various ML algorithms, and the presentation of case studies demonstrating the practical application of these techniques. Furthermore, it also discusses the ethical considerations and challenges associated with the use of ML in dermatology and outlines prospective avenues for upcoming investigations and therapeutic uses.

Chapter 13: A Comparative Study of Different Machine Learning Techniques for Skin Disease Detection

For effective treatment and improved patient outcomes, early identification and precise detection of skin cancer are critical. However, subjective skin cancer diagnosis can be time-consuming and inconsistent among individuals, necessitating the development of automated and objective diagnostic techniques. Different models like CNN and Alexnet architecture are implemented on HAM10000 database and compared in Chap. 13.

Chapter 14: Leveraging MLP-Mixer for Improved Melanoma Diagnosis Using Skin Lesion Images

Chapter 14 makes use of the MLP mixer to address some of the shortcomings of CNN and transformers in the image classification task—that is, the identification of melanoma. To give the model a variety of images to learn from, the three datasets have been mixed. This proposed effort has concentrated on developing a model without complex computing modules, such as convolution or attention, and aims to achieve performance comparable to that of cutting-edge CNN and Transformer models.

Chapter 15: Application of AI to Detect Brain Tumors

Chapter 15 deals in exploring the role of AI in detecting brain tumor, importance of early detection, limitation of traditional methods, and accuracy and efficiency of diagnosis.

Chapter 16: Revolutionizing Brain Tumor Detection: Unleashing the Power of Artificial Intelligence

Chapter 16 underscores the transformative potential of AI, emphasizing the urgent need to surmount the challenges faced in the detection of brain tumor. It serves as a beacon, guiding the exploration of innovative approaches crucial for the seamless integration of AI into brain tumor diagnosis. In navigating the present landscape, addressing challenges, and envisioning future applications, this chapter provides a concise yet comprehensive roadmap for the future of brain tumor detection.

Chapter 17: Disease Detection and Diagnosis Using Artificial Intelligence Techniques for Sustainable Economic Growth

Chapter 17 covers in detail about the use of AI-based algorithms in disease detection and diagnosis, methodology for disease detection, role of AI in early detection of diseases, AI interventions in the detection of rare diseases, and contribution of AI in medical diagnostic market. The findings of the chapter hold significance for the policy makers, medical practitioners, and health technology enthusiasts.

Chapter 18: Developing a COVID-19 Prediction Kit Using Machine Learning

Chapter 18 gives an insight into the innovation, difficulties, and prognosis for the invention and wide implementation of the ML-based COVID-19 prediction kits. The chapter analyzes and focuses on the present methods where the application of ML has been proven to contribute to the accuracy of prediction rather than the classical procedures.

Chapter 19: Plant Disease Detection: Comprehensive Review of Methods and Techniques

Chapter 19 explores the popular machine learning (ML) and deep learning approaches that have proven their strength in the classification and detection of various plant and leaf diseases in the realm of agriculture. The chapter highlights the benefits of these computational models by showing how well they can handle large datasets and precisely identify diseases. The chapter also discusses the drawbacks of these models, such as the need for large datasets for model training and potential issues with generalize ability across different crop species or environmental conditions.

We sincerely hope our efforts are appreciated and that readers find value and benefit in this book.

Dehradun, Uttarakhand, India Tanupriya Choudhury
Dehradun, Uttarakhand, India Avita Katal

Acknowledgments

The editorial team would like to express their sincere thanks to their institution, UPES, Dehradun, Uttarakhand, India, for creating a supportive research environment that served as the foundation for this project. We extend our sincere gratitude to the wide range of authors and in particular to the international reviewers who have painstakingly examined every chapter in order to maintain the high caliber of this book. Their insightful input on how to make the chapters' content presentation, coherence, and quality better has been invaluable. We sincerely thank everyone who were engaged for their commitment and willingness to take on challenges that were outside of their comfort zones.

Contents

About the Editors

Tanupriya Choudhury completed his undergraduate studies in Computer Science and Engineering at the West Bengal University of Technology, Kolkata, India (2004–2008), followed by a master's degree in the same field from Dr. MGR University, Chennai, India (2008–2010). In 2016, he successfully obtained his PhD degree from Jagannath University, Jaipur. With a total of 15 years of experience in both teaching and research, Dr. Choudhury holds the position of visiting professor at Daffodil International University, Bangladesh. Currently, he is working as Professor in School of Computer Science, UPES, Dehradun.

Prior to this role, he served as a Professor and Associate Dean of Research at Graphic Era Deemed to be University, Dehradun, India; Professor at Symbiosis International Deemed University. Pune; Graphic Era Hill University, Dehradun (Research Professor); UPES, Dehradun (Professor and Research Head Informatics); Amity University, Noida (Assistant Professor Grade 3 and International Dept. Head); and other prestigious academic institutions (Dronacharya College of Engineering, Gurgaon; Lingaya's University, Faridabad; Babu Banarsi Das Institute of Technology, Ghaziabad; Syscon Solutions Pvt. Ltd., Kolkata; etc.). He was recently recognized for his outstanding contributions to education with the Global Outreach Education Award for Excellence in Best Young Researcher Award at GOECA 2018.

His areas of expertise encompass Human Computing, Soft Computing, Cloud Computing, and Data Mining, among others. Notably accomplished within his field thus far is filing 25 patents and securing copyrights for 16 software programs from MHRD (Ministry of Human Resource Development). He has actively participated as an attendee or speaker at numerous national and international conferences across India and abroad.

With more than 150 research papers (Scopus) authored to date on record, Dr. Choudhury has also been invited as a guest lecturer or keynote speaker by esteemed institutions such as Jamia Millia Islamia University, Maharaja Agersen College (Delhi University), Duy Tan University Vietnam, etc. He has also contributed significantly to various conferences throughout India serving roles like Technical Program Committee (TPC) member and session chairperson.

As an active professional within the technical community; Dr. Choudhury holds lifetime membership with IETA (International Engineering & Technology Association) along with being affiliated with Institute of Electrical and Electronics Engineers (IEEE), Institution of Engineering & Technology (IET), UK and other reputable technical societies. Additionally, he is associated with corporate entities and serves as a Technical Adviser for Deetya Soft Pvt. Ltd., Noida, IVRGURU, and Mydigital360. He is also serving as Editor in reputed journals.

He currently serves as the Honorary Secretary of IETA (Indian Engineering Teacher's Association, India), alongside his role as the Senior Advisor Position in INDO-UK Confederation of Science, Technology and Research Ltd., London, UK, and International Association of Professional and Fellow Engineers, Delaware, USA.

Avita Katal is working as an Assistant Professor (Selection Grade) in the School of Computer Science, UPES, Dehradun, Uttarakhand, India. She has received her BE degree from University of Jammu, India, in Computer Science Engineering in 2010 and MTech degree from Graphic Era University, Dehradun, India, in 2013. She has Postgraduate Certificate in Academic Practice (PGCAP). She received her PhD in the area of Cloud Computing from UPES, Dehradun, India. Her research interest is in the area of Cloud Computing, IOT, and Artificial Intelligence. She has published various research papers and has also served as reviewer in renowned conferences and journals. She has a keen interest in improving the teaching pedagogies and has participated in various faculty development programs. She is actively involved in all areas of education including research, curriculum development, teacher mentoring, student career preparation, and community work with a genuine interest in student's cognitive and social growth.

Chapter 1
Introduction to Machine Learning and Image Processing for Disease Detection

Pradeep Kumar Kaushik, Mahesh Manchanda, and Raj Kishor Bisht

1.1 Introduction

Machine learning is a powerful tool and it is frequently used for data-driven prediction and analysis, as well as image analysis and observation, including space exploration. Nowadays, there are several examples of the independent or combined application of machine learning and image processing-based algorithms for the recognition and prediction of sickness.

Figure 1.1 shows the relationship between machine learning, image processing and prediction, analysis, and detection of disease.

Machine learning is a subdomain of artificial intelligence, and intelligent health or patient care systems can benefit from the assistance of artificial intelligence. Artificial intelligence methodologies, such as machine learning and deep learning, are extensively employed in health management to recognize patient health problems and also find medications and assess sickness. To fully identify disorders using artificial intelligence algorithms, a diversity of medical data sources is required, involving ultrasonography, genetics, computed tomography, mammography, and magnetic resonance imaging. Furthermore, artificial intelligence has significantly improved the dispensary experience and expedited the process of getting patients ready to go home and continue their therapy.

Artificial intelligence in the health sector usually concentrates on developing ideas and procedures to conclude whether a system is successful in diagnosing sickness. The medical approach uses the patient's symptoms and other clues to determine the illness or other health condition. Usually, the patient's physical examination

P. K. Kaushik (✉) · M. Manchanda
Department of Computer Science and Engineering, Graphic Era Hill University, Dehradun, Uttarakhand, India

R. K. Bisht
School of Computing, Graphic Era Hill University, Dehradun, Uttarakhand, India

T. Choudhury, A. Katal (eds.), *Machine Learning for Disease Detection, Prediction, and Diagnosis*, https://doi.org/10.1007/978-981-96-4241-0_1

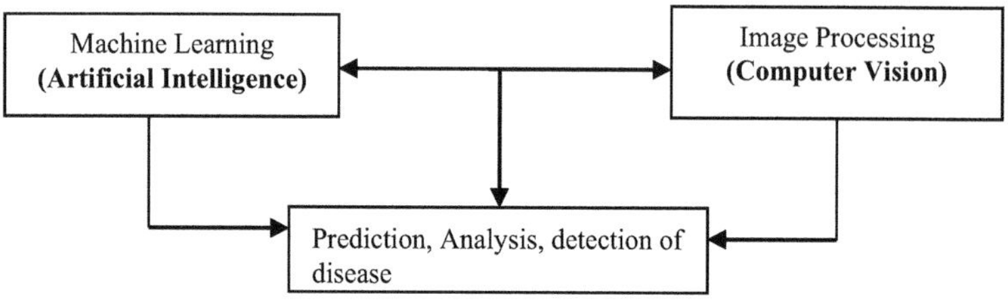

Fig. 1.1 Relation between machine learning, image processing, and prediction, analysis, and detection of disease

and previous medical records provide important facts [1]. Because a lot of signs and indicators are ambiguous and should only be recognized by licensed physicians, it is continually difficult. Thus, developing nations like Bangladesh and India, which have a dearth of medical professionals relative to their population, find it challenging to provide a suitable diagnostic procedure for a sizable patient base [2]. Furthermore, medical examinations are typically required as part of the evaluation process, although many low-income individuals consider them to be unneeded and costly.

People make mistakes during the diagnosis of a disease and do diagnosis again and again, even different doctors prescribe the same diagnosis again, and it is not unexpected that a patient could receive a more generalized overdiagnosis. Overdiagnosis can lead to issues such as unnecessary medical care, which can affect an individual's health and finances [3]. The research conducted by Balogh et al. in 2015 states that most people will receive a false detection at least once in their lives [4]. Various factors can influence an inaccurate diagnosis such as

- Lack of appropriate symptoms, which are typically undetectable.
- Uncommon disease.
- Illness is unintentionally left out of the examination [5]

The chapter is organized as follows: Sect. 1.1 covers the introduction; Sect. 1.2 covers related work; Sect. 1.3 discusses the role of machine learning in disease detection; Sect. 1.4 discusses the importance of disease detection; Sect. 1.5 covers the application of machine learning and image processing; Sect. 1.6 covers various machine learning techniques; Sect. 1.7 focuses on the various data sources used for disease detection; Sect. 1.8 discusses medical images; Sect. 1.9 covers discussion; Sect. 1.10 covers conclusion; and Sect. 1.11 addresses future scope.

1.2 Related Work

We searched various resources like Scopus, Google Scholar, PubMed, Web of Science, Excerpta Medical Database, and Psychology Information for artificial intelligence-based disease early detection. Here, we discuss some of the exploration work related to the present study. When it comes to blood examination, machine learning-based image recognition technology is quite helpful. Research has shown that this methodology can identify white blood cells with 90% accuracy, even in cases when they are difficult to notice with the human eye [6]. For the detection of a certain disease using image processing, the results of distinct machine learning methodologies may be different. A study to observe the working of different machine learning algorithms using image processing for rice disease detection is conducted [7]. Machine learning methods are time-saving and cost-effective, particularly when dealing with non-native diseases and attempting to identify disease patterns and colors. A clever and efficient technique for identifying agricultural diseases that leverages machine learning and computer vision features is proposed [8]. In their research, researchers showed that their suggested method could identify 20 distinct plant diseases with 93% accuracy, affecting five common plant species.

In the case of plants, damaged plant leaves indicate infections or presence of a certain disease. The images of such damaged leaves are quite useful for the detection of diseases through machine learning algorithms. The study conducted in [9] considers samples of damaged tomato leaves to diagnose the infections on the basis of early indicators. The tomato samples are of higher quality when the histogram is employed; equalization is done after the leaf samples have been scaled to 256 × 256 pixels. The data was divided using the K-means clustering algorithm into Voronoi cells. Using contour tracing, the margins of leaf samples are identified. The authors in [10] examined the most recent developments in deep learning and machine learning for the diagnosis of plant diseases. The study's publications span the years 2015–2022. The difficulties and restrictions associated with applying machine learning and deep learning to the diagnosis of plant diseases are also covered in this study. These include problems with data access, imaging quality, and the ability to differentiate between healthy and diseased plants.

The integration of advance deep learning models into a mobile application for skin cancer identification is presented in this study [11]. Focusing on a single powerful model DenseNet169, the achieved accuracy is 92.25%, recall is 93.59%, and F1-score is 93.27% on the HAM10000 dataset. The mobile application not only classifies skin lesions but also provides sun exposure information (UV radiation and sunscreen index) based on skin photographs. However, Inception-V3 and InceptionResnet-V2 were employed in the study [12] to enhance melanoma diagnosis. After fine-tuning the feature extraction layers, new additional upper layers are trained. Data augmentation was used on the HAM10000 dataset, which involves details on seven distinct types of skin cancer, to address the issues with visualization. For Inception-V3 and InceptionResnet-V2, the melanoma detection accuracy is 0.89 and 0.91, respectively. Then, the U-Net approach [13], a kind of CNN, is used for precise skin lesion segmentation. U-Net combines network designs from convolutional and deconvolutional layers. U-Net is designed to segment images used in biological applications. Because of its lightweight network architecture, the MobileNetV2 model is employed in the work [14] to classify melanoma pictures into benign and malignant categories. A deep neural network approach was trained using 220,680 pictures of 174 disorders in a distinct study [15]. It was then validated using datasets from Edinburgh (1300 pictures; 10 diseases) and SNU (2201 pictures; 134 diseases). This framework has the ability to classify 134 diseases into multiple classes, identify the most crucial treatments, anticipate cancers, and enhance physician performance.

Next, image processing is one of the most fascinating fields at present. This approach is typically used to enhance images obtained from a variety of sources. An overview of image processing methods is provided in [16], and a basic understanding of this work is to characterize the numerous methodologies used in different periods of image processing. The study [17] reviews the use of eye-tracking and facial emotion identification for image processing in cognitively healthy but impaired people with communication impairments.

Currently, a lot of data is easily accessible. This study [18] presents the use of machine learning across several application areas. To extract some crucial information from the data and create an algorithm based on the findings, analysis is essential. Artificial intelligence relies heavily on machine learning, which generates algorithms based on data trends and historical connections. Machine learning is used in many areas, such as bioinformatics, marketing, image deconvolution, intrusion detection, virus detection, information retrieval, and gaming. In case of interpretability of machine learning models, a recurrent challenge is addressed by studies [19, 20]. So, understanding the rationale behind algorithmic decisions is important for a successful unification of these techniques in clinical practice. Therefore, in the study [21], the shortcomings in the field of recent disease prediction are discussed and some current solutions provided. In order to predict diseases, this study presents a number of deep learning methods, including FM-Deep Learning, ANN, CNN, and RNN.

In case of transfer learning, the study [22] highlights the relevance in increasing model generalization, enabling the application of pretrained models to novel dermatological conditions. However, transfer learning is a crucial tactic for overcoming the limitations of dermatological datasets, as discussed in the article [23].

1.3 Role of Machine Learning in Disease Detection

Machine learning can significantly improve healthcare outcomes and plays a significant role in disease identification. The following are a few significant applications of machine learning in illness identification.

1.3.1 Initial Identification

To find patterns and spot early warning indicators of illness, large volumes of patient data such as images, scans, and genetic information, including lab test results, can be examined using machine learning algorithms. This makes it possible for prompt therapy and intervention; it can greatly improve the health of the patient.

1.3.2 Predictive Analysis

Machine learning methods can analyze patient data to predict the probability that a verified condition will manifest or predict the risk of a disease worsening. Through the integration of multiple factors such as genetics, lifestyle, environment, and biomarkers, these models are able to present accurate risk assessments and promote focused preventive initiatives.

1.3.3 Image Analysis

Medical image interpretation in particular has greatly benefited from deep learning. Because the algorithms are trained on large databases of annotated medical images, they can accurately detect anomalies and help diagnose diseases such as cancer, congestive heart failure, and neurological disorders.

1.3.4 Clinical Decision Support Software

Clinical decision support software can incorporate machine learning approaches to help medical professionals make well-informed decisions. The tool can evaluate patient data, suggest appropriate diagnostic procedures and treatments, and provide real-time notifications for any drug interactions or adverse effects.

1.3.5 Research and Manufacturing of Medicine

As machine learning methods examine large amounts of biological and molecular data, medical research can proceed more quickly. These models can be used to find potential new drug candidates, predict interactions between target drugs, repurpose current drugs to find new warning signals, and improve the properties of drugs.

1.3.6 Real-Time Monitoring and Early Recognition

Both real-time monitoring and early disease progression detection can benefit from machine learning. Machine learning techniques can analyze a variety of data sources, including search engine results, social media posts, and electronic health data, to detect patterns and anomalies that may point to the onset or development of infectious diseases. This enables quick public health measures to be taken.

It is important to remember that machine learning models require constant validation and enhancements as they are trained on real data. Effective collaboration between regulatory agencies, data scientists, and physicians is critical for the ethical and effective implementation of the above-discussed techniques in the diagnosis of medical conditions.

1.4 Importance of Disease Detection

For several reasons, disease detection is important in the medical field. Here are some key ideas that emphasize how important it is to identify diseases.

1.4.1 Early Intervention

Early detection of the disease allows prompt treatment and action. Early diagnosis improves the prognosis of many diseases, including infections, heart problems, and cancer. Timely diagnosis can reduce the severity of the disease, prevent it from getting worse, and increase the chances that the treatment strategy will work.

1.4.2 Improved Patient Outcomes

Early disease identification allows medical professionals to implement the best treatment plans and protocols. This can lead to better patient outcomes, such as increased survival rates, reduced morbidity, and improved quality of life. For example, early diagnosis and treatment of chronic conditions, such as diabetes or high blood pressure, can help avoid problems and support good health.

1.4.3 Disease Avoidance and Control

To determine who is susceptible to an infectious disease or who is prone to an infectious disease, one must have a thorough understanding of the diseases. Public health experts can prevent the infection from spreading throughout communities by identifying and isolating affected individuals. This is important, especially for the management of epidemics and pandemics.

1.4.4 Resource Assignment

Healthcare organizations can allocate resources effectively with the help of early detection. Recognizing the prevalence and frequency of diseases helps medical workers more effectively allocate staff, supplies, and medicines to areas with higher rates of disease. This optimization guarantees resource efficiency, while improving the efficiency of healthcare delivery.

1.4.5 Cost Savings

Detection of disease in the early stage can save a lot of money in the medical industry. Medical costs are usually lower when a condition is detected early and treated later. Early disease detection allows healthcare systems to potentially save costs by reducing hospitalization rates, the need for intensive care, and the overall burden on health services.

1.4.6 Public Health Observation

Public health surveillance systems cannot function without the identification of diseases. Public health workers can identify trends, anticipate unexpected events, and implement necessary control measures by tracking the prevalence and incidence of diseases. Preventive intervention planning, resource allocation, and policy development all depend on these observational data.

1.4.7 Personal Training

Individuals who are aware of diseases are better equipped to take care of their health. People can detect potential health problems and make educated decisions about their lifestyle, complementary and alternative medicine, and preventive care by having regular screenings and diagnostic tests. If people get early attention, they may be able to actively participate in their healthcare and significantly improve their well-being.

It enables early intervention, enhances patient health, prevents the spread of disease, updates resource allocation, reduces costs, enables public health surveillance, and trains people.

1.5 Applications of Machine Learning and Image Processing for the Detection of Disease

Machine learning and image processing are mostly used together to identify and diagnose diseases. For example:

1.5.1 *Medical Image Analysis*

Machine learning algorithms can be used to evaluate medical images such as MRI or CT scans and X-rays to recognize patterns and abnormalities associated with particular diseases. For example, they may be useful in identifying tumors, lesions, or anomalies in multiple organs.

1.5.2 *Computer-Aided Diagnosis (CAD)*

The creation of machine learning methods benefits from the availability of large medical imaging datasets along with diagnostics. Later, these models can be used to automatically diagnose patients, help doctors interpret medical images, and pinpoint underlying conditions or symptoms.

1.5.3 *Disease Classification*

Photographs can be classified by machine learning algorithms using patterns and visual features. They can learn to differentiate between different diseases or sickness based on labels using labeled datasets. It has wide application in the field of histological studies, skin diseases, and eye diseases.

1.5.4 *Early Disease Detection*

It is possible to train machine learning algorithms to identify biomarkers or early warning signs of particular diseases. They can help in early detection of diseases when examining photographs can make medical treatment options more beneficial.

1.5.5 Risk Assessment and Prognosis

Machine learning techniques require only the analysis of an image and additional relevant data to predict the likelihood of a person suffering from a particular disease or not.

1.5.6 Remote Monitoring and Telemedicine

The use of imaging techniques for remote disease monitoring has become possible with advances in machine learning and image processing. Medical specialists can review medical images remotely through telemedicine, improving access to healthcare in remote areas.

1.5.7 Automated Tumor Identification

Machine learning algorithms trained on medical images are able to identify and classify deadly diseases. This makes it easier to accurately detect tumors, diagnose them early, and track their growth over time.

1.5.8 Retinal Disorder Recognition

To quickly recognize retinal images and detect eye diseases associated with age-related macular degeneration and diabetic retinopathy, for this purpose image processing and machine learning models are used. These techniques make it possible to identify and classify particular retinal lesions or anomalies, facilitating prompt intervention and treatment.

1.5.9 Skin Cancer Recognition

Machine learning methods can be used to identify melanoma, a dangerous type of skin cancer, using photographs of lesions and moles. These algorithms use patterns and features from vast stores of skin images to determine the condition of skin lesion. It aids doctors in diagnosing skin lesions.

1.5.10 Neurological Disease Recognition

Diffusion tensor imaging and functional magnetic resonance imaging data have been processed using machine learning and image processing methods. These methods can be used to recognize patterns and biomarkers linked to neurological diseases.

1.5.11 Recognition of Cardiovascular Disorders

Angiograms and echocardiograms are examples of cardiac snapshots that machine learning algorithms can discover irregularities in the heart, assess heart function, and predict cardiovascular risk factors. These methods are helpful in the initial detection of cardiac disorders as well as the planning of therapeutic interventions.

It should be recognized that while machine learning and image processing methods might be useful tools, their goal is not to replace the functions played by medical personnel.

1.6 Different Machine Learning Techniques for Diseases Detection

The detection and analysis of diseases have proven to be remarkably effective when using machine learning techniques. The following are some commonly used algorithms in this field.

1.6.1 Supervised Learning

In this method, a model is trained on labeled data, with each sample associated with a recognized disease label. To forecast results for new, unidentified samples, the model discovers patterns and linkages in the data. Typical supervised learning models for illness detection consist of

- Decision trees
- Random forests
- Logistic regression
- Neural network
- Support vector machines (SVM)

1.6.2 Unsupervised Learning

Unsupervised learning methods can be used when labeled data is hard to come by or not available. These methods evaluate data to look for patterns, clusters, or anomalies without prior knowledge of labels associated with diseases. For example

- Hierarchical clustering
- Gaussian mixture models (GMM)
- K-means clustering
- Autoencoders
- Principal component analysis (PCA)

1.6.3 Deep Learning

Deep learning methodologies have attracted much interest in the field of disease diagnosis because they are a subclass of neural networks and can automatically generate hierarchical representations from complex data. Deep learning algorithms can extract high-level features from huge stores of data.

1.6.4 Transfer Learning

Transfer learning improves the accuracy of pretrained models by employing smaller, domain-specific datasets to learn from their experiences with larger datasets. This method can be useful when there is a lack of labeled data for a certain situation. Transfer learning can increase the effectiveness of disease detection algorithms by employing insights from datasets or diseases similar to each other.

1.6.5 Ensemble Learning

Ensemble learning produces more accurate predictions by combining multiple models. Techniques such as majority voting and weighted averaging are used to combine important outputs after training different models on the same or different datasets. Better generalization, bias reduction, and overall performance are made possible through holistic learning.

In most cases, experts and researchers experiment with multiple models before selecting the most effective model for a particular disorder detection project.

1.7 Different Data Sources Used for Disease Detection

The following are some typical data sources for disease identification.

1.7.1 Electronic Health Data (EHD)

EHDs store a lot of patient data, such as lab results, treatment plans, diagnoses, and more. Because EHD data collects a person's medical history over time, it can be a valuable tool for detecting disease.

1.7.2 Medical Imaging Data

Data from medical images such as ultrasound, CT scan, MRI, and X-ray are especially important in applications involving radiology to diagnose or detect specific diseases. These images can be used to train deep learning algorithms to diagnose pneumonia, identify diabetic retinopathy, and detect tumors.

1.7.3 Hereditary and Genomic Information

This category includes details about an individual's genetic makeup, including expression patterns, mutations, and gene variants. This type of information can be used to develop personalized disease diagnosis algorithms or identify genetic markers associated with disease.

1.7.4 Wearable Devices and Sensor Data

As the popularity of wearable technology and the Internet of Things continues to grow, specific health information can be obtained, such as heart rate, activity level, and sleep patterns. Two examples of how examining this data can help monitor and detect disease are identifying irregular heart rhythms and predicting the worsening of a condition.

1.7.5 Public Health Record

Public health records collect information about the incidence, prevalence, risk factors, and outcomes of diseases at the population level. These records may take the form of disease lists, health surveys, and observation systems. The epidemiological information from this data is important for tracking and identifying diseases.

1.7.6 Social Media and Online Health Societies

A significant amount of user-generated health data is produced by social media sites and online health communities. Mining and analyzing this data can reveal trends in symptoms and disease patterns beyond traditional data sources.

1.7.7 Mobile Health Applications

These applications, which are sometimes called "mHealth apps," ask users to self-report their health-related activities, symptoms, and prescription compliance. The combined and anonymized data from these apps can support efforts aimed at identifying diseases, especially those related to mental health and lifestyle-associated disorders.

1.7.8 Biobanks and Research Datasets

Biobanks are repositories of biological samples, such as blood, tissue, and DNA, along with the clinical and demographic data related to them. Another advantage of research datasets that are made publicly available or through partnerships is that they provide large-scale data for examination, which makes them indispensable for studies on illness detection.

1.8 Medical Images

Medical imaging is crucial to the healthcare sector for the detection, diagnosis, and planning of sickness. Through their ability to make the human body's internal workings visible, they help medical practitioners to identify ailments, evaluate abnormalities, and monitor treatment success. A few common kinds of medical pictures that are used to identify disorders are as follows.

1.8.1 X-Rays

Ionizing radiation is utilized in X-ray imaging to provide images of the bones and soft tissues. X-rays are commonly used to diagnose dental problems, pneumonia, lung infections, and fractures.

1.8.2 Computed Tomography (CT) Scans

CT scans use X-ray equipment and computer tools to create extremely detailed cross-sectional images of the body. Computed tomography scans can be used to evaluate and diagnose a variety of medical conditions, including malignancies, heart conditions, severe trauma, and abnormalities of the abdomen, pelvis, brain, and chest.

1.8.3 Magnetic Resonance Imaging (MRI)

In MRI, radio waves and strong magnetic fields are used to provide detailed images of systems inside the body. It works best when evaluating soft tissues such as the brain, spinal cord, muscles and joints, as well as organs such as the kidneys, liver, and heart. MRI can detect a wide variety of medical diseases, such as cancer, neurological disorders, musculoskeletal injuries, and vascular anomalies.

1.8.4 Ultrasound

Ultrasound imaging creates real-time images of organs, tissues, and blood flow using high-frequency sound waves. It is widely used for evaluating the heart, blood arteries, abdomen, pelvis, and other body components, as well as for obstetric imaging during pregnancy. Ultrasound is a very useful tool for identifying disorders such as kidney and gallstones, as well as monitoring prenatal development.

1.8.5 Positron Emission Tomography (PET) Scans

During a PET scan, positrons are released from the body by radioactive tracers, which are detected by the scanner. A wide variety of diseases, including cancer, cardiovascular problems, and neurological disorders, can be evaluated and diagnosed using imaging equipment. In addition, information regarding metabolic activities is provided.

1.8.6 Mammography

A specific type of X-ray imaging called mammography is used to detect and screen breast cancer. It is needed to detect anomalies in the breast, such as calcification or malignancy.

1.8.7 Nuclear Medicine Imaging (NMI)

A special camera records a very small amount of radioactive material (radiopharmaceutical) given to the patient as part of NMI. This method can be used to identify diseases such as cancer, infection, or anomalies in organ function, as well as problems in the function of multiple organs, such as the thyroid, heart, and bones.

1.8.8 Radiology Images

Images from radiology were collected with various imaging techniques and used to plan treatment and make diagnoses. These images are typically taken by radiologists, who are physicists, specialized in medical imaging. Examples of radiological imaging include CT, MRI, and X-ray scans. Radiology images are essential medical imaging resources that provide a comprehensive understanding of the internal architecture of the human body. Radiologists are medical professionals who have specific training in interpreting and examining images. It is their duty to interpret images obtained from radiological examinations such as X-ray, CT scan, and MRI. Radiologists examine the images for anomalous signs, cross-reference them with previous imaging investigations, and prepare comprehensive reports that support patient care and treatment options. Artificial intelligence (AI) systems and computer-aided diagnosis (CAD) frameworks are two examples of image analysis technological advancements that have been created to help radiologists find and explain anomalies in radiological imaging, thereby increasing accuracy and efficiency. When performing a thorough evaluation, it is essential to keep in mind that radiological images must be evaluated in light of the patient's symptoms, medical history, and other clinical data. The integration of additional clinical data with qualified radiologists and radiologic results is essential for effective disease identification, diagnosis, and therapy planning.

1.8.9 Histopathology Images

Histopathology snapshots are microscopic pictures of tissue samples taken from surgical specimens or biopsies. Pathologists are able to investigate and diagnose a variety of diseases and conditions using these pictures. Changes in cells and tissues can also be detected using these images. This is a list of pictures related to histopathology.

1.8.9.1 Formation of Histopathology Slides

Several processes are used to prepare histopathology slides. First, tissue samples are obtained through surgical excision or biopsy. The specimens are then processed, which includes thin sectioning and soaking in paraffin wax. These slices are placed on glass slides and stained in various ways to highlight specific features and architecture of the cell.

1.8.9.2 Hematoxylin and Eosin (H-E) Staining

H-E staining is the most commonly used staining method in histopathology. Hematoxin turns the cell nucleus blue-violet, while eosin turns the extracellular elements and cytoplasm pink. Because this staining integration provides a comparison and reveals important cellular features, pathologists are better able to evaluate the shape of the tissue and detect abnormalities.

1.8.9.3 Microscopic Diagnosis

Histopathology slides are examined by a pathologist using a light microscope. They identify patterns, distorted features or structures, and the architecture of the cell. Pathologists examine tissue and cell characteristics for patterns to diagnose diseases and determine their presence, severity, and potential predictors.

1.8.9.4 Disease Identification and Examination

Histopathology images are important for the investigation of various diseases, including cancer, autoimmune diseases, inflammatory diseases, and viral infections. When examining tissue samples, pathologists look for abnormal cell proliferation, changes in tissue architecture, the presence of certain cells, and other pathological features. Decisions about patient care, diagnosis, and therapy are made using this data.

1.8.9.5 Immunohistochemistry (IHC)

In histopathology, immunohistochemistry is a technique used to identify specific proteins or biomarkers in tissue samples. Target proteins are bound to antibodies tagged with fluorescent or chromogenic markers, which provide information about the amount, location, and presence of certain molecules. IHC is useful for determining tissue origin, classifying malignant diseases, and assessing potential treatment targets.

1.8.9.6 Electronic Pathology

Histopathology slides are scanned in electronic pathology to create digital pictures that may be examined and studied on computer monitors. Histopathology images can be stored, retrieved, and shared more easily using digital pathology. It also makes it easier to combine artificial intelligence algorithms for automatic identification and examination, as well as computer-aided picture examination approaches.

These represent just a small selection of the medical imaging modalities used in disease diagnosis. Based on clinical parameters, the body under investigation, and the suspected condition, healthcare practitioners select the most appropriate imaging strategy. Every imaging method has advantages and disadvantages. Medical image interpretation requires expertise and is usually carried out by skilled medical professionals with training in image analysis, such as radiologists. Additionally, automated image analysis and computer-aided diagnosis systems that help medical personnel understand and detect diseases from medical images are now possible because of advancements in the above-discussed fields.

1.9 Discussion

The uses of machine learning and image processing methods for illness recognition in the plant and human domains are comprehensively summarized in this chapter. Apart from delineating several uses of machine learning and image processing, it emphasizes the need of early sickness identification in enhancing agricultural productivity and healthcare outcomes. Through a survey of the literature, the chapter highlights the advancements in machine learning-based sickness detection systems and investigates their applicability in many fields. The importance of disease detection is highlighted, with special focus on agricultural production, population health, and individual health. Machine learning is crucial to illness detection because it enables the examination of complex data patterns and the automation of decision-making procedures. The chapter covers many machine learning techniques used in disease detection, including reinforcement learning, supervised learning, and unsupervised learning algorithms. Depending on the circumstance, each technique offers unique advantages. The chapter also looks at a range of data sources that are utilized in disease detection, such as medical imaging, including histopathological images, and radiology images (X-rays, CT scans, and MRI) for plant or human disease detection.

1.10 Conclusion

The combination of machine learning and image processing holds great potential to revolutionize the diagnosis of illnesses in a variety of domains. By applying contemporary computational methodologies, researchers and practitioners can increase the scalability, accuracy, and effectiveness of disease diagnosis systems, which will enhance diagnostic outcomes and resource allocation. However, other challenges remain, such as varied datasets, the stability and interpretability of algorithms, and ethical issues, including bias and data privacy. To fully leverage image processing and machine learning for illness identification, these challenges should be the focus of future exploration efforts. In the end, this will enhance agricultural sustainability and human health.

1.11 Future Scope

All things considered, the future using image processing and machine learning in the diagnosis of illness will be determined by consistent advancements in algorithm development, data integration, real-time monitoring, tailored medicine, and ethical considerations. Collaboration between scientists, physicians, legislators, and business stakeholders will be required to completely fulfill the promise of these technologies to advance human health and well-being.

References

1. McPhee SJ, Papadakis MA, Rabow MW, editors. Current medical diagnosis & treatment 2024. New York: McGraw-Hill Medical. ISBN 978-1-2655-5603-7, MHID 1-2655-5603-2, ISSN 0092-8682.
2. Ahsan MM, Ahad MT, Soma FA, Paul S, Chowdhury A, Luna SA, Yazdan MMS, Rahman A, Siddique Z, Huebner P. Detecting SARS-CoV-2 from chest X-ray using artificial intelligence. IEEE Access. 2021;9:35501–13. https://doi.org/10.1109/ACCESS.2021.3061621.
3. Coon ER, Quinonez RA, Moyer VA, Schroeder AR. Overdiagnosis: how our compulsion for diagnosis may be harming children. Pediatrics. 2014;134:1013–23. https://doi.org/10.1542/peds.2014-1778.
4. Balogh EP, Miller BT, Ball JR. Improving diagnosis in health care. Washington: National Academic Press; 2015.Bookshelf ID: NBK338596,PMID: 26803862, https://doi.org/10.17226/21794.
5. Ahsan MM, Luna SA, Siddique Z. Machine-learning-based disease diagnosis: a comprehensive review. Healthcare (Basel). 2022;10(3):541. https://doi.org/10.3390/healthcare10030541.
6. Huang J, Li J, Li Z, Zhu Z, Shen C, Qi G, Yu G. Detection of diseases using machine learning image recognition technology in artificial intelligence. Computational Intelligence of Neuroscience. 2022;2022:5658641. https://doi.org/10.1155/2022/5658641.
7. Hasan MM, Uddin AFMS, Akhond MR, Uddin MJ, Hossain MA, Hossain MA. Machine learning and image processing techniques for rice disease detection: a critical analysis. Int J Plant Biol. 2023;14:1190–207. https://doi.org/10.3390/ijpb14040087.

8. Kulkarni P, Karwande A, Kolhe T, Kamble S, Joshi A, Wyawahare M. Plant disease detection using image processing and machine learning. arXiv preprint arXiv:2106.10698. https://doi.org/10.48550/arXiv.2106.10698.
9. Harakannanavar SS, Rudagi JM, Puranikmath VI, Siddiqua A, Pramodhini R. Plant leaf disease detection using computer vision and machine learning algorithms. Global Trans Proc. 2022;3(1):305–10. https://doi.org/10.1016/j.gltp.2022.03.016.
10. Shoaib M, Shah B, Ei-Sappagh S, Ali A, Ullah A, Alenezi F, Gechev T, Hussain T, Ali F. An advanced deep learning models-based plant disease detection: a review of recent research. Front Plant Sci. 2023;14 https://doi.org/10.3389/fpls.2023.1158933.
11. Kousis I, Perikos I, Hatzilygeroudis I, Virvou M. Deep learning methods for accurate skin cancer recognition and mobile application. Electronics. 2022;11:1294. https://doi.org/10.3390/electronics11091294.
12. Alwakid G, Gouda W, Humayun M, Jhanjhi NZ. Diagnosing melanomas in dermoscopy images using deep learning. Diagnostics (Basel). 2023;13(10):1815. https://doi.org/10.3390/diagnostics13101815.
13. Seeja RD, Suresh A. Deep learning based skin lesion segmentation and classification of melanoma using support vector machine (SVM). APJCP. 2019;20(5):1555. https://doi.org/10.31557/APJCP.2019.20.5.1555.
14. Indraswari R, Rokhana R, Herulambang W. Melanoma image classification based on MobileNetV2 network. Proc Comput Sci. 2022;197:198–207. https://doi.org/10.1016/j.procs.2021.12.132.
15. Han SS, Park I, Eun Chang S, Lim W, Kim MS, Park GH, Chae JB, Huh CH, Na J-I. Augmented intelligence dermatology: deep neural networks empower medical professionals in diagnosing skin cancer and predicting treatment options for 134 skin disorders. J Invest Dermatol. 2020;140(9):1753–61. https://doi.org/10.1016/j.jid.2020.01.019.
16. Gupta S, Mahajan S, Pandit AK. A review on image processing techniques. In: 2020 12th international conference on computational intelligence and communication networks (CICN); 2020. p. 20–4. https://doi.org/10.1109/CICN49253.2020.9242606.
17. Kaushik PK, Pandey S, Rauthan SS. Facial emotion recognition and eye-tracking based expressive communication framework: review and recommendations. Int J Comput Appl. 2022;184:20–8. https://doi.org/10.5120/ijca2022922494.
18. Angra S, Ahuja S. Machine learning and its applications: a review. In: 2017 international conference on big data analytics and computational intelligence (ICBDAC); 2017. p. 57–60. https://doi.org/10.1109/ICBDACI.2017.8070809.
19. Caruana R, Lou Y, Gehrke J, Koch P, Sturm M, Elhadad N. Intelligible models for healthcare: predicting pneumonia risk and hospital 30-day readmission. In: Proceedings of the 21th ACM SIGKDD international conference on knowledge discovery and data mining; 2015. p. 1721–30. https://doi.org/10.1145/2783258.2788613.
20. Ribeiro MT, Singh S, Guestrin C. "Why should i trust you?" Explaining the predictions of any classifier. In: Proceedings of the ACM SIGKDD international conference on knowledge discovery and data mining; 2016. p. 1135–44. https://doi.org/10.1145/2939672.2939778.
21. Yu Z, Wang K, Wan Z, et al. Popular deep learning algorithms for disease prediction: a review. Clust Comput. 2023;26:1231–51. https://doi.org/10.1007/s10586-022-03707-y.
22. Qureshi, A.S., Roos, T. Transfer Learning with Ensembles of Deep Neural Networks for Skin Cancer Detection in Imbalanced Data Sets. Neural Process Lett 55, 4461–4479 (2023). https://doi.org/10.1007/s11063-022-11049-4.
23. Gulshan V, Peng L, Coram M, Stumpe MC, Wu D, Narayanaswamy A, Venugopalan S, Widner K, Madams T, Cuadros J, Kim R, Raman R, Nelson PC, Mega JL, Webster DR. Development and validation of a deep learning algorithm for detection of diabetic retinopathy in retinal fundus photographs. JAMA. 2016;316(22):2402–10. https://doi.org/10.1001/jama.2016.17216.

Chapter 2
Comparative Study of Various Deep Learning-Based Methods for the Prediction of Disease

Shaweta Sachdeva, Aleem Ali, and Nawaf R. Alharbe

2.1 Introduction

It is anticipated that artificial intelligence would improve computers' computing capabilities, and it continuously increases machine intelligence. A kind of artificial intelligence called machine learning is used to diagnose different medical disorders. Professionals from many fields agree that intelligence cannot advance in the absence of learning. Various machine learning methodology categories are shown in Fig. 2.1, including reinforcement, unsupervised, semi-supervised, supervised, and deep learning methods.

The following is a definition of various machine learning techniques.

Supervised Learning Using a dataset of different inputs and the intended outputs for each, algorithms are taught using this technique. The learning of regression and classification are two instances of supervised learning.

Unsupervised Learning In this method, no predetermined goals or right answers are given. This method seeks to identify common patterns in the input data and then uses those patterns to classify the data. Density estimation is another name for this method [1].

S. Sachdeva (✉)
Yamuna Nagar's Guru Nanak Khalsa College, Yamunanagar, Haryana, India

A. Ali
Chandigarh University, Mohali, Punjab, India

N. R. Alharbe
Saudi Arabia's Taibah University in Medinah (KSA), Medina, Saudi Arabia

T. Choudhury, A. Katal (eds.), *Machine Learning for Disease Detection, Prediction, and Diagnosis*, https://doi.org/10.1007/978-981-96-4241-0_2

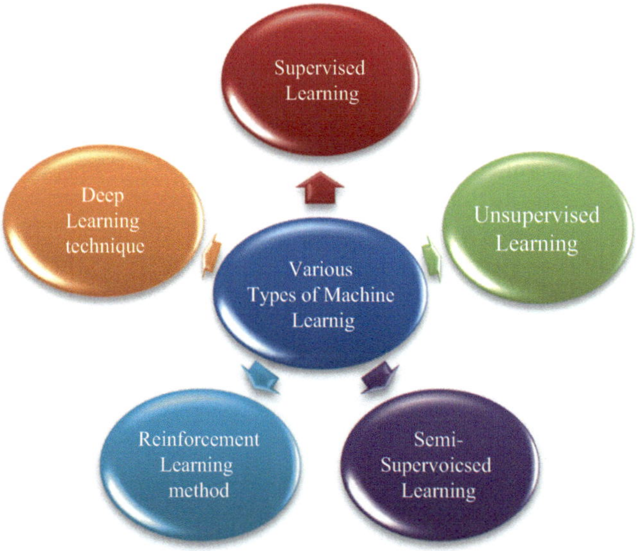

Fig. 2.1 Various machine learning techniques

Semi-supervised Learning A subset of supervised learning methods known as semi-supervised learning makes use of some unlabeled data in addition to a lesser quantity of labeled data. By using unlabeled data, it functions as a bridge between supervised learning, which depends on labeled data, and unsupervised learning.

Reinforcement Learning This method tells the user when their reaction is wrong but does not tell them how to fix it. Before determining the best answer, it investigates and tests a wide range of options. It is also referred to as "learning with a critic" and does not provide well-defined suboptimal actions or particular input–output combinations. It also stresses the importance of real-time performance.

Deep Learning This area of machine learning is the foundation of deep learning. These techniques use a sophisticated graph structure with multiple processing layers that include both linear and nonlinear transformations to imitate high levels of data abstraction.

Data classification and pattern recognition have long been established as useful techniques. The ability to sense their environment and react to what they see is innate in human nature [2]. Processes that combine machine learning, databases, and statistics with other fields are essential for handling large volumes of data. In the modern field of medical science, a precise sickness diagnosis is crucial. Clinical examinations and assessments are essential for gaining a thorough grasp of patients' problems. Accurate diagnosis and effective patient care may require computer-based decision support systems. Big data is produced by the healthcare sector in large quantities from a variety of data sources, including clinical evaluations, patient records, treatments, follow-ups, and pharmaceutical information.

It can be difficult to organize data correctly, and poor data management has led to a decline in the caliber of data organization. The increasing amount of data makes it necessary to use appropriate technologies for data extraction and processing [3]. Several machine learning tools are used to create a classifier that can divide data into groups according to their properties. These classifiers are used in the analysis and detection of diseases in medical data. The data collection is split into two or more categories.

Initially, medical datasets were analyzed using ML algorithms. Currently, there are several methods for expedited data processing available thanks to machine learning. Data storage and collecting are now more affordable and widely available because of the digital revolution. In particular, contemporary hospitals are outfitted with instruments for data collection and analysis, which enables them to trade and gather data inside of vast information networks.

The application of ML technology used for analysis of medical data has been shown to be quite successful, leading to notable advancements in the field of medical diagnosis. Contemporary medical facilities have developed effective methods for producing accurate diagnostic data via reports, medical records, or specialized data departments. Precise patient diagnostic data are entered into a computer to run an algorithm. Based on cases that have already been settled, the system can automatically produce results. These built classifiers are used by doctors to quickly and reliably diagnose certain cases. These classifiers are also useful resources for teaching problem detection to persons or students lacking specialist knowledge.

2.2 Review of Literature

An artificial intelligence (ML)-based approach to forecast common medical diseases was proposed by Dahiwade et al. [4]. They made use of a dataset of symptoms from the UCI ML repository, which comprised symptoms and indicators of several common illnesses. The system predicted several diseases by using CNN and KNN as classification methods. Additional lifestyle-related data about the patients under examination was included in the therapy suggestions, which helped to improve the evaluation of risk connected with the anticipated medical disorders. They performed a comparison of the KNN and CNN algorithms' processing speeds and output accuracy.

A comparison analysis was carried out using the KFT dataset, that is, kidney function test, which was provided by Serek et al. [5], to assess the performance of classifiers in the diagnosis of chronic kidney disease (CKD). Random forest (RF), naive Bayes (NB), and K-nearest neighbors (KNN) were the classifiers used in this investigation. Each classifier's F-measure, accuracy, and precision were used to evaluate performance. The results of the investigation showed that, although NB showed greater precision, RF performed better in terms of accuracy and F-measure.

Research on renal disease diagnosis using naive Bayes and support vector machine was done by Vijayarani [6]. Finding the classification method that performed best in terms of accuracy and processing time was another important goal. The results showed that SVM greatly surpassed NB in terms of accuracy.

Several data mining techniques, such as random forest (RF) and support vector machine (SVM), were used in the study by Yao [7] to identify the best algorithm for breast cancer prediction. The findings showed that the random forest approach showed classification rates of 96.27, 96.78, and 94.57% for accuracy, sensitivity, and specificity, respectively, whereas the SVM algorithm obtained accuracy rates of 95.85, 95.95, and 95.53%. Yao concluded that in terms of prediction accuracy the random forest technique performs better than SVM because it yields more accurate estimations of feature attributes.

Marimuthu et al.'s [8] goal was to use supervised machine learning techniques to forecast heart disorders. The authors categorized the following data attributes: gender, age, target, slope, and chest pain. Decision trees (DT), K-nearest neighbors (KNN), logistic regression (LR), and naive Bayes (NB) were the four machine learning algorithms they used. With an accuracy score of 86.89%, the investigation showed that the logistic regression algorithm outperformed the other techniques.

Shubair [9] used ML algorithms, such as random forest (RF), Bayesian networks, and support vector machine (SVM), in a study aimed at diagnosing breast cancer. Utilizing key metrics, including ROC graph area, accuracy, recall, and precision, the researchers evaluated the learning models using the original Wisconsin breast cancer dataset that they had obtained from the UCI Repository. After establishing a value of K = 10, the classifiers were assessed using the K-fold validation procedure.

Khan R [10] uses methods that are modeled after the composition and operations of the brain. ANN, that is, artificial neural networks with an extra hidden layer are used in deep learning (DL), while cascading neural networks are used to handle nonlinear datasets. DL entails examining several buried data layers in order to find crucial information. This approach is not limited to diagnostics and has significantly improved the healthcare industry's capacity for accurate and timely data analysis.

2.3 Methodology

The disease prognosis system uses a variety of data mining classifier techniques, such as random forest, decision tree, and naive Bayes. These algorithms' functionality and workings are explained next.

2.3.1 Decision Tree

The decision tree approach produces a number of categorization models that are arranged in a tree-like manner. To anticipate a result, it divides the data into smaller subsets and obtains a set of particular if-then rules based on feature values, like symptoms. Decision nodes and leaf nodes make up this structure.

Decision Node: All symptoms in this study are regarded as decision nodes, which are nodes with numerous branches.

Leaf Nodes: These nodes show the choice or classification made for a specific branch. They line up with the leaf node units.

2.3.2 The ID3 Algorithm

This strategy, created by J.R. Quinlan, is a key research approach that we use. Through a top-down, greedy search among the available characteristics, ID3, or Iterative Dichotomiser3, assesses each attribute, such as symptoms, at every node and chooses those that are most effective in categorizing a particular dataset. ID3 uses the ideas of entropy and information gain to identify which symptom is most suited for a decision tree.

Entropy is a measure of a particular occurrence occurring that is similar to uncertainty or unpredictability. Using frequency tables, we attempt to compute two distinct kinds of entropy for each attribute to build a decision tree.

Entropy is expressed as E(C), where C is the current state, or the results. Entropy is the frequency table of a particular property. P(h) is the probability that an event h will occur in that condition.

$$\mathrm{E}(\mathrm{C}) = \sum \mathrm{h}\varepsilon\mathrm{H} - \mathrm{P}(\mathrm{h})\log 2\,\mathrm{P}(\mathrm{h}) \tag{2.1}$$

Information gain, also called Kullback–Leibler divergence and represented as IG(C, A) for a given state C, is the large change in entropy that results from choosing attribute A for the final decision. The equation is as follows:

$$\mathrm{IG}(\mathrm{CA}) = \mathrm{E}(\mathrm{C}) — \mathrm{E}(\mathrm{C,A}) \tag{2.2}$$

2.3.3 Random Forest

The decision tree method is often flexible and easy to use, and it frequently produces good outcomes when left unchecked. However, as the decision tree's nature suggests, a major disadvantage of decision trees is their propensity to overfit.

Random forest, an ensemble learning technique, is used to overcome this difficulty. Applying a new algorithm or the same method repeatedly is known as ensemble learning. To reduce overfitting, random forest uses a group of decision trees cooperating with one another.

The way random forest functions is as follows:

1. From a dataset of m symptoms extracted from a patient's medical records, k symptoms (where k equals m) are chosen at random. These selected k symptoms are then used to form a decision tree.
2. Iteratively, different sets of k symptoms chosen at random (or, alternatively, by generating unique random subsets of the data, known as bootstrap samples) are used to create multiple decision trees, each with a number n.
3. A random variable is run through each of the n decision trees in order to forecast the disease. It stores the anticipated course of the disease in a way that makes it possible to utilize n decision trees to forecast the course of a total of n diseases.
4. The random forest algorithm uses the mode, or the disease predicted most frequently, as its final forecast after computing the votes for each projected sickness.

2.3.4 *Naive Bayes*

In this case, every attribute influences the result in an equal and autonomous manner. Its ability to efficiently use fewer computing resources enables it to function quickly, even on massive datasets.

The algorithm naive Byes is provided by

$$P(s/h) = P(s)/P(h) \tag{2.3}$$

2.3.5 *Model of Prediction*

As seen in Fig. 2.2, our primary goal is to create a prediction model that analyzes different user symptoms and forecasts the most likely ailment.

(a) Inputs: Symptoms: We assumed that the user was fully aware of the symptoms they were now experiencing when we created the model. The 95 symptoms that the prediction algorithm takes into account as input are available for the user to choose from.
(b) Data Preprocessing: Also referred to as data mining, this method entails converting or encoding raw data in order to speed up the algorithm's execution. The following preprocessing techniques were used in the study that is being presented.

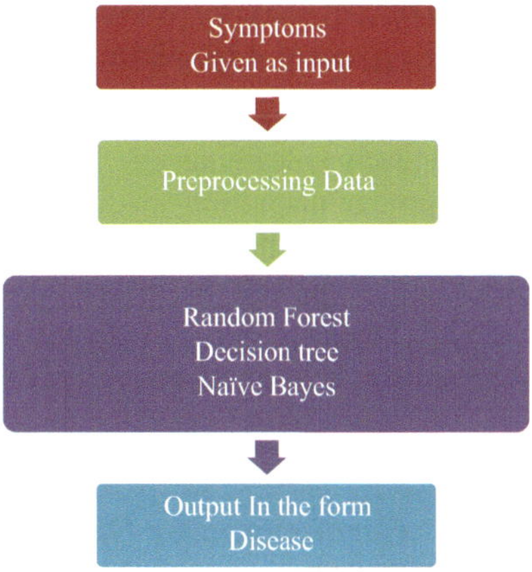

Fig. 2.2 The prediction model

(c) Data cleaning involves removing any anomalies from the data, which may involve activities like adding missing information.
(d) Data Reduction: There are particular difficulties when analyzing large databases. As such, we eliminate independent variables (symptoms) that are probably not going to have a big influence on the dependent variable (disease). Out of the 132 symptoms, 95 were chosen for this analysis because they showed strong relationships with the disorders.

2.3.6 Models Selected

Three methods are used to train the system to forecast the diseases.

(i) Random forest.
(ii) Decision tree.
(iii) Naive Bayes.

2.3.7 Diseases as an Output

The system generates a set of rules after it has been trained on the training dataset using the previously described processes. The symptoms are quickly evaluated by the model using the pre-established rule set to produce classifications and a prediction for the most likely diagnosis when a user enters their symptoms.

2.4 Diagnosing Illnesses via Various Machine Learning Algorithms

Several machine learning algorithms have been investigated in research studies for the purpose of diagnosing medical disorders. Scholars have exhibited the effectiveness of ML algorithms in diagnosing a broad spectrum of medical conditions. A conceptual framework for the disorders identified by ML approaches is depicted in Fig. 2.3. Diabetes, hepatitis, liver disorders, heart disease, dengue, and diabetes were among the medical issues that MLT found in this survey report.

2.4.1 Cardiovascular Disease

A monitoring and analysis system for the identification and tracking of coronary artery disease was presented by Otoom et al. [11]. They utilized the UCI Cleveland Heart Dataset, which has 76 traits or features and 303 samples. Also, 13 of the 76 functionalities are being used right now. Using the WEKA tool for detection, the study used three algorithms—Bayes net, support vector machine, and functional trees FT—to find patterns as shown in Table 2.1. The SVM approach produces an accuracy of 88.3% in the Holdout test. Both the SVM and the Bayes net demonstrated an accuracy of 83.8% in the cross-validation test.

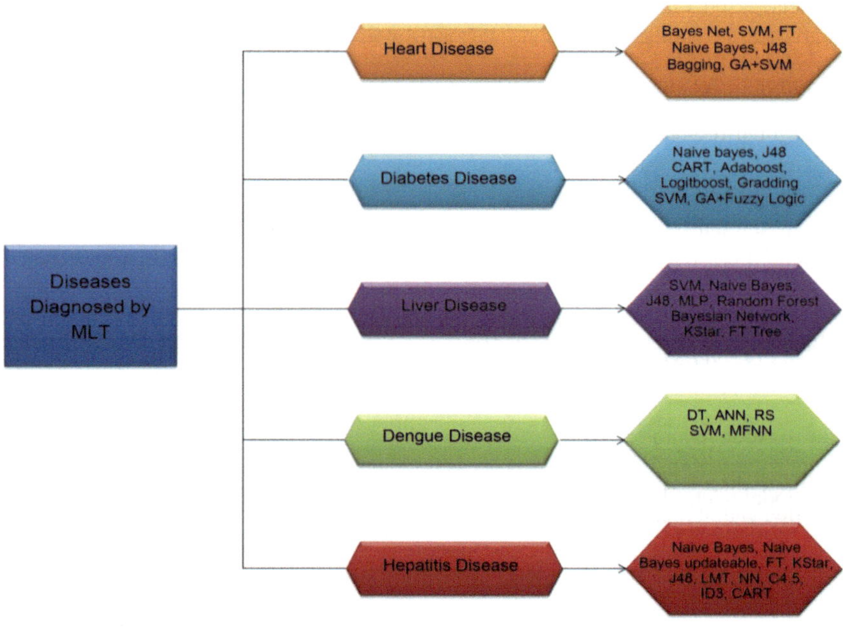

Fig. 2.3 Illnesses that MLT diagnoses

Table 2.1 A comparative analysis of machine learning methods for the detection of heart disease

Techniques of machine learning	Year	Disease	Resources of data set	Tool	Accuracy value (%)
Bayes net					84.5
SVM		Coronary artery disease	UCI	WEKA	85.1
J48	2013	Heart disease	UCI	WEKA	84.35

Table 2.2 A comparative analysis of machine learning methods for the diagnosis of diabetes

Machine learning techniques	Year	Disease	Resources of dataset	Tool	Accuracy value (%)
Naive Bayes	2015	Diabetic disease	Pima Indian diabetes dataset	WEKA	79.562
SVM	2013	Diabetic disease	UCI	MATLAB	78
GA + fuzzy logic	2011	Diabetic disease	UCI	MATLAB	87

2.4.2 Diabetes

The goal of Iyer et al.'s work [12] was to use decision trees and naive Bayes to forecast the beginning of diabetes, which is frequently brought on by problems with insulin synthesis. They used diabetes data from the Pima Indians and ran a number of experiments using the WEKA data mining program. They discovered that, in contrast to cross-validation, a 70:30 percentage split in the dataset produced more accurate predictions. J48 specifically obtained an accuracy of 76.9565% with the percentage split and 74.8698% in cross-validation. At 79.5652%, naive Bayes demonstrated the highest accuracy while utilizing the percentage split. The most accurate approach was the one that employed the % split test.

Sen and Dash [13] used a dataset from the UCI machine learning lab to examine diabetes diagnosis using meta-learning methods, with a particular focus on Pima Indians with the disease. They used algorithms, including CART, Adaboost, Logiboost, and gradient learning, to diagnose patients with diabetes, and used WEKA for analysis. They ascertained the algorithms' correctness by comparing the outcomes of their experiments. Gradient learning had a categorization accuracy of 66.406%, Adaboost scored 77.864%, CART reached 78.646%, and Logiboost exhibited 77.479% accuracy as shown in Table 2.2. With the best accuracy of 78.646% and the lowest misclassification rate of 21.354% among them, CART had the lowest misclassification rate [14–18].

2.5 Conclusion

Many statistical models have taken the position of the evaluation field, frequently failing to produce high-quality outcomes. Missing values, large numbers of data points, and categorical data are difficult for statistical models to handle. This emphasizes how important machine learning techniques (MLT) are. Machine learning is widely used in many applications, including data mining, picture identification, natural language processing, and medical diagnosis. It provides possible solutions in each of these fields. An overview of the numerous machine learning algorithms used in the diagnosis of heart disease, diabetes, hepatitis, dengue fever, and liver illness is provided in this chapter. Numerous algorithms have demonstrated encouraging outcomes in precisely recognizing the characteristics of diseases. Previous studies have shown that naive Bayes produces accurate diabetic diagnoses, whereas SVM improves the accuracy of cardiac illness diagnosis. The survey provides comprehensive performance graphs related to machine learning models used for disease prediction, along with an overview of the benefits and drawbacks of these various algorithms. According to the analysis, these algorithms help in more accurate disease diagnosis. A variety of AI community-developed tools are also presented in the survey article, providing useful resources and the chance to make well-informed decisions.

References

1. Marshland S. An algorithmic perspective on machine learning. Boca Raton: CRC Press; 2009. p. 6–7.
2. Kaur M, Sharma P. A review on classification in pattern recognition. IJARCSSE. 2013;3:298.
3. Deepanker W, Rambhajani M, et al. An overview of the application of machine learning techniques for the classification of diseases in dermatology. Int J Eng Technol Adv. 2015;8:194–5.
4. Patle DG, Dahiwade E, Meshram G. Using machine learning to design a disease prediction model. In: Proceedings of the 3rd International conference on computing methodologies and communication (ICCMC 2019); 2019. p. 1211–5.
5. Amirgaliyev Y, Shamiluulu S, Serek A. Applying machine learning techniques to analyze a dataset on chronic kidney disease. In: IEEE 12th international conference on application of information and communication technologies (AICT); 2018. p. 1–4.
6. Vijayarani S, Dhayanand S. Data mining classification algorithms for kidney disease prediction. Int J Cybernetics Inform. 2015;4(4):13–25.
7. Yao D, Yang J, Zhan X. A novel approach to disease prediction: combining multivariate adaptive regression splines with random forests. J Comput. 2013;8(1):170–7.
8. Marimuthu M, Abinaya M, Hariesh KS, Madhankumar K, Pavithra V. A review on heart disease prediction using machine learning and data analytics approach. Int J Comput Appl. 2018;181(18):20–5.
9. Sengar PP, Gaikwad MJ, Nagdive AS. Comparative study of machine learning algorithms for breast cancer prediction. In: 2020 third international conference on smart systems and inventive technology (ICSSIT). IEEE; 2020. p. 796–801.
10. Khan R, Srivastava AK, Gupta M, Kumari P, Kumar S. [Retracted] Medicolite-machine learning-based patient care model. Comput Intell Neurosci. 2022;2022(1):8109147.

11. Otoom AF, Abdallah EE, Kilani Y, Kefaye A, Ashour M. Effective diagnosis and monitoring of heart disease. IJSEA. 2015;9(1):143–56.
12. Sasipriya R, Vembandasamy K, Deepa E. Naive Bayes algorithm for the detection of heart disorders. JISET. 2015;2:441–4.
13. Iyer A, Jeyalatha S, Sumbaly R. Diabetes diagnosis employing classification mining methods. IJDKP. 2015;5:1–14. https://doi.org/10.5121/ijdkp.2015.5101.
14. Sen SK, et al. Application of meta learning algorithms for the prediction of diabetes disease. IIJARCSMS. 2014;2:396–401.
15. Sachdeva S, Ali A. A hybrid approach using digital forensics for attack detection in a cloud network environment. IJFGCN. 2021;14(1):1536–46.
16. Sachdeva S, Ali A. Machine learning with digital forensics for attack classification in cloud network environment. Int J Syst Assur Eng Manag. 2022;13(Suppl 1):156–65.
17. Sachdeva S, Ali A, Khan S. Secure and privacy issues in telemedicine: issues, solutions, and standards. In: Telemedicine: the computer transformation of healthcare. Cham: Springer International Publishing; 2022. p. 321–31.
18. Sachdeva S, Ali A, Khalid S. Telemedicine in healthcare system: a discussion regarding several practices. In: Telemedicine: the computer transformation of healthcare. Cham: Springer International Publishing; 2022. p. 295–310.

Chapter 3
Introduction to Deep Learning for Disease Prediction

Abhinay Thakur and Ashish Kumar

3.1 Introduction

With broad repercussions for both patients and healthcare organizations, disease prognosis has become a crucial aspect of research in the medical domain. One of the main tenets of contemporary medicine is the capacity to forecast illnesses, whether they affect a group or an individual. It promises improved healthcare, earlier diagnosis, more efficient therapy, and cost reductions for healthcare systems. Conventional illness prediction techniques, which are based on statistical algorithms, historical information, and clinical skills, have been very important [1–3]. Disease prognosis has many applications and is very important in the healthcare industry. Several advantages come with disease prognosis, such as early detection, better treatment strategies, preventative interventions, cost savings, effective allocation of resources, and patient engagement. For example, accurate diagnosis is essential because it enables prompt and focused therapies that greatly enhance clinical outcomes. Healthcare is benefiting greatly from deep learning, especially in the area of illness prediction [4, 5]. One prominent use is in the diagnosis of breast cancer. CNNs, one sort of deep learning model, are utilized to scan mammograms and detect potentially malignant tumors, helping doctors diagnose patients. Within the field of diabetes risk assessment, these models analyze a variety of patient data to forecast a person's chance of acquiring diabetes, such as blood glucose measurements, medical information in the family, and lifestyle habits. This strengthens customized prevention techniques and early intervention [6–8]. Moreover, deep

A. Thakur
Department of Chemistry, School of Chemical Engineering and Physical Sciences, Lovely Professional University, Phagwara, Punjab, India

A. Kumar (✉)
Nalanda College of Engineering, Bihar Engineering University, Science, Technology and Technical Education Department, Government of Bihar, Patna, Bihar, India

T. Choudhury, A. Katal (eds.), *Machine Learning for Disease Detection, Prediction, and Diagnosis*, https://doi.org/10.1007/978-981-96-4241-0_3

learning is essential for the prognosis of cardiovascular disease since it analyzes electrocardiograms (ECGs) to identify abnormal heartbeats and determine the risk of heart failure. Moreover, it has the ability to forecast cardiac problems, which permits prompt diagnosis and individualized therapeutic plans. Deep learning models are employed in the cure of Alzheimer's disease to examine brain imaging data, including MRI and PET scanning, in order to find morphological changes associated with the disease. This allows for an early diagnosis and continuous follow-up. Deep learning also performs exceptionally well in the identification of skin cancer. Dermatologists use these models to reliably categorize skin lesions as benign or malignant by examining photographs of the lesions, like moles. This lessens the requirement for invasive biopsies and helps identify skin cancer. Deep learning algorithms are used in lung cancer diagnosis to evaluate CT scans in order to identify lung nodules and determine whether they are malignant [9–11]. Prompt therapy and better patient results depend on this prompt exploration. By evaluating data on patient stride, stiffness, and other motor functions, deep learning is also used in the diagnosis of Parkinson's disease. As the illness worsens, this helps with the creation of treatment programs and the required modifications. Deep learning benefits the subject of public health transcend individual diagnosis, especially when it comes to Epidemiological Outbreak Forecasting. These models forecast and track contagious disease epidemics by examining health data, route maps, and external conditions. This allows for quick reactions and confinement measures during medical emergencies. Deep learning (DL) is utilized in the field of psychological health to identify mental health disorders by looking at text data, like posts on social media or patient records [12, 13]. This is essential for earlier treatment and helping people who are struggling with psychological health issues. Finally, genetic disease diagnosis is made possible by deep learning, which analyzes genomic data to find gene variations and offer clues into the underlying reasons for various disorders. These findings can inform therapy choices. These varied uses demonstrate the revolutionary possibilities of deep learning in the healthcare industry, enhancing patient care through individualized treatment plans, prompt detection, and improved diagnostic reliability.

In the healthcare system, illness containment methods and immunization program development are guided by prevention methods predicated on disease projections. Moreover, effective disease forecasting may result in the best possible distribution of medical resources, improving the effectiveness and economy of healthcare provision. In the end, knowledgeable illness prognoses enable people to take charge of their health and adopt preventive lifestyle decisions [14–16]. However, these approaches have their drawbacks, especially when it comes to handling the immensely complicated and varied world of medical data. In the past, scientific findings and instinct were frequently used to forecast illness. Early medical professionals made assumptions about the causes of particular diseases based on trends they saw in groups or groups, attributing them to lifestyle, nutrition, or external conditions. Although these insights provided insightful information, they failed the scientific validity necessary to make accurate individual forecasts. Clinical competence gained prominence in the twentieth century when doctors began to forecast patients' wellness based on physiological examination results, past medical history, and symptomatic expertise. In parallel, epidemiology research and mathematical analysis were developed, offering a more methodical, data-driven method of illness forecasting. Risk assessment became an essential element that made it possible to determine a person's probability of contracting a certain disease depending on lifestyle, genetics, and age [17–19]. Even with these improvements, there were still drawbacks to using old techniques. These included their inability to manage high-dimensional data, reliance on human feature extraction, and challenges in recording complex illness trends.

Approaches for predicting diseases in the past had their share of difficulties. Comprising a variety of healthcare data formats, high-dimensional datasets presented challenges in identifying intricate trends and regularities. It takes a lot of work for features engineers to actively choose and create pertinent characteristics for forecasting. Different interpretations led to different diagnoses and outcomes [20, 21]. Furthermore, the main emphasis of old models was on organized data, which hindered their capacity to fully utilize unorganized sources of data, including genomics information, patient records, and imaging techniques. These drawbacks highlighted the necessity for even more sophisticated, data-driven methods of illness prognosis. The introduction of electronic health data, wearable tech, genetic decoding, diagnostic imaging, and digitalization of medical records all contributed to the "data revolution" that shattered the healthcare industry. The advent of several, varied healthcare datasets presented obstacles as well as possibilities for illness forecasting. Patient data were digitalized via electronic health records (EHRs), making data more organized and easily available for research and forecasting. Image data rich in detail was produced by medical imaging technology for the purpose of detecting and diagnosing diseases [22, 23]. Genetic susceptibilities to illnesses were revealed by genomic data. Patient data is continuously streamed in real time thanks to wearables technology and the Internet of Things (IoT). To create thorough patient assessments, initiatives to improve information sharing and interchange made it easier to aggregate data from different sources. The possibility for more precise and customized illness forecasts was made possible by the information revolution, but there were also difficulties with data preparation, analytics, and modeling.

The artificial intelligence area of deep learning has arisen as a transformational influence in the healthcare industry in relation to the hurdles encountered by conventional disease forecasting models in this data-rich context. The capacity of deep learning systems, especially neural networks, to automatically extract complex structures and descriptions from data brought about a paradigm change. The necessity for time-consuming classifiers was removed by these models' exceptional autonomous showcase learning capabilities. They might efficiently utilize the full capacity of heterogeneous healthcare data by handling unorganized datasets like clinical records and medical photographs. Deep learning adopted a data-centric strategy, adapting to trends seen in vast amounts of labeled and unstructured information. A model of the healthcare system based on Friedreich Ataxia is shown in Fig. 3.1. This model is prepared to meet the unique healthcare requirements and

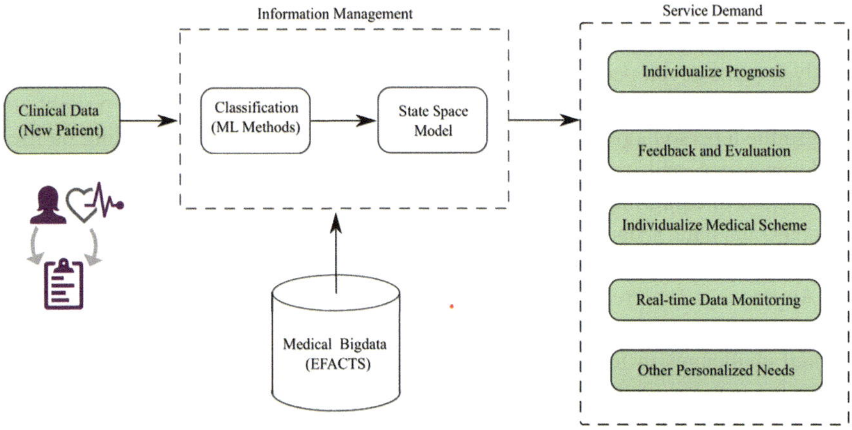

Fig. 3.1 Friedreich Ataxia-based healthcare system model. (Adapted from Ref. [24] under CCBY 4.0)

difficulties faced by people with Friedreich Ataxia, a degenerative neurological illness that is genetic. The model is designed to offer a thorough foundation for comprehending and treating the illness, accounting for many facets of assistance and treatment. It offers a comprehensive perspective on the procedure of managing a condition and emphasizes the value of a multidisciplinary team effort, including medical experts, patients, and their relatives.

These models addressed the shortcomings of conventional techniques by demonstrating the ability to recognize complicated structures and nuanced connections. Furthermore, their flexibility in real time matched the dynamic nature of patient characteristics and health information. Including its neural networks, deep learning ushered in a revolutionary age in prediction and diagnosis, changing the face of healthcare and creating opportunities for more precise, data-driven forecasts. The backdrop of illness prognosis and the critical function of deep learning are explained in the following sections:

- Historical Disease Prediction Methods: In the past, epidemiological research and empirical findings were frequently used to forecast sickness. Although useful, these techniques could not manage huge and complicated datasets. The core components of these early forecasting attempts were manual data processing and human judgment [25].
- Traditional Statistical Models: A major contribution of statistical models to disease prognosis in the twentieth century was made. These models were applied to data analysis in order to find trends that would suggest a higher chance of a disease developing. However they frequently needed specific characteristic engineering, and they had trouble utilizing the high-dimensional, complex data that is becoming more and more accessible in contemporary healthcare.
- The Data Revolution: A data revolution has recently occurred in the healthcare sector. Numerous sources of healthcare data have been produced, including

wearable technology, genetic testing, EHRs, and diagnostic imaging [26, 27]. Although there are never-before-seen possibilities for disease prognosis with this data, there are also difficulties with data preparation, extraction of features, and the requirement for more sophisticated modeling methods.

- Challenges of Traditional Methods: The capacity to identify intricate trends from a variety of data sources, analyze high-dimensional information, and adjust to the ever-changing nature of health-related material was a limitation of conventional illness prediction techniques. Furthermore, the objectivity present in forecasts made by humans created questions regarding consistency and accuracy.
- Emergence of DL: Neural networks, a feature of DL, enable a fundamental change in disease prognosis. The necessity for human feature engineering is eliminated by these models' ability to dynamically understand and describe complex characteristics and connections inside data. They are flexible instruments for healthcare purposes because of their significant degree of adaptability to many data kinds, such as text, pictures, and series data [28, 29].
- Unstructured Data Handling: Deep learning models exhibit extraordinary accuracy when handling complex data, including genome information, medical imaging, and physician's comments. This capacity is essential for obtaining important information from various healthcare sources of data.
- Data-Driven Approach: In the field of disease prognosis, deep learning signals a move toward a more data-driven methodology. Deep learning models use enormous volumes of labeled and unstructured data to provide estimates based on learned patterns rather than just depending on preset constraints and assumptions.
- Rise of AI in Healthcare: A surge of progress in patient care, medication development, and diagnostic performance has been spurred by employing AI, especially DL, in the healthcare industry. AI-powered illness prediction models could greatly enhance the efficiency and quality of healthcare.

3.2 Fundamentals of Deep Learning

Contemporary illness prediction relies heavily on deep learning, which has revolutionized the way we handle and evaluate complicated medical data [30, 31]. An in-depth discussion of the training algorithm, components, and the key algorithms and tools supporting deep learning's use in illness prediction is provided in this section.

3.2.1 Neural Networks and Their Building Blocks

Neural networks, which are computer models based on the framework and operation of the human brain, are crucial to deep learning. Different building elements of neural networks are essential for interpreting and responding to data.

3.2.1.1 Neurons and Activation Functions

Neural networks are primarily composed of neurons, which are computational elements that perform data processing and transformation. In order for neurons to function, they must first acquire weighted stimuli, which are signals from other neurons in the input level or, in the instance of the previous layers, from outside data sources. A weight is applied to each input, which indicates the importance or strength of the link between the neurons. Then, the neuron aggregates, or sums, these weighted inputs [32, 33]. This accumulation stage simulates how a biological neuron integrates incoming data, giving each signal a different weight according to its related relevance. Nevertheless, an activation function—a key component of non-linearity—is introduced by neural networks. The neural network's ability to recognize intricate connections in the input is largely dependent on its nonlinear activation function. Without it, the network's capacity to efficiently learn and reflect complex data patterns would have been significantly confined to modeling only linear correlations. There are several sorts of activation functions, and choosing the right one can have a big effect on how well the network learns from and interprets the data. Activation functions that are frequently utilized include

- Rectified Linear Unit (ReLU): One of the most well-liked activation mechanisms is ReLU. It produces zero if the input is negative and the input if it is positive. ReLU is a popular option in deep learning since it is computationally effective and helps with gradient disappearing problems.
- Sigmoid: Compressing input values into a spectra of 0–1 is accomplished via the sigmoid activation function, which is especially helpful in situations involving binary classification. But in deep networks, it may run into gradient vanishing issues [34].
- Hyperbolic Tangent (tanh): Tanh, like the sigmoid, reduces some of the gradient vanishing problems related to the sigmoid function by mapping input values to a range across -1 and 1.

The type of data being used and the particular issue being tackled determine which activation function is best. Finding the best activation function for a particular neural network architecture usually requires testing. Activation functions are vital to the performance of contemporary DL algorithms because they shape the network's capacity to represent complex data relationships.

3.2.1.2 Layers and Architectures

The framework of a neural network consists of the layers that are created when neurons are arranged in a particular way. The input layer and the output layer are the three primary types of layers that make up the feedforward neural network, which is the most basic type of neural network. Every layer has a distinct function in information processing and adds to the network's capacity for forecasting and data-driven learning. The neural network first accepts outside inputs at the input layer

[35–37]. Anything from text to pictures to numerical data may be included in this. All that the input layer does is transfer this data to the network so that it can be processed subsequently. The number of neurons in the input layer is determined by the dimensionality of the input data. The hidden layers serve as the computational engines of the neural network, responsible for analyzing and transforming the input data. Within these layers, complex mathematical operations such as weighted accumulation and activation function implementation process the input layer's data. The configuration of the hidden layers, including the number of layers and the quantity of neurons within each layer, can be adjusted based on the complexity of the problem and the availability of data. Designing effective neural network architectures entails customizing both the depth (number of layers) and width (number of neurons per layer) of the network to suit the specific task at hand.

Ultimately, the neural network generates its projections or outputs at the output layer. The kind of forecast or task that the network is intended for determines how many neurons are present in the output layer. For example, in binary classification, a single-output neuron that produces a probability score ranging from 0 to 1 might exist [38, 39]. There could be several output neurons in a multiclass classification, each of which would indicate the likelihood of falling into a certain class. The output layer in regression tasks could generate numerical results immediately. The way these layers are organized and created, known as network architecture, has a big impact on how well the network can recognize intricate patterns and forecast the future. Neural network topologies have been developed to support various jobs and data kinds. For instance, because convolutional neural networks (CNNs) can recognize spatial patterns, they are ideally suited for handling image data. Natural language or time-series data are examples of sequential data that are well suited for recurrent neural networks (RNNs). More sophisticated designs, such as the Transformer, have completely changed tasks involving natural language processing and demonstrated the flexibility and versatility of neural network topologies in resolving a broad array of real-world matters.

3.2.2 Backpropagation

Neural networks use a basic process called learning to enable them to identify complex patterns in data and make precise forecasts. The process of backpropagation is one of the main ways that this learning takes place. Backpropagation is an iterative refinement process that reduces loss or error—the inconsistency among the real recorded results and the framed outcomes produced by the neural network. It is a way to measure how much the model's forecasts deviate from the actual or intended results. The backpropagation works by modifying the network's individual neurons' weights and biases [40–42]. The output that each neuron produces is significantly influenced by these weights and biases, which also establish the intensity of connections between neurons. Reducing the inaccuracies in the model's projections by continuously fine-tuning these parameters is the aim of backpropagation. The computation of gradient with regard to the forfeiture function is the fundamental procedure of backpropagation. The rate of increase of the error with respect to the distinct weights and biases is represented by these gradients. Following their computation, these gradients are methodically propagated backward through the network, from the output layer to the concealed layers and, finally, back to the input layer, via backpropagation. The algorithm can determine which weights and biases ought to be adjusted because they are mostly causing prediction mistakes thanks to this backward propagation.

These gradients are then used via backpropagation to modify the network's parameters. By making this change, the weights and biases are positioned to reduce error. Through numerous iterations of this process, the model eventually coheres to a point where the error is much decreased and the neural network can produce more precise forecasts [43, 44]. A key component of deep learning model training is backpropagation. DL models possess the capability to adapt to complex, high-dimensional data structures, rendering them valuable across a spectrum of applications, including natural language understanding and image recognition. The effectiveness of DL and artificial intelligence largely stems from their ability to learn from data and refine predictions through iterative processes like backpropagation.

3.2.3 Deep Learning Frameworks and Libraries

Developing, testing, and distributing deep learning models require infrastructure, which is provided by deep learning web frameworks. They provide pre-built functions and algorithms, which cut down on the amount of effort and intricacy needed to construct a model. Numerous important deep learning frameworks and tools have become more well-known:

- TensorFlow: It is an open-source DL framework initially developed by Google and has become widely adopted in various fields. It provides an adaptable framework for developing and honing different kinds of DL models.
- PyTorch: A more open-source deep learning framework is PyTorch, which was created by Facebook's AI Research group. It is well-known for having a dynamic computation graph, which greatly increases its flexibility for study and model testing [45, 46].
- Keras: For the purpose of creating and retraining neural networks, Keras is an open-source DL library that offers a high-level, user-friendly API. It operates on top of Theano and TensorFlow, two more deep learning frameworks.
- Caffe: Berkeley Vision and Learning Center created the deep learning framework Caffe. It is renowned for being quick and effective, and it works especially well for computer vision applications.
- MXNet: It is an open-source DL framework formed by Apache with an emphasis on usability, stability, and performance. It is widely used in industrial and scientific applications.

3.3 Models and Techniques

To evaluate and understand complicated medical data, the discipline of prediction and diagnosis in healthcare uses a range of deep learning models and approaches. The main models and approaches that are frequently employed for illness prediction will be covered in detail in this part. These include feedforward neural networks, CNNs, RNNs, transfer learning, and evolutionary algorithms [47, 48].

3.3.1 Feedforward Neural Networks

Multilayer perceptrons (MLPs), another name for feedforward neural networks, are the basic models used in deep learning. An input layer, one or more hidden layers, and an output layer make up these networks. Each neuron in a single layer is linked to every other layer's neuron. The following are the feedforward neural networks' main characteristics:

- Universal Function Approximators: Because feedforward neural networks can estimate any functional form, they are particularly versatile for challenges involving the prediction of disease. This property makes them ubiquitous function approximators [49, 50].
- Nonlinearity: The model gains nonlinearity from the activation functions employed in hidden layers, which allows it to grasp intricate correlations in the data.
- Structured Learning: Because correlations between different health measures could be complex, these networks are useful for discovering organized patterns and linkages in data.

In order to forecast disease, feedforward neural networks are used for tasks, including risk analysis, processing of medical images, and clinical prognosis.

3.3.2 Convolutional Neural Networks (CNNs)

CNNs are specially made for applications that require grid-like data, like photographs. They have completely changed how healthcare uses imaging techniques. Important characteristics of CNNs are

- Convolution Layers: Convolutional layers are used by CNNs to automatically recognize and learn from features in photos. These layers are essential for identifying patterns and trends in medical images, which allows for accurate illness identification [51, 52].
- Spatial Hierarchies: CNNs are able to identify complicated structures at various scales by creating spatial hierarchies inside images.

- Transfer Learning: Fine-tuning pretrained CNN models, including those that rely on ImageNet, has the potential to significantly reduce the quantity of associated medical data needed for medical image analysis.

CNNs are frequently employed in the medical field to do tasks, including classifying skin lesions for the forecast of dermatological diseases, detecting tumors in medical pictures, and spotting anomalies in radiological scans.

3.3.3 Recurrent Neural Networks (RNNs)

Sequential data processing is a good fit for recurrent neural networks (RNNs). Time-series data, electronic health, and natural language processing jobs in the healthcare industry have all found use for them. Key characteristics of RNNs include

- Sequential Processing: Because of its memory system, RNNs can handle sequential data while retaining historical knowledge. Tasks requiring time-series patient data and clinical notes benefit from this capacity.
- Long Short-Term Memory (LSTM) and Gated Recurrent Unit (GRU): RNN variants that deal with vanishing gradients and long-range dependencies, including LSTMs and GRUs, improve the efficiency of RNNs at identifying intricate patterns in sequential data.

RNNs are utilized for illness prognosis in tasks like medication adherence evaluation, expression evaluation of clinical notes, and client prognosis.

3.3.4 Transfer Learning

Using pretrained models to improve a model's efficiency on a particular illness prediction challenge is known as transfer learning. Important aspects of transfer learning consist of

- Model Reusability: Pretrained models could be adjusted for particular healthcare purposes. These models are frequently developed on huge datasets for generic activities. This method lessens the requirement for comprehensively labeled medical data, which could be hard to come by and costly to obtain [53, 54].
- Domain Adaptation: It is the procedure of transferring a model's subject expertise to another. Models learned on generic images, for instance, could be adjusted for medical image processing.
- In disease prognosis, transfer learning is especially helpful when there is a lack of labeled healthcare data, and it is necessary to draw on expertise from related fields.

3.3.5 Ensemble Methods

To increase general forecast robustness and accuracy, ensemble approaches aggregate forecasts from several models. Important features of group techniques include

- Diversity: By integrating their forecasts, ensemble techniques take advantage of the variability of different models. This diversity frequently results in improved generalization and accuracy rates.
- Bagging and Boosting: Using a different set of data, bagging techniques like random forests build various models and afterward aggregate their forecasts. Boosting techniques, like AdaBoost, train models one after the other by concentrating on evidence that earlier models incorrectly identified.
- When predicting diseases, ensemble approaches are employed to decrease overfitting, increase model efficiency, and make forecasts more robust—especially when dealing with complicated healthcare datasets.

3.4 Evaluation and Performance Metrics

It is critical to evaluate illness predictive model in the medical industry in order to determine their efficacy and reliability. This section explores the approaches and measurements utilized to assess these models' effectiveness. These include the utilization of testing and training sets, measurements for assessing models such as accuracy, recollect, and F1-score; ROC curves and the area under the curve (AUC); cross-validation; and regularization techniques to mitigate overfitting [55, 56].

3.4.1 Training and Validation Sets

Separating the data into training and validation is a basic step in the model assessment process. This is necessary to access the generalization ability of a model. Following this procedure, important points are

- Training Set: The model is proficient in utilizing the training set. By exposing the model to previous data, it can pick up trends and connections.
- Validation Set: Unlike the training set, the validation set is a different subset of the data. It is employed to assess how well the model performed throughout training. The validation set aids in tracking the model's ability to generalize to fresh, untested data.
- Holdout Method: Using a randomization technique, the dataset is sectioned into strength and conditioning and validation sets. 80–20 and 70–30 splits are popular, although it is a split ratio.

3.4.2 Model Evaluation Metrics

The efficacy of illness forecast models is measured utilizing a range of measures during evaluation. These measures shed light on the reliability, accuracy, memory, and other aspects of the model's forecasting capabilities.

3.4.2.1 Accuracy, Precision, Recall, and F1 Score

- Accuracy: A simple metric called accuracy counts the percentage of cases that are properly categorized out of all the occurrences. It may not work well with imbalanced data when one class predominates over the other, even though it offers a straightforward indicator of the general effectiveness of the model.
- Precision: The ration of precise anticipated positive occurrences to all positively anticipated instances is measured as accuracy [57, 58]. When false positives might have serious repercussions, like in illness forecasting where expensive misdiagnoses occur, it is crucial.
- Sensitivity (Recall): The proportion of all immediate valid occurrences to all accurately projected positive occurrences is known as recall. It is especially crucial in situations where omitting a positive instance is crucial, like in the prognosis of sickness.
- F1 score: The F1 score, a metric used in binary classification tasks, represents the harmonic average of precision and recall. It provides a balanced measure that considers both precision (the proportion of true positive predictions among all positive predictions) and recall (the proportion of true positive predictions among all actual positives). This harmonic average encapsulates the trade-off between precision and recall, offering a single numerical value that reflects the model's performance in capturing both true positives and minimizing false positives and false negatives. It is particularly useful in scenarios where there is an imbalance between the classes or when both precision and recall are equally important. It is frequently employed in predictive modeling, where it is crucial to strike a balance between reducing false positives and false negatives.

3.4.2.2 ROC Curves and AUC

- Receiver Operating Characteristic (ROC) curves: These diagrams show how well a model performs over a range of classification thresholds. As the threshold varies, it displays the real positive rate (recall) versus the false-positive rate. In general terms, a model with a larger ROC curve performed best.
- Area Under the Curve (AUC): By figuring out the region underneath the ROC slope, the AUC measures a model's overall efficiency. AUC values between 0.5 and 1.0 denote randomized behavior and perfect classification, respectively. It

offers a solitary metric to evaluate a model's capacity to differentiate between good and bad occurrences.

3.4.3 *Cross-Validation*

A crucial method for more reliably evaluating a model's generalization ability is cross-validation. The dataset is divided into several subsets or folds, and the model is continuously trained and assessed on various permutations of these folds [59, 60]. Key aspects of cross-validation:

- K-Fold Cross-Validation: This technique divides the data into K sections of equal size (folds). K-1 folds are used for learning and the leftover fold is used for verification during each of the K training cycles of the model. During the K-fold cross-validation procedure, each fold takes turns acting as the validation set exactly once within the K iterations. This means that the entire dataset is partitioned into K equally sized subsets, and in each iteration, one of these subsets is designated as the validation set while the remaining K-1 subsets are used for training the model. This process ensures that every data point is included in the validation set exactly once throughout the K iterations, allowing for a comprehensive evaluation of the model's performance across different subsets of the data. K-fold cross-validation helps to mitigate potential biases and variability in model performance that may arise from using a single fixed validation set.
- Leave-One-Out Cross-Validation (LOOCV): This variation uses the remaining data for training and each data point as the validation data once. Although this method requires a lot of calculation, it offers a reliable assessment. Cross-validation aids in reducing problems with dataset bias and variation and yields more accurate efficiency predictions.

3.4.4 *Overfitting and Regularization*

In machine learning, overfitting occurs when a model fits the training set of data too accurately, which leads to poor generalization of new information. Approaches for regularization aid in addressing overfitting. The following are some important regularization points:

- Ls. 1 and 2 Regularization: By introducing penalty words to the model's loss function, L1 and L2 regularization help prevent the model from giving undue weight to particular features. They lessen the complexity of the model, which helps to avoid overfitting.

- Early Termination: One way to keep an eye on a model's progress during learning on the validation set is to use premature pausing. To avoid overfitting, training is halted early when efficiency on the validation set begins to decline.

Dropout: A random subset of the model's neurons is "dropped out" after each training iteration as part of the regularization process known as dropout. This makes the model less dependent on any one neuron and promotes it to be more resilient overall.

An important part of disease prediction is assessing a model's effectiveness and correcting overfitting. Effective evaluation criteria, cross-validation, and regularization strategies guarantee that the models function effectively on both fresh, untested data and training data. This is crucial for precise illness prognostications and efficient patient treatment in the medical domain.

3.5 Deep Learning in Disease Prediction

Deep learning has had a profound impact on the field of disease prediction in healthcare, offering more accurate and data-driven approaches to diagnosis, prognosis, and risk assessment [61–63]. In this section, we elaborate on how DL is applied in various disease prediction domains, including cancer diagnosis, heart disease risk assessment, neurodegenerative disorders, and other disease prediction applications.

3.5.1 Cancer Diagnosis

When it comes to identifying different types of cancer, deep learning is leading the way in a revolutionary shift in the field of diagnostics, offering significant improvements in accuracy, economy, and timeliness. This technological advancement has the potential to completely change how medical professionals identify and treat cancer. In this regard, let us examine in more depth how deep learning is helping with cancer diagnosis across a number of different types.

3.5.1.1 Breast Cancer Diagnosis

Mammography and breast tissue image analysis have shown great success with deep learning models, particularly CNNs. These models are capable of independently identifying minute abnormalities in the images, such as irregular lumps or microcalcifications, which could be signs of breast cancer. Even more astounding is deep learning's capacity to further categorize these abnormalities into distinct cancer kinds and distinguish between benign and malignant ones. The use of deep learning models, especially convolutional neural networks, has significantly changed the way that breast cancer is diagnosed (CNNs). The precision and effectiveness of identifying possible cancers in mammography and breast tissue pictures are changing as a consequence of these models. The influence of DL on breast cancer diagnosis is significant and can be attributed to multiple factors. One such factor is the ability of deep learning models, particularly CNNs, to detect microcalcifications in mammography pictures autonomously. These tiny calcium deposits, or microcalcifications, can act as early warning signs of breast cancer. They can be difficult to find using conventional techniques due to their small size and complicated nature. These tiny abnormalities are exceptionally well-identified by deep learning algorithms, allowing for early identification and action. Images of breast tissue can be used to identify irregular masses with great accuracy by deep learning algorithms. These lumps can indicate cancers, benign or malignant, and early detection is essential for prompt and efficient treatment. The ability of deep learning to identify these abnormalities exhibits a vital role in enhancing patient findings by aiding in the early identification of possible cases of breast cancer. The capacity of DL algorithms to recognize abnormalities and classify them into distinct cancer kinds is what makes them unique. These models are able to distinguish between benign and malignant tumors, which provides important information for developing accurate and customized treatment plans. Medical specialists are able to identify the type of anomaly and choose the best plan of action, which may include vigorous treatment or close observation. The DL usage in the diagnosis of breast cancer greatly improves diagnostic precision. Large datasets of pictures of breast tissue from mammograms are used to train models like CNNs. They can often identify complex patterns and relationships in the photos that the human eye cannot due to their vast training. Their diagnostic performance can therefore equal or even surpass that of skilled medical professionals, guaranteeing reliable and consistent outcomes. In situations of breast cancer, early identification plays a crucial role in enhancing patient outcomes. Patients can undergo therapy when it is most likely to be effective because deep learning models are excellent at spotting possible cancers early on. By reducing the severity of the illness and increasing the likelihood of a favorable outcome, this prompt action ultimately saves lives. In an experiment, Ragab et al. [64] discussed the clinical decision support systems (CDSS) that offer an effective means to detect diseases like breast cancer utilizing ultrasound images (USIs). Tumors in USIs are sometimes identified and categorized utilizing CAD models. CADs are painstakingly made to make suggestions that help radiologists diagnose breast cancers and provide insightful information about the prognosis of the condition. The caliber of the photographs and the radiologist's expertise are both crucial for the arrangement procedure's accuracy. The classification of breast cancer has been shown to be a highly efficient use of DL models. In this study, a novel ensemble deep learning-enabled clinical decision support system for breast cancer diagnosis and classification (EDLCDS-BCDC) was developed using ultrasound images (USIs). The primary objective of this multistep approach, EDLCDS-BCDC, is to accurately detect the presence of breast cancer. The methodology involves several preprocessing steps applied to the USIs. Initially, Wiener filtering and contrast enhancement techniques are employed to improve the quality of the images.

Subsequently, a refined image segmentation technique is implemented using Kapur's entropy (KE) coupled with the Chaotic Krill Herd Algorithm (CKHA). Furthermore, for feature extraction, three deep learning models—namely, VGG-16, VGG-19, and SqueezeNet—are utilized. These models are proficient in extracting meaningful features from the USIs, which are crucial for accurate classification. In the final classification phase, a multilayer perceptron (MLP) model is employed in conjunction with cat swarm optimization (CSO). This combination aids in effectively identifying whether breast cancer is present or absent based on the extracted features. Using benchmark databases, a thorough series of simulations was carried out, and the extensive findings show that the suggested EDLCDS-BCDC technique performs better than current approaches. Naranjo et al. [65] studied consecutive patients from the Memorial Sloan Kettering Cancer Center (January 2018–March 2020) and the Medical University Vienna (January 2011–August 2014) who presented with suspicious enhancing breast tumors on breast MRI classified as BI-RADS 4 and subsequently underwent image-guided biopsy. The 93 patients in the study, all of whom were female, had a mean age of 49 ± 12 years. There were 104 lesions in these patients, ranging in size from 7 to 99 mm and with an average size of 22.8 mm; 46 of the lesions were malignant and the remaining 58 were benign. These lesions' radiomic features were calculated, and a robust machine learning model that could distinguish between benign and malignant lesions was created using the five most important features. Two distinct models were developed based on radiomic features extracted from dynamic contrast-enhanced imaging (DCEI) and diffusion-weighted imaging (DWI) modalities. The DCEI-based model achieved an area under the curve (AUC) of 0.83 (95% confidence interval: 0.75–0.91), indicating strong discriminatory performance in distinguishing between classes. Conversely, the DWI-based model exhibited a slightly lower AUC of 0.79 (95% confidence interval: 0.70–0.88), suggesting a slightly less robust discriminatory capability compared to the DCEI-based model. These findings underscore the potential of utilizing radiomic features derived from both DCEI and DWI modalities to enhance the accuracy and effectiveness of diagnostic models for various medical imaging applications. Combining these modalities in a multiparametric radionics model produced the most promising findings, with the best diagnostic accuracy of 81.7% (95% CI: 73.0–88.6) and AUC of 0.85 (95% CI: 0.77–0.92). To sum up, our study showed that using machine learning and radiomics analysis to multiparametric MRI greatly improves the assessment of suspicious breast tumors, aids the precise recognition of breast cancer, and minimizes the necessity for needless benign breast biopsies.

3.5.1.2 Lung Cancer Diagnosis

One of the world's primary contributors to cancer-related deaths, lung cancer is a powerful foe. Improving patient results hinges on early diagnosis, and deep learning has become an essential component of this effort. Its function in the diagnosis of lung cancer consists of carefully examining lung scans, especially computed tomography (CT) pictures. These photos may be examined closely by deep learning models, which are particularly good at spotting lung nodules, that may be early signs of lung cancer. These nodules in the complex lung architecture are frequently inconspicuous and difficult to find. Nevertheless, deep learning models—in particular, CNNs—perform exceptionally well at identifying these minute irregularities, making early diagnosis possible—a critical step toward enhancing treatment outcomes. Deep learning models are not only capable of detecting nodules but also of determining whether or not these nodules are malignant, which can provide information on the probable severity of the cancer. When creating treatment regimens that are customized to each patient's condition, healthcare experts find this classification to be quite helpful. Furthermore, tumor segmentation within CT images is another area in which deep learning excels. CNNs perform remarkably well in lung tumor segmentation, providing information about the exact location and limits of the tumors. Radiation therapy and surgical planning both benefit from this. Deep learning models can also forecast these tumors' future growth, which is a useful feature that enables medical professionals to decide on the best course of action for their patients promptly and intelligently. Deep learning is revolutionizing the field of lung cancer diagnosis overall by focusing on tumor segmentation, growth prediction, early detection, and malignancy assessment. In the continuous fight against lung cancer, these developments help to improve patient outcomes by facilitating quicker and more precise diagnoses. By addressing a major global health concern and revolutionizing healthcare, deep learning is being incorporated into the diagnosis of lung cancer, providing patients with this deadly disease with hope for improved outcomes and survival rates. In an experiment, Riquelme et al. [66] presented a review article where they studied the detection of malignant lung nodules from computed tomography (CT) scans. CAD systems have been presented as a way to reduce this strain. DL techniques have shown impressive outcomes recently, outperforming conventional methods across a range of disciplines. Researchers are currently investigating several DL methods to improve the efficacy of CAD arrangements in computed tomography-based lung tumor detection. The most recent cutting-edge DL algorithms created for CAD aimed at lung tumor diagnosis are thoroughly reviewed in this chapter. There are two categories for these systems: (1) systems for nodule detection identify potential nodules during the initial CT scan. (2) Systems for reducing false positives classify prospective nodules as benign or malignant tumors. The report does a comprehensive examination of the performance of the different approaches and explains their salient aspects. It also provides an overview of the CT lung datasets that are accessible for research and compares and contrasts the various methods. This work contributes to the ongoing efforts to enhance early diagnosis and patient outcomes by providing insightful information about the use of DL in the classification and detection of lung tumor nodules in CT scans. Shafi et al. [67] introduced a cancer diagnostic model that leverages deep learning in conjunction with a support vector machine (SVM). In cross-sectional lung cancer lesions, the suggested computer-aided design (CAD) model is intended to detect pathological and physiological alterations in soft tissues. First, by detecting and analyzing specific profile values in CT scans obtained from patients with confirmed lung cancer as well as control patients, the model is trained to identify lung cancer. CT scans from patients and control participants who were not involved in the training phase are then used to test and validate the model. Three CT scans are shown in Fig. 3.2: the top two show early-stage cancerous lung disease; the top one is malignant, while the bottom one is benign. It is difficult to distinguish cancer nodules from blood arteries at this stage. The last picture shows advanced-stage cancer, which is distinguished by big nodules that are easier to find but also have a decreased chance of survival. This study examined 888 CT scans with annotations that were taken from the LIDC/IDRI database, which is accessible to the general public. When it came to identifying pulmonary nodules linked to early-stage lung cancer, the DL-assisted SVM-based model had a 94% precision rate. When applied to lung CT images for nodule diagnosis, it performs better than the current approaches, which include hybrid techniques, simple machine learning, and complicated DL. The empirical outcomes highlight the potential of the suggested method to greatly assist radiologists in the early diagnosis of lung cancer, allowing for timely patient management.

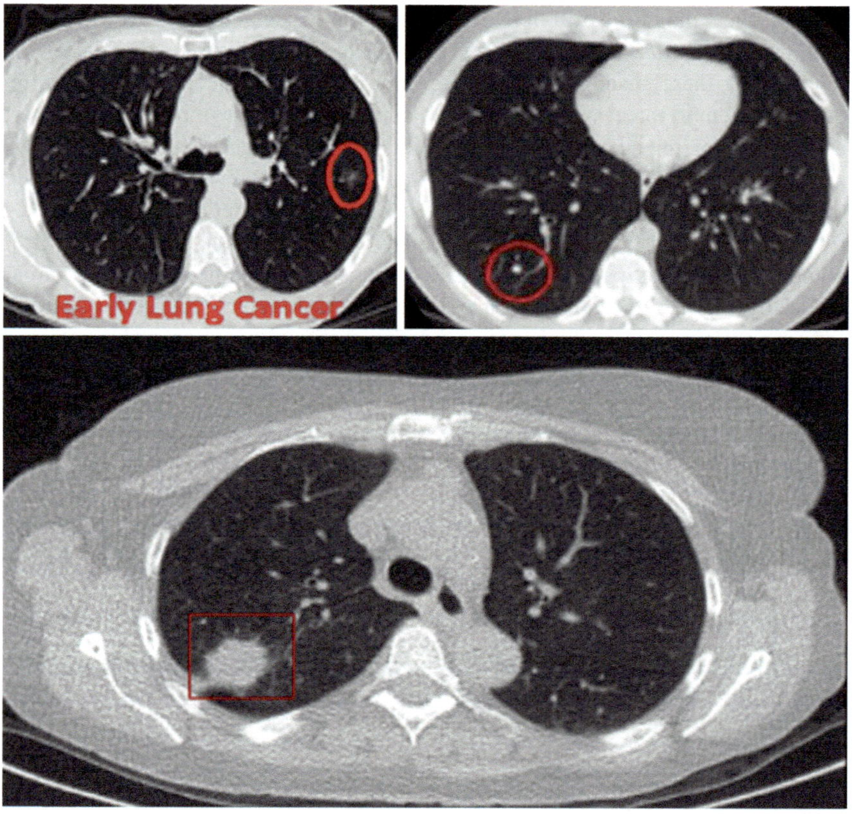

Fig. 3.2 CT scans created using the maximum intensity projection technique. (Adapted from Ref. [67] under CCBY 4.0)

3.5.1.3 Skin Cancer Diagnosis

With the usage of DL, particularly the use of CNNs, skin cancer diagnosis is experiencing a revolutionary makeover. These models have been used by dermatology, a discipline that has historically relied on clinical judgment, in order to aid in the early diagnosis of skin cancer. DL models are particularly good at identifying minute irregularities that can be signs of skin cancer. They carefully analyze photos of skin lesions, moles, and other dermatological traits. Early cancer detection is crucial because it allows for prompt intervention, which greatly improves patient outcomes and may even save lives in certain situations. Additionally, deep learning algorithms automatically classify skin lesions as benign or malignant, giving doctors a crucial decision-support tool. Healthcare practitioners can make well-informed judgments regarding patient management with the help of this self-classifying system, which provides insights into the origin of skin lesions. In difficult and sometimes fatal instances, dermatologists find these simulations to be extremely helpful tools for prompt and precise diagnosis. The reliability of skin cancer diagnosis is greatly improved by the use of DL. These models could identify complex patterns and correlations in the photos since they have been trained on large databases of dermatological images. As such, their diagnostic abilities can be on par with or even better than those of board-certified dermatologists. This guarantees accurate and consistent outcomes in various healthcare settings. In conclusion, skin cancer diagnostics and dermatology are changing as a result of deep learning, especially with CNNs. These models improve the precision and promptness of skin cancer diagnosis by facilitating early identification, automating lesion classification, and providing decision support for medical professionals. This not only helps patients receive better care but also solves the urgent need for a prompt and precise diagnosis in skin cancer cases, especially in difficult and sometimes fatal situations. The purpose of DL in the diagnosis of skin tumor demonstrates the technology's potential to transform healthcare and enhance the prognosis for those with the disease. In an experiment, Jinaai et al. [68] curated a dataset of 5725 clinical visuals featuring pigmented skin lesions from 3543 patients. These lesions comprised benign tumors and malignant tumors. They chose 666 patients at random and used those results to construct a test dataset with one photograph for each patient. To create the training dataset, bounding boxes were added to the other images (4732 images from 2885 patients). They then utilized the training dataset to train a faster region-based CNN (FRCNN) and evaluated its performance using the test dataset. Ten dermatologic trainees (including BCDs) and ten board-certified dermatologists (BCDs) were given identical tests, and they compared the diagnostic accuracy of each group using FRCNN. FRCNN obtained 86.2% reliability for a six-class classification, whereas BCDs and TRNs obtained 79.5%. FRCNN showed 91.5% accuracy, 83.3% sensitivity, and 94.5% specificity for two-class categorization (benign or malignant), comparable to 87.4, 84.7, and 85.9% for BCDs and 84.2, 81.9, and 88.9% for TRNs, correspondingly. For FRCNN, the false positive rates and positive predictive values were 5.5 and 84.7%, respectively; for BCDs, the corresponding figures were 13.4 and 70.5%, and for TRNs, 14.1 and 67.6%. In comparison to the classification performance of 20 dermatologists, FRCNN exhibited superior performance. Their primary goal is to enhance skin cancer prognosis, and therefore, they aim to integrate this system into society and make it accessible to the general public. Tahir et al. [69] introduced a DL-based skin cancer CNN, known as DSCC_Net, which relies on a CNN. ISIC 2020, HAM10000, and DermIS are 3 well-known benchmark data on which the model has been applied and evaluated. With an area under the curve (AUC) of 99.43%, an accuracy of 93.75%, a recall rate of 95.87%, a precision of 94.57%, and an F1 value of 91.35%, DSCC Net produced remarkable results for classifying the four different sorts of skin tumor disorders. In contrast, precision of 89.68, 92.51, 91.46, 89.12, 89.46, and 91.82% was obtained from other well-known models, including Vgg-19, ResNet-152, Vgg-16, MobileNet, Inception-V3, and EfficientNet-B0. These results highlight the enhanced efficacy of the recommended DSCC Net model over baseline methods, thereby offering dermatologists and other medical professionals invaluable assistance in the precise detection of skin cancer.

3.5.1.4 Colorectal Cancer Diagnosis

Deep learning models are used in the field of colorectal cancer detection to assess colonoscopy pictures. These models are very good at identifying precancerous lesions like adenomas and polyps. Given that these lesions are precursors to colon cancer and that early treatment could save lives, it is critical to identify these lesions as soon as possible. Deep learning offers a revolutionary effect on cancer diagnosis because of its unmatched ability to evaluate medical imagery and identify complex patterns that might be invisible to human observers. Deep learning increases diagnosis reliability, economy, and uniformity in a variety of healthcare situations by automation the detection method. This technology has the ability to reduce the workload for medical practitioners, which is particularly beneficial in areas where accessibility to specialist knowledge is scarce. Deep learning has several benefits, including early diagnosis, personalized treatment strategies, and improved patient results and survival percentages. Deep learning offers the efficacy to progress healthcare by making cancer diagnosis more accurate, affordable, and fast, thereby advancing the general goal of improving public health. In an experiment, Attallah et al. [70] presented a framework utilizing several lightweight deep learning models for the early detection of lung and colon cancers. In order to decrease feature dimensionality and offer a more useful data presentation, the framework integrates a variety of transformation techniques. ShuffleNet, MobileNet, and SqueezeNet models were used in this context to process histopathological scans. Fast Walsh-Hadamard transform (FHWT) and principal component analysis (PCA) methods are used to further minimize the deep features that are derived from these models. Furthermore, the discrete wavelet transform (DWT) is employed to integrate the attributes derived from the FHWT reduction, and the PCA features from the 3 DL models are blended. Four several machine learning algorithms are then fed the reduced features obtained from PCA, FHWT-DWT reduction, and feature fusion. This results in an amazing accuracy rate of 99.6%. The outcomes of the suggested framework, which is built on lightweight deep learning models, show that it can differentiate between variations of colon and lung cancer with smaller attributes set and less computational complexity than other approaches. This method also emphasizes how important it is to use feature reduction approaches to improve data interpretation, which will ultimately increase the precision of cancer detection. In Nakajima et al. [71], liquid chromatography spectrometry was utilized to analyze seven polyamines, including spermine and spermidine and their acetylated forms. This study examined urine specimens from 201 patients with colorectal cancer (CRC) and 31 non-CRC patients. Among the polyamines, N1, N12-diacetylspermine exhibited the highest area under the receiver operating characteristic curve (AUC) at 0.794 (95% confidence interval (CI): 0.704–0.885, $p < 0.0001$) for discriminating between CRC and healthy or benign controls. To assess the stability of measured quantities, 59 samples from healthy controls were evaluated three times over 3 days, confirming the consistency of observed quantitative values. Utilizing combinations of polyamines in a machine learning approach resulted in a higher AUC value of 0.961 (95% CI: 0.937–0.984, $p < 0.0001$). Computational validations confirmed the models' ability to generalize.

3.5.2 *Heart Disease Risk Assessment*

Deep learning algorithms are transforming the way that heart disease risk is assessed and providing tailored suggestions that can significantly affect a person's health [72, 73]. These models use their ability to collect and analyze various forms of healthcare data to produce precise risk levels and suggestions for preventing cardiac disease. Here is a closer look at the role deep learning plays in this situation:

3.5.2.1 Electrocardiograms (ECGs)

Electrocardiograms (ECGs) are analyzed using deep learning models, which are essential for keeping an eye on cardiovascular health. ECGs monitor the electrical cardiac activities and therefore may indicate aberrant cardiac beats and other relevant concerns. Deep learning models are very good at detecting alterations in the ST-segment of the ECG and diagnosing abnormal heart rhythms (arrhythmias), particularly when they use CNNs or RNNs. Early detection of these abnormalities is essential for averting cardiac events and guaranteeing prompt intervention.

3.5.2.2 Risk Prediction Models

The creation of thorough risk prediction models for cardiovascular disease also makes use of deep learning. These models consider a multitude of factors that influence an individual's risk, such as genetic information, lifestyle decisions, and medical history. DL algorithms could possess this abundance of data and produce extremely customized risk evaluations [74–85]. The generated risk assessments enable the creation of individualized guidelines for the prevention of cardiovascular disease by assisting consumers and healthcare professionals in understanding their unique risk factors [86, 87]. These suggestions may involve medication, dietary changes, or other preventative actions.

3.5.2.3 Early Detection

Deep learning algorithms can analyze cardiac imaging data, including angiograms and echocardiography, to identify early indicators of heart illness. These models may locate problem areas, assess heart function, and detect structural anomalies. The ability of healthcare professionals to act quickly and possibly stop or slow the evolution of heart disease makes early identification essential.

The capability of these deep learning programs to collect and interpret the large and diverse amount of healthcare data that is currently accessible makes them significant for assessing the risk of heart disease. This technology-driven method allows for earlier identification and more focused preventative measures in addition to improving risk assessment reliability [88, 89]. Additionally, it gives people the ability to take charge of their heart health by giving them tailored advice on how to live a safer lifestyle and make better medical decisions.

3.5.2.4 Neurodegenerative Disorders

With deep learning at the vanguard, neurodegenerative illnesses may now be detected sooner and more accurately, revolutionizing the observation and diagnosis of these crippling conditions. Deep learning models are demonstrated to be beneficial in the following manner by interpreting and analyzing a wide variety of data sources, comprising medical pictures, patient motion and gait characteristics, and even language and voice trends.

3.5.2.5 Alzheimer's Disease

Numerous brain imaging data, such as those from magnetic resonance imaging (MRI) and positron emission tomography (PET), are analyzed using deep learning models. These models are incredibly good at identifying anatomical alterations in the brain linked to Alzheimer's disease. They are able to identify minute changes in brain morphology that are indicative of Alzheimer's disease, including the existence of neurofibrillary and amyloid plaques. Because of this skill, medical personnel can diagnose patients earlier and start therapy and monitoring earlier. Additionally, deep learning is essential for monitoring the course of Alzheimer's disease, offering insightful information about how the condition is changing and supporting the evaluation of treatment efficacy [90, 91].

3.5.2.6 Parkinson's Disease

In order to evaluate how Parkinson's disease is progressing, deep learning is essential. These models assess movement, gait, convulsions, and other motor symptoms in patients by examining data from wearables and sensors. The planning and modification of treatment plans are aided by this study. With the use of deep learning, patients' motor performance may be continuously monitored. This allows medical professionals to make well-informed judgments on medication and therapy regimens by using objective metrics. Early detection of abnormalities in motor function could also help with timely treatments and modifications to enhance the quality of life for the patient.

3.5.2.7 Amyotrophic Lateral Sclerosis (ALS)

A decrease in muscle control results from the progressive damage that ALS causes to motor neurons. Deep learning is used to analyze voice and speech characteristics to track the development of ALS. These models are able to monitor minute alterations in the patient's voice and speech patterns, which are suggestive of the course of the illness. Keeping an eye on speech and voice patterns makes it possible to evaluate the disease's course more frequently and can reveal information about how well interventions or therapies are working [92–94]. This technique offers a constant, noninvasive way to evaluate the patient's status.

Deep learning's revolutionary influence in the field of neurodegenerative illnesses stems from its capacity to handle and evaluate a wide range of data kinds, making early management and tracking possible. For tasks like monitoring voice changes, evaluating the motor function, or analyzing complex brain imaging data, deep learning models offer a level of precision, speed, and consistency that was previously unachievable. Deep learning helps to better patient outcomes, more effective treatment planning, and an elevated quality of life for those with neurodegenerative disorders by identifying these conditions in their early stages.

3.5.3 *Other Disease Prediction Applications*

Deep learning is influencing a wide range of disease prediction technologies, from common health issues to uncommon and complicated diseases, thanks to its amazing versatility and data-driven methodology. Deep learning's adaptability has completely changed the healthcare sector by making it possible to anticipate diseases more precisely and promptly, which has enhanced health results in the end. Here is a closer look at a few of these uses.

3.5.3.1 Diabetes Prediction

Diabetes risk prediction relies heavily on deep learning algorithms. These algorithms produce highly individualized risk evaluations by evaluating patient data such as blood glucose, health information, lifestyle habits, and genetic susceptibility [95–101]. The forecasts play a crucial role in early intervention as well as providing tailored advice for illness prevention and efficient treatment. Deep learning gives people the ability to make educated decisions about their health by giving them personalized recommendations [102].

3.5.3.2 Infectious Disease Outbreak Prediction

Deep learning has become a powerful tool in epidemiology to forecast and track infectious disease epidemics. To find possible epidemics, these algorithms examine trends in traveler information, health data, environmental variables, and even social media data. Timely identification of infectious diseases may facilitate prompt public health measures, such as immunization programs and travel restrictions, which can eventually aid in the prevention of disease transmission and the preservation of human life.

3.5.3.3 Mental Health Assessment

Models of DL have been applied to the evaluation of mental health issues. To determine a person's mental health condition, they could analyze text data from a variety of sources, such as posts on social media, patient records, and patient surveys. This technology-driven method makes it possible to detect mental health problems earlier, which opens the door to prompt help and treatment. Deep learning-powered mental health evaluations could have a significant influence on both individual well-being and psychiatric care.

3.5.3.4 Rare Diseases

Through the analysis of genomic data, deep learning also aids in the detection of uncommon and hereditary disorders. This could identify morphological variances, sequence anomalies, and gene abnormalities, giving important information about the underlying causes of these uncommon illnesses. Early detection of rare diseases could result in better-targeted management and treatment plans, which will ultimately enhance the standard of life for those who are afflicted [103, 104].

Deep learning excels in these various illness prediction fields because of its capacity to process high-dimensional data, adjust to various data kinds, and pick up intricate patterns. With the use of this information, medical personnel can now make better judgments, provide individualized suggestions, and, in many cases, identify diseases early on, when treatment is most successful.

3.6 Ethical Considerations and Challenges

To guarantee the legitimate and impartial employment of these technologies in healthcare, a number of moral issues and obstacles related to the implementation of deep learning in disease prognosis should be resolved. We go into more detail about three important ethical issues in this section: bias and injustice, data privacy and security, and compliance and ethical guidelines.

3.6.1 Data Privacy and Security

- Data Privacy: Strict privacy regulations must be followed while collecting and using patient data to forecast illness. Patients should have the privacy of their data respected, and their private medical records must be safeguarded and anonymized. It is imperative to comply with data privacy provisions, such as the Health Insurance Portability and Accountability Act (HIPAA) in the USA.
- Security: Because healthcare data is so important and sensitive, hackers attack it frequently. To protect patient information, deep learning professionals and healthcare institutions need to put strong security mechanisms in place. This covers network access, decryption, and ongoing surveillance to find and stop breaches [105–107].
- Informed Consent: Patients must give their aware permission after being fully briefed about how their data will be utilized to forecast illness. People ought to be able to revoke their consent or opt out at any moment.
- Data-Sharing: Clear data-sharing rules should be established by healthcare organizations and researchers to make sure that data is only provided when needed and that it is always done so in a secured and private method.

3.6.2 Bias and Fairness

- Data Bias: Deep learning models are greatly influenced by the data that was utilized to create them. The model may acquire and perpetuate biases found in the training set, including the under-representation of specific demographic groups. This could lead to unfair and discriminatory forecasts.

- Algorithmic Bias: Models have the ability to include bias due to the way they extract patterns from data. Examining models for bias is vital, especially in the healthcare sector where equality is vital [108–110].
- Fairness: In order to ensure fairness in sickness prediction, models need to operate consistently throughout different demographic subgroups. Fairness-aware training, impartiality metrics, and postprocessing techniques could all be used to reduce bias and promote equitable healthcare outcomes.

3.6.3 Regulatory Compliance and Ethical Guidelines

- Regulatory Compliance: Based on the area, deep learning systems in healthcare must abide by different regulations. It is imperative to adhere to standards including HIPAA, the European General Data Protection Regulation (GDPR), and the standards set forth by the U.S. Food and Drug Administration (FDA).
- Ethical Rules: Prioritizing patient welfare, healthcare facilities, and professionals must adhere to ethical standards [111, 112]. This entails being open about the application of deep learning models, subjecting them to thorough validation and verification, and abiding by an ethical code that supports patient-centered treatment.
- Transparency: It is critical to maintain transparency during the creation and use of DL models. Patients attain the opportunity to know the methodology used to make forecasts and should be educated about this process.
- Accountability: Those using deep learning techniques need to take responsibility for the choices their models make. This entails accepting accountability for any mistakes, biases, and unfavorable results that could arise from model projections [113, 114].
- Patient Autonomy: Patients ought to be in charge of making their own healthcare judgments. Deep learning models ought to be employed as instruments, not as a substitute for, clinical judgment.

3.7 Future Directions and Conclusion

Deep learning for illness predictions has a bright future ahead of it, with new developments that could change the face of healthcare. In this section, we go into further detail about these new developments as well as the expected development of healthcare. We also offer some closing thoughts and important lessons learned.

3.7.1 Emerging Trends in Deep Learning for Disease Prediction

- Explainable AI (XAI): Transparency and interpretability are becoming increasingly important as deep learning models get more complicated. It is anticipated that XAI approaches will be crucial in the prognosis of diseases, enabling medical professionals to comprehend and have confidence in the choices these models make.
- Multimodal Data Integration: Wearable technology, genetics, electronic health data, medical imaging, and wearable data are just a few of the sources from which disease prognosis models are progressively incorporating data. Predictions will be more accurate and comprehensive when these many data kinds are integrated.
- Personalized Medicine: The creation of individualized treatment regimens and preventative measures will be made possible by deep learning. To customize healthcare solutions, models will take into account an individual's genetic, behavioral, and environmental data.
- Real-Time Monitoring: The healthcare industry will see a rise in the use of wearable technology and remote monitoring. In order to improve proactive healthcare actions, deep learning algorithms will constantly evaluate data from these devices to deliver real-time illness evaluations.
- Applications in Global Health: Deep learning can help with issues related to global health, such as disease epidemics and resource distribution in underdeveloped areas. Models will help in earlier identification, healthcare crisis management, and avoidance.

3.7.2 The Future Landscape of Healthcare

The application of deep learning in healthcare will drastically alter the current state of the industry:

- Data-Driven Healthcare: A proactively, data-driven strategy will gradually replace the current reactive paradigm in healthcare. In addition to lowering healthcare expenditures and increasing patient experiences, deep learning models may help with prompt identification, surveillance, and prophylactic interventions.
- Empowered Patients: People will be more actively involved in choosing their medical treatment and will have more accessibility to their health information. Users will be empowered with tailored health information and advice attributable to deep learning algorithms.

- Better Resource Allocation: Predictive analytics will be used by healthcare organizations to optimize resource allocation. Their ability to deploy resources more effectively will improve, resulting in fewer bottlenecks and improved patient care.
- Cooperation and Assimilation: There will be an increase in the frequency of interaction between data analysts, deep learning specialists, and healthcare practitioners. Improved AI integration in health coverage and the creation of creative remedies are the outcomes of this multidisciplinary strategy.
- Advances in Ethics and Regulation: As artificial intelligence (AI) becomes more widely used in healthcare, ethical questions and legal requirements will change accordingly. To guarantee the responsible and ethical application of deep learning models, regulations will be created.

3.7.3 Concluding Remarks and Key Takeaways

To summarize, deep learning for illness predictions is transforming the medical field. Enhancing diagnosis, prediction, and tailored treatment has the ability to improve patient results. But at the vanguard of this change needs to be the moral issues of data privacy, justice, and governmental conformity. Important lessons to learn from this chapter are

- Medical data, particularly photographs, electronic health information, and genetic information, can be effectively analyzed by deep learning models like CNN and RNNs.
- Metrics, including precision, accuracy, recollection, and cross-validation methods, are used in the assessment of these models in order to guarantee reliable functioning.
- Privacy protection, impartiality and prejudice, and compliance with legal and moral requirements are examples of moral issues.
- Explainable AI, multimodal processing, personalized medicine, real-time monitoring, and applicability in global health are some of the recent developments in deep learning for illness prognosis.
- It is anticipated that data-driven, patient-centered healthcare will be characterized by cooperation, resource efficiency, and changing moral and legal requirements in the long term.

In summary, there is a tremendous chance to revolutionize patient care, treatment, and illness prognosis using the integration of DL in the healthcare industry. Through moral and societal responsibility, we may fully utilize the possibilities of new technologies to improve worldwide health while tackling the accompanying issues.

References

1. Sakr AS, Soliman NF, Al-Gaashani MS, Pławiak P, Ateya AA, Hammad M. An efficient deep learning approach for colon cancer detection. Appl Sci. 2022;12:1–13.
2. Mengash HA, Alamgeer M, Maashi M, Othman M, Hamza MA, Ibrahim SS, Zamani AS, Yaseen I. Leveraging marine predators algorithm with deep learning for lung and colon cancer diagnosis. Cancers (Basel). 2023;15 https://doi.org/10.3390/cancers15051591.
3. Carrillo-Perez F, Morales JC, Castillo-Secilla D, Gevaert O, Rojas I, Herrera LJ. Machine-learning-based late fusion on multi-omics and multi-scale data for non-small-cell lung cancer diagnosis. J Pers Med. 2022;12 https://doi.org/10.3390/jpm12040601.
4. Huang CH, Zeng C, Wang YC, Peng HY, Lin CS, Chang CJ, Yang HY. A study of diagnostic accuracy using a chemical sensor array and a machine learning technique to detect lung cancer. Sensors (Switzerland). 2018;18 https://doi.org/10.3390/s18092845.
5. Alfi IA, Rahman MM, Shorfuzzaman M, Nazir A. A non-invasive interpretable diagnosis of melanoma skin cancer using deep learning and ensemble stacking of machine learning models. Diagnostics. 2022;12:1–18.
6. Alam TM, Shaukat K, Khan WA, Hameed IA, Almuqren LA, Raza MA, Aslam M, Luo S. An efficient deep learning-based skin cancer classifier for an imbalanced dataset. Diagnostics. 2022;12 https://doi.org/10.3390/diagnostics12092115.
7. Dildar M, Akram S, Irfan M, Khan HU, Ramzan M, Mahmood AR, Alsaiari SA, Saeed AHM, Alraddadi MO, Mahnashi MH. Skin cancer detection: a review using deep learning techniques. Int J Environ Res Public Health. 2021;18 https://doi.org/10.3390/ijerph18105479.
8. Quoc N, Le K, Kha QH, Nguyen VH, Chen Y, Cheng S. Machine learning-based radiomics signatures for EGFR and KRAS mutations prediction in non-small-cell lung cancer. Int J Mol Sci. 2021;22:9254.
9. Park JH, Cho HE, Kim JH, Wall MM, Stern Y, Lim H, Yoo S, Kim HS, Cha J. Machine learning prediction of incidence of Alzheimer's disease using large-scale administrative health data. NPJ Digit Med. 2020;3:46. https://doi.org/10.1038/s41746-020-0256-0.
10. Ketu S, Mishra PK. Internet of healthcare things: a contemporary survey. J Netw Comput Appl. 2021;192:103179.
11. Kukar M, Gunčar G, Vovko T, Podnar S, Černelč P, Brvar M, Zalaznik M, Notar M, Moškon S, Notar M. COVID-19 diagnosis by routine blood tests using machine learning. Sci Rep. 2021;11:1–9.
12. Sajal MSR, et al. Telemonitoring Parkinson's disease using machine learning by combining tremor and voice analysis. Brain Inform. 2020;7:1–11.
13. Roy SS, Mallik A, Gulati R, Obaidat MS, Krishna PV. A deep learning based artificial neural network approach for intrusion detection. Commun Comput Inf Sci. 2017;655:44–53.
14. Van Ginneken B, Schaefer-Prokop CM, Prokop M. Computer-aided diagnosis: how to move from the laboratory to the clinic. Radiology. 2011;261:719–32.
15. Singh AJ, Kumar M. Comparative analysis on prediction of software effort estimation using machine learning techniques. SSRN Electron J. 2020; https://doi.org/10.2139/ssrn.3565822.
16. Shiokawa Y, Date Y, Kikuchi J. Application of kernel principal component analysis and computational machine learning to exploration of metabolites strongly associated with diet. Sci Rep. 2018;8:1–8.
17. May M. Eight ways machine learning is assisting medicine. Nat Med. 2021;27:2–3.
18. Qadri YA, Nauman A, Bin ZY, Vasilakos AV, Kim SW. The future of healthcare internet of things: a survey of emerging technologies. IEEE Commun Surv Tutor. 2020;22:1121–67.
19. Suzuki K. A review of computer-aided diagnosis in thoracic and colonic imaging. Quant Imaging Med Surg. 2012;2:163–76.
20. Rampisela TV, Rustam Z. Classification of schizophrenia data using support vector machine (SVM). J Phys Conf Ser. 2018; https://doi.org/10.1088/1742-6596/1108/1/012044.

21. Javed AR, Fahad LG, Farhan AA, Abbas S, Srivastava G, Parizi RM, Khan MS. Automated cognitive health assessment in smart homes using machine learning. Sustain Cities Soc. 2021;65:102572.
22. Zemel RS, Pitassi T. A gradient-based boosting algorithm for regression problems. In: Advances in neural information processing systems; 2001.
23. Miao Q, Cao Y, Xia G, Gong M, Liu J, Song J. RBoost: label noise-robust boosting algorithm based on a nonconvex loss function and the numerically stable base learners. IEEE Trans Neural Netw Learn Syst. 2016;27:2216–28.
24. Yue W, Wang Z, Chen H, Payne A, Liu X. Machine learning with applications in breast cancer diagnosis and prognosis. Designs. 2018;2:1–17.
25. Krishnamurthy S, Kapeleshh KS, Dovgan E, Luštrek M, Gradišek Piletič B, Srinivasan K, Li YC, Gradišek A, Syed-Abdul S. Machine learning prediction models for chronic kidney disease using national health insurance claim data in Taiwan. Healthc. 2021;9:1–13.
26. Kim TS, Sohn SY. Multitask learning for health condition identification and remaining useful life prediction: deep convolutional neural network approach. J Intell Manuf. 2021;32:2169–79.
27. Bhavsar KA, Abugabah A, Singla J, AlZubi AA, Bashir AK, et al. A comprehensive review on medical diagnosis using machine learning. Comput Mater Contin. 2021;67:1997–2014.
28. Mavroforakis ME, Theodoridis S. A geometric approach to support vector machine (SVM) classification. IEEE Trans Neural Netw. 2006;17:671–82.
29. Mishra M. Machine learning techniques for structural health monitoring of heritage buildings: a state-of-the-art review and case studies. J Cult Herit. 2021;47:227–45.
30. Gao Y, Geras KJ, Lewin AA, Moy L. New frontiers: an update on computer-aided diagnosis for breast imaging in the age of artificial intelligence. Am J Roentgenol. 2019;212:300–7.
31. Verma C, Quraishi MA, Rhee KY. Present and emerging trends in using pharmaceutically active compounds as aqueous phase corrosion inhibitors. J Mol Liq. 2021;328:115395.
32. Rajesh N, Maneesha T, Hafeez S, Krishna H. Prediction of heart disease using machine learning algorithms. Int J Eng Technol. 2018;7:363–6.
33. Nilashi M, Ibrahim O, et al. An analytical method for diseases prediction using machine learning techniques. Comput Chem Eng. 2017;106:212–23.
34. Polat K, Güneş S. Detection of ECG arrhythmia using a differential expert system approach based on principal component analysis and least square support vector machine. Appl Math Comput. 2007;186:898–906.
35. Hosseinzadeh M, Ahmed OH, Ghafour MY, Safara F, et al. A multiple multilayer perceptron neural network with an adaptive learning algorithm for thyroid disease diagnosis in the internet of medical things. J Supercomput. 2021;77:3616–37.
36. Osareh A, Shadgar B. Machine learning techniques to diagnose breast cancer. In: 2010 5th international symposium on health informatics and bioinformatics. IEEE; 2010. p. 114–20.
37. Nishat MM. Performance assessment of different machine learning algorithms in predicting diabetes mellitus. Biosci Biotechnol Res Commun. 2021;14:74–82.
38. Nithya B, Ilango V. Predictive analytics in health care using machine learning tools and techniques. In: 2017 international conference on intelligent computing and control systems (ICICCS). IEEE; 2017. p. 492–9.
39. Kwekha-Rashid AS, Abduljabbar HN, Alhayani B. Coronavirus disease (COVID-19) cases analysis using machine-learning applications. Appl Nanosci. 2021;13:2013. https://doi.org/10.1007/s13204-021-01868-7.
40. Balakrishnan M, Arockia Christopher AB, Ramprakash P, Logeswari A. Prediction of cardiovascular disease using machine learning. J Phys Conf Ser. 2021;1767:1–7.
41. Park C, Took CC, Seong JK. Machine learning in biomedical engineering. Biomed Eng Lett. 2018;8:1–3.
42. Ghosh P, et al. Optimization of prediction method of chronic kidney disease using machine learning algorithm. In: 2020 15th international joint symposium on artificial intelligence and natural language processing (iSAI-NLP). IEEE; 2020. https://doi.org/10.1109/iSAI-NLP51646.2020.9376787.

43. Ancochea J, Lumbreras S, Soriano JB. Evidence of gender differences in the diagnosis and management of coronavirus disease 2019 patients: an analysis of electronic health records using natural language processing and machine learning. J Women's Heal. 2021;30:393–404.
44. Kim J, Lee D, Park E. Machine learning for mental health in social media: bibliometric study. J Med Internet Res. 2021;23:1–17.
45. Kishor A, Chakraborty C. Artificial intelligence and internet of things based healthcare 4.0 monitoring system. Wirel Pers Commun. 2021; https://doi.org/10.1007/s11277-021-08708-5.
46. Peyvandi A, Majidi B, Peyvandi S, Patra J. Computer-aided-diagnosis as a service on decentralized medical cloud for efficient and rapid emergency response intelligence. N Gener Comput. 2021;39:677–700.
47. Gürfidan R, Ersoy M. Classification of death related to heart failure by machine learning algorithms. Adv Artif Intell Res. 2021;1:13–8.
48. Castiglioni I, Rundo L, Codari M, Di Leo G, Salvatore C, Interlenghi M, Gallivanone F, Cozzi A, D'Amico NC, Sardanelli F. AI applications to medical images: from machine learning to deep learning. Phys Med. 2021;83:9–24.
49. Alizadehsani R, Roshanzamir M, Hussain S, et al. Handling of uncertainty in medical data using machine learning and probability theory techniques: a review of 30 years (1991–2020). Ann Oper Res. 2021:1. https://doi.org/10.1007/s10479-021-04006-2.
50. van der Schaar M, Alaa AM, Floto A, Gimson A, Scholtes S, Wood A, McKinney E, Jarrett D, Lio P, Ercole A. How artificial intelligence and machine learning can help healthcare systems respond to COVID-19. Mach Learn. 2021;110:1–14.
51. Wang C, Zhai H. Machine learning of frustrated classical spin models (II): kernel principal component analysis. Front Phys. 2018;13:1–7.
52. Mahesh Kumar SV, Gunasundari R. Computer-aided diagnosis of anterior segment eye abnormalities using visible wavelength image analysis based machine learning. J Med Syst. 2018;42:128. https://doi.org/10.1007/s10916-018-0980-z.
53. Yassin NIR, Omran S, El Houby EMF, Allam H. Machine learning techniques for breast cancer computer aided diagnosis using different image modalities: a systematic review. Comput Methods Prog Biomed. 2018;156:25–45.
54. Vu QV, Truong VH, Thai HT. Machine learning-based prediction of CFST columns using gradient tree boosting algorithm. Compos Struct. 2021;259:113505.
55. Choudhury K, Mohd Nawi N, et al. Training artificial neural network using particle swarm optimization algorithm Studying the effect of training Levenberg Marquardt neural net work by using hybrid meta-heuristic A… hybrid learning enhancement of RBF net work with part icle swarm opt. Int J Adv Res Comput Sci Softw Eng. 2013;3:2277.
56. Somvanshi M, Chavan P, Tambade S, Shinde SV. A review of machine learning techniques using decision tree and support vector machine. In: 2016 international conference on computing communication control and automation (ICCUBEA). IEEE; 2017. https://doi.org/10.1109/ICCUBEA.2016.7860040.
57. Gunčar G, Kukar M, Notar M, Brvar M, Černelč P, Notar M, Notar M. An application of machine learning to haematological diagnosis. Sci Rep. 2018;8:1–12.
58. Khan B, Shukla PK, Ahirwar MK. Strategic analysis in prediction of liver disease using different classification algorithms. Int J Comput Sci Eng. 2019;7:71–6.
59. Kudo T, Matsumoto Y. A boosting algorithm for classification of semi-structured text. In: Proceedings of the 2004 conference on empirical methods in natural language processing; 2004. p. 301–8.
60. Xia Y, Wang J. A one-layer recurrent neural network for support vector machine learning. IEEE Trans Syst Man Cybern Part B Cybern. 2004;34:1261–9.
61. Yang H, Zhang Z, Zhang J, Zeng XC. Machine learning and artificial neural network prediction of interfacial thermal resistance between graphene and hexagonal boron nitride. Nanoscale. 2018;10:19092–9.
62. Xue JH, Titterington DM. Comment on "on discriminative vs. generative classifiers: a comparison of logistic regression and naive bayes". Neural Process Lett. 2008;28:169–87.

63. Tsai JC, Leu JS, Prakosa SW, Hsiao LC, Huang PC, Yang SY, Huang YT. Design and implementation of an internet of healthcare things system for respiratory diseases. Wirel Pers Commun. 2021;117:337–53.
64. Ragab M, Albukhari A, Alyami J, Mansour RF. Ensemble deep-learning-enabled clinical decision support system for breast cancer diagnosis and classification on ultrasound images. Biology (Basel). 2022;11 https://doi.org/10.3390/biology11030439.
65. Naranjo ID, Gibbs P, Reiner JS, et al. Radiomics and machine learning with multiparametric breast mri for improved diagnostic accuracy in breast cancer diagnosis. Diagnostics. 2021;11:1–13.
66. Riquelme D, Akhloufi MA. Deep learning for lung cancer nodules detection and classification in CT scans. AI. 2020;1:28–67.
67. Shafi I, Din S, Khan A, Díez IDLT, et al. An effective method for lung cancer diagnosis from CT scan using deep learning-based support vector network. Cancers (Basel). 2022;14:1–18.
68. Jinnai S, Yamazaki N, Hirano Y, Sugawara Y, Ohe Y, Hamamoto R. The development of a skin cancer classification system for pigmented skin lesions using deep learning. Biomol Ther. 2020;10:1–13.
69. Tahir M, Naeem A, Malik H, Tanveer J, Naqvi RA, Lee SW. DSCC_Net: multi-classification deep learning models for diagnosing of skin cancer using dermoscopic images. Cancers (Basel). 2023;15 https://doi.org/10.3390/cancers15072179.
70. Attallah O, Aslan MF, Sabanci K. A framework for lung and colon cancer diagnosis via lightweight deep learning models and transformation methods. Diagnostics. 2022;12 https://doi.org/10.3390/diagnostics12122926.
71. Nakajima T, Katsumata K, Kuwabara H, et al. Urinary polyamine biomarker panels with machine-learning differentiated colorectal cancers, benign disease, and healthy controls. Int J Mol Sci. 2018;19:1–14.
72. Ha T, Jung Y, Kim JY, Park SY, Kang DK, Kim TH. Comparison of the diagnostic performance of abbreviated MRI and full diagnostic MRI using a computer-aided diagnosis (CAD) system in patients with a personal history of breast cancer: the effect of CAD-generated kinetic features on reader performance. Clin Radiol. 2019;74:817.
73. Anzanpour A, Amiri D, Azimi I, Levorato M, Dutt N, Liljeberg P, Rahmani AM. Edge-assisted control for healthcare internet of things: a case study on PPG-based early warning score. ACM Trans Internet Things. 2021;2:1–21.
74. Thakur A, Kumar A. Recent trends in nanostructured carbon-based electrochemical sensors for the detection and remediation of persistent toxic substances in real- time analysis. Mater Res Express. 2023;10:034001.
75. Thakur A, Kumar A, Sharma S, Ganjoo R, Assad H. Materials today: proceedings computational and experimental studies on the efficiency of Sonchus arvensis as green corrosion inhibitor for mild steel in 0.5 M HCl solution. Mater Today Proc. 2022;66:609–21.
76. Thakur A, Sharma S, Ganjoo R, Assad H, Kumar A. Anti-corrosive potential of the sustainable corrosion inhibitors based on biomass waste: a review on preceding and perspective research. J Phys Conf Ser. 2022;2267:012079.
77. Thakur A, Kumar A. Sustainable inhibitors for corrosion mitigation in aggressive corrosive media: a comprehensive study. J Bio- Tribo-Corrosion. 2021;7:1–48.
78. Thakur A, Kumar A. Recent advances on rapid detection and remediation of environmental pollutants utilizing nanomaterials-based (bio)sensors. Sci Total Environ. 2022;834:155219.
79. Thakur A, Kaya S, Abousalem AS, Kumar A. Experimental, DFT and MC simulation analysis of Vicia Sativa weed aerial extract as sustainable and eco-benign corrosion inhibitor for mild steel in acidic environment. Sustain Chem Pharm. 2022;29:100785.
80. Kaya S, Lgaz H, Thakkur A, Kumar A, Özbakır D, Karakuş N, et al. Molecular insights into the corrosion inhibition mechanism of omeprazole and tinidazole: a theoretical investigation. Mol Simul. 2023;49:1–15.

81. Kaya S, Thakur A, Kumar A. The role of in Silico/DFT investigations in analyzing dye molecules for enhanced solar cell efficiency and reduced toxicity. J Mol Graph Model. 2023;124:108536.
82. Kumar A, Thakur A. Encapsulated nanoparticles in organic polymers for corrosion inhibition. Corros Prot Nanoscale. 2020; https://doi.org/10.1016/b978-0-12-819359-4.00018-0.
83. Thakur A, Kumar A, Kaya S, Vo DVN, Sharma A. Suppressing inhibitory compounds by nanomaterials for highly efficient biofuel production: a review. Fuel. 2022;312:122934.
84. Thakur A, Kumar A, Kaya S, Marzouki R, Zhang F, Guo L. Recent advancements in surface modification, characterization and functionalization for enhancing the biocompatibility and corrosion resistance of biomedical implants. Coatings. 2022;12:1459.
85. Thakur A, Kaya S, Kumar A. Recent innovations in nano container-based self-healing coatings in the construction industry. Curr Nanosci. 2021;18:203–16.
86. Shah SA, Ren A, Fan D, et al. Internet of things for sensing: a case study in the healthcare system. Appl Sci. 2018;8:1–16.
87. Eun JC, Gong YJ, Young MH, Young SL, Keun SK. Solitary pulmonary nodule on helical dynamic CT scans: analysis of the enhancement patterns using a computer-aided diagnosis (CAD) system. Korean J Radiol. 2008;9:401–8.
88. Bakkouri I, Afdel K. Computer-aided diagnosis (CAD) system based on multi-layer feature fusion network for skin lesion recognition in dermoscopy images. Multimed Tools Appl. 2020;79:20483–518.
89. Okazaki Y, Takenaga T, Miyake T, Iwabuchi M, Okubo T, Yokoi N. Design and implementation of smartphone-based tear volume measurement system. In: Internet of things technologies for healthcare: third international conference, HealthyIoT 2016; 2016. https://doi.org/10.1007/978-3-319-51234-1_27.
90. Pramanik PKD, Solanki A, Debnath A, Nayyar A, El-Sappagh S, Kwak KS. Advancing modern healthcare with nanotechnology, nanobiosensors, and internet of Nano things: taxonomies, applications, architecture, and challenges. IEEE Access. 2020;8:65230–66.
91. Benzakoun J, Bommart S, Coste J, Chassagnon G, Lederlin M, Boussouar S, Revel MP. Computer-aided diagnosis (CAD) of subsolid nodules: evaluation of a commercial CAD system. Eur J Radiol. 2016;85:1728–34.
92. Bhatia M, Kaur S, Sood SK, Behal V. Internet of things-inspired healthcare system for urine-based diabetes prediction. Artif Intell Med. 2020;107:101913.
93. Küçük H, Eminoılu I. Classification of ALS disease using support vector machines. In: 2015 23nd signal processing and communications applications conference (SIU); 2015. p. 1664–7.
94. Hazra A, Choudhary P, Sheetal Singh M. Recent advances in deep learning techniques and its applications: an overview. In: Advances in biomedical engineering and technology: select proceedings of ICBEST 2018; 2021. p. 103–22.
95. Bashir S, Lgaz H, Chung IM, Kumar A. Effective green corrosion inhibition of aluminium using analgin in acidic medium: an experimental and theoretical study. Chem Eng Commun. 2020:1–10.
96. Bashir S, Thakur A, Lgaz H, Chung I-M, Kumar A. Corrosion inhibition performance of Acarbose on mild steel corrosion in acidic medium: an experimental and computational study. Arab J Sci Eng. 2020;45:4773–83.
97. Bashir S, Thakur A, Lgaz H, Chung IM, Kumar A. Corrosion inhibition efficiency of bronopol on aluminium in 0.5 M HCl solution: insights from experimental and quantum chemical studies. Surf Interfaces. 2020;20:100542.
98. Dhonchak C, Agnihotri N. Computational insights in the Spectrophotometrically 4H-chromen-4-one complex using DFT method. Biointerface Res Appl Chem. 2023;13:357.
99. Thakur A, et al. Recent trends in the characterization and application progress of nano – modified coatings in corrosion mitigation of metals and alloys. Appl Sci. 2023;13:730.
100. Bashir S, Thakur A, Lgaz H, Chung I-M, Kumar A. Computational and experimental studies on phenylephrine as anti-corrosion substance of mild steel in acidic medium. J Mol Liq. 2019;293:111539.

101. Thakur A, et al. Recent trends in the characterization and application progress of nano-modified coatings in corrosion mitigation of metals and alloys. Appl Sci. 2023;13:730.
102. Kandemir M, Hamprecht FA. Computer-aided diagnosis from weak supervision: a benchmarking study. Comput Med Imaging Graph. 2015;42:44–50.
103. Dreiseitl S, Ohno-Machado L. Logistic regression and artificial neural network classification models: a methodology review. J Biomed Inform. 2002;35:352–9.
104. Pranckevičius T, Marcinkevičius V. Comparison of naive Bayes, random forest, decision tree, support vector machines, and logistic regression classifiers for text reviews classification. Balt J Mod Comput. 2017;5:221–32.
105. Alizadehsani R, Abdar M, Roshanzamir M, Khosravi A, Kebria PM, Khozeimeh F, Nahavandi S, Sarrafzadegan N, Acharya UR. Machine learning-based coronary artery disease diagnosis: a comprehensive review. Comput Biol Med. 2019;111:103346.
106. Itoo F, Meenakshi SS. Comparison and analysis of logistic regression, naïve Bayes and KNN machine learning algorithms for credit card fraud detection. Int J Inf Technol. 2021;13:1503–11.
107. Tanaka H, Chiu SW, Watanabe T, Kaoku S, Yamaguchi T. Computer-aided diagnosis system for breast ultrasound images using deep learning. Phys Med Biol. 2019;64 https://doi.org/10.1088/1361-6560/ab5093.
108. Karamizadeh S, Abdullah SM, Manaf AA, Zamani M, Hooman A. An overview of principal component analysis. J Signal Inf Process. 2013;04:173–5.
109. Lahmiri S, Shmuel A. Performance of machine learning methods applied to structural MRI and ADAS cognitive scores in diagnosing Alzheimer's disease. Biomed Signal Process Control. 2019;52:414–9.
110. Huang W, Wu J, Mao Y, Zhu S, Huang GF, Petritis B, Huang RP. Developing a periodontal disease antibody array for the prediction of severe periodontal disease using machine learning classifiers. J Periodontol. 2020;91:232–43.
111. Amarappa S, Sathyanarayana SV. Data classification using support vector machine (SVM), a simplified approach. Int J Electron Comput Sci Eng. 2011;3:435–45.
112. Niitsu K. Energy-autonomous biosensing platform using supply-sensing CMOS integrated sensor and biofuel cell for next-generation healthcare internet of things. Jpn J Appl Phys. 2018;57:1–14.
113. Souri A, Ghafour MY, Ahmed AM, Safara F, Yamini A, Hoseyninezhad M. A new machine learning-based healthcare monitoring model for student's condition diagnosis in internet of things environment. Soft Comput. 2020;24:17111–21.
114. Lee YW, Choi JW, Shin EH. Machine learning model for diagnostic method prediction in parasitic disease using clinical information. Expert Syst Appl. 2021;185:115658.

Chapter 4
A Survey of Image Classification Techniques for the Prediction of Diseases

Manjeet Kaur, Ankita Pandey, and Arun Kumar

4.1 Introduction

Among individuals aged 20–30 years, primary spontaneous pneumothorax (PSP) is notably prevalent. In the United States, PSP manifests at an annual incidence rate of 7 cases per 100,000 among men and 1 case per 100,000 among women. Recurrences predominantly occur within the initial year, with recurrence rates ranging from 25% to 50%, with the highest risk observed within the initial 30 days. Conversely, secondary spontaneous pneumothorax (SSP) is more frequently observed in elderly patients, particularly those aged 60–65 years. SSP affects 6.3 out of 100,000 men, 2 out of 100,000 women, with a male-to-female ratio of 3 to 1. Chronic obstructive pulmonary disease (COPD) is associated with a pneumothorax incidence of 26 per 100,000 patients, with heavy smokers facing a substantially elevated risk, 102 times greater than nonsmokers, for spontaneous pneumothorax development. The primary cause of iatrogenic pneumothorax is transthoracic needle aspiration, frequently conducted for biopsies, followed by central venous catheterization. These occurrences surpass the frequency of spontaneous pneumothorax and are increasing alongside advancements in intensive care techniques, transpiring once per 10,000 hospital admissions. Assessing tension pneumothorax incidence proves challenging due to many cases receiving decompressive needle thoracostomies prior to hospital arrival, with not all necessitating treatment for tension pneumothorax. Pneumomediastinum is identified at a rate of one case per 10,000 hospital admissions. Within the pleural space, situated between the lungs and chest wall, abnormal air accumulation initiates pneumothorax. This trapped air exerts pressure on the lungs, leading to partial

M. Kaur · A. Kumar
Department of Mathematics, Statistics and Computer Science, G. B. Pant University of Agriculture and Technology, Pantnagar, Uttarakhand, India

A. Pandey (✉)
Institute of Applied Sciences and Humanities, GLA University, Mathura, Uttar Pradesh, India

T. Choudhury, A. Katal (eds.), *Machine Learning for Disease Detection, Prediction, and Diagnosis*, https://doi.org/10.1007/978-981-96-4241-0_4

or complete lung collapse. Pneumothorax can stem from various origins, such as blunt or penetrating chest traumas, specific medical procedures, or damage caused by underlying pulmonary conditions. Occasionally, it may occur without discernible causation. Typical symptoms encompass the sudden onset of acute chest pain on one side and breathing difficulties. The severity of these symptoms often correlates with the degree of lung collapse. Figure 4.1 shows a chest radiograph featuring a pneumothorax extracted from a dataset.

There are several potential causes of pneumothoraxes, including

1. Chest Trauma: Lung collapse can ensue from either blunt or penetrating thoracic trauma, originating from sources such as physical altercations, vehicular accidents, or invasive medical procedures involving thoracic needle punctures.
2. Pulmonary Ailments: The occurrence of lung collapse is widespread, especially in cases where the pulmonary parenchyma is compromised. Various underlying conditions, such as chronic obstructive pulmonary disease, cystic fibrosis, lung tumors, or pneumonia, can lead to damage to lung tissue.
3. Ruptured Pulmonary Blisters: In the vicinity just above the lungs, diminutive air-filled blisters known as blebs may form. On occasion, these blebs can rupture, facilitating the ingress of air into the pleural space surrounding the lungs.
4. Ventilator-Associated Pneumothorax: Individuals reliant on mechanical ventilation for respiratory support are susceptible to experiencing a severe manifestation of pneumothorax. The utilization of a ventilatory apparatus carries the potential to induce imbalances in thoracic air pressure, which may culminate in complete lung collapse.

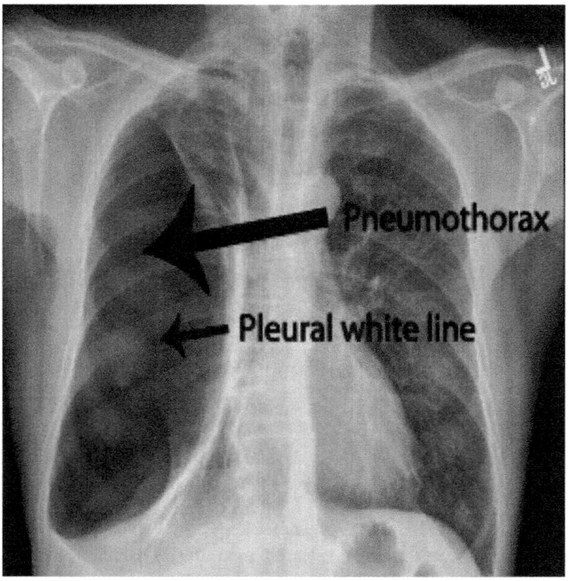

Fig. 4.1 A chest radiograph featuring a pneumothorax extracted from a dataset

Diagnosis
Pneumothorax can be diagnosed using several techniques:

- Diagnosis can be accomplished through chest X-ray imaging.
- For more detailed visuals, a computed tomography (CT) scan may be required.
- Ultrasound is another viable method for detecting pneumothorax.

Types of pneumothorax:

- Primary pneumothorax
- Secondary pneumothorax
- Traumatic pneumothorax
- Tension or nontension pneumothorax

Treatment
The main goal of treating a pneumothorax is to release the lung's pressure so that it may expand once again. Depending on the underlying etiology of the pneumothorax, avoiding recurrences may be a secondary objective. The specific therapeutic interventions employed are contingent upon the degree of pulmonary collapse and the overall well-being of the patient. Therapeutic modalities include vigilant monitoring, thoracentesis via needle aspiration, the insertion of a chest tube, noninvasive medical procedures, and surgical interventions. On certain occasions, adjunctive oxygen therapy may be requisite to expedite the reabsorption of air and promote lung re-expansion.

Pneumothorax may present with subtle or faint radiographic signs, making it challenging for doctors to identify in chest X-rays, especially in its early stages. Small pneumothoraxes or those occurring in specific locations within the thoracic cavity may be more difficult to detect on chest X-rays. Pneumothorax may coexist with other pulmonary abnormalities or be mistaken for other conditions, such as pleural effusion or pneumonia, complicating the diagnostic process. The anatomy of the chest involves overlapping structures, and distinguishing between normal anatomy and the presence of pneumothorax requires a keen eye and experience.

In this investigation, we have seamlessly fused the capabilities of deep learning (DL) and machine learning (ML) methodologies to detect pneumothorax in chest X-ray images. Our model structure is anchored by ResNet-50, a sophisticated deep neural network consisting of 50 pretrained layers specialized in extracting intricate image features. We utilize a support vector machine (SVM) as the discerning classifier. The dataset is meticulously divided into three distinct subsets for training, testing, and validation, with a batch size carefully set at 20 throughout 40 epochs. The primary focus of this study is to enhance pneumothorax classification accuracy and enable early prognosis through comprehensive feature extraction. The main objective of this research is to anticipate the presence or absence of pneumothorax within chest X-ray images.

The main contribution of this study can be summed up as follows:

- Construct an automated detection and classification system.
- Examine the fusion of deep feature extractor and classifier and develop a model for detection and binary classification of pneumothorax.
- Develop a dependable, superior, less time-consuming, and single-click classifier.
- Compare the optimal model's performance in terms of accuracy, specificity, sensitivity, and F1 score to that of state-of-the-art models.

The study is structured into several sections. Section 4.2 explores previous research on pneumothorax detection. Section 4.3 presents information about the dataset and tools used in the study. Section 4.4 offers a comprehensive overview of the experiments conducted. Section 4.5 delves into the details of the experimental results. Section 4.6 focuses on discussing these results. Finally, Sect. 4.7 wraps up the chapter with a conclusion.

4.2 Related Works

Identifying pneumothorax in chest X-ray images presents a daunting obstacle, often demanding the expertise of radiologists. Moreover, delayed diagnoses can precipitate critical conditions in patients, underscoring the imperative for the development of computer-aided detection (CAD) systems. He et al. [1] expounded that the training of increasingly intricate neural networks presents considerable hurdles. To render the training of substantially deeper networks more manageable, they introduced a novel residual learning framework. Rather than learning uncorrelated functions, they intentionally adjusted the network layers to obtain residual functions concerning the inputs of the layers. An abundance of empirical evidence is presented, underscoring the ease with which these residual networks can be fine-tuned and the enhanced accuracy that can be derived from significantly increased network depth.

In Wang et al. [2], the researcher conducted an experiment utilizing the ChestX-ray8 database. The database consists of 108,948 chest X-ray images captured from 32,717 different patients, all in frontal view. These images are associated with eight different medical conditions, and their labels were extracted from radiological reports using natural language processing techniques. The study demonstrates a unified approach for identifying diseases and categorizing images with weak supervision, allowing for the efficient detection and even identification of various common thoracic ailments. This validation is carried out using the dataset proposed in the study.

Then, Chen et al. [3] introduced a novel approach using SVM for the detection of frequent cases of pneumothorax. They employed a technique called local binary pattern (LBP) to extract relevant features from lung images. Subsequently, SVM was employed to distinguish cases of pneumothorax from non-pneumothorax instances. Additionally, their method involved segmenting abnormal lung regions. This automatic pneumothorax detection technique relied on multiscale intensity texture segmentation to eliminate background noise from chest images. The texture information was computed from numerous overlapping blocks, which were used to delineate the areas of the lung with anomalies. Their method's effectiveness was assessed by measuring accuracy, precision, and sensitivity across different patch sizes. Interestingly, the (5,5) patch size produced the best outcomes, achieving an accuracy rate of 85.8%, precision of 83.6%, and sensitivity of 87.4%.

In an independent investigation, as elucidated in Cheng et al. [4], a team of researchers introduced a convolutional neural network (CNN) framework designed to integrate distinctive attributes for pneumothorax detection in chest X-ray images. In the initial stages, they harnessed two methodologies to refine the preprocessed image specimens. Subsequently, the classification process was executed by means of a feature fusion CNN model, which amalgamated Gabor characteristics with the enriched information extracted from the images. Through a comprehensive range of qualitative and quantitative experiments, they demonstrated that their proposed model demonstrated superior performance, especially when faced with chest X-ray images obtained from various angles or viewpoints.

Gooßen et al. [5] proposed a method for identifying and localizing pneumothorax in X-ray images. They methodically analyzed the attributes of various deep learning methodologies, including convolutional neural network (CNN), multiple instance learning (MIL), and fully convolutional network (FCN). They conducted a fivefold cross-validation on a dataset containing 1003 chest X-ray images. Through receiver operating characteristic (ROC) analysis, they assessed the performance of these three techniques. The results showed that CNN achieved the highest area under the curve (AUC) at 0.96, indicating its superior overall detection capability. However, MIL and FCN provided a higher level of confidence in terms of localization, allowing radiologists to identify the specific image region responsible for the network's decision. This enhancement in localization increased trust in the proposed deep learning architecture.

In the same year, another group of researchers led by Wang et al. [6] introduced a technique for the high-resolution analysis of medical images depicting pneumothorax in X-rays. Their methodology utilized a deep convolutional neural network (DCNN) for image classification. This DCNN model integrated several crucial elements, including a network-in-network (NIN) module for data refinement, random histogram equalization for data augmentation, and the core DCNN architecture itself. Through their experiments, they showcased the efficacy of their approach in significantly enhancing the accuracy of pneumothorax diagnosis. Their method achieved an impressive area under the curve (AUC) of 0.9844 on the ZJU-2 test dataset and an even higher AUC of 0.9906 on the ChestX-ray14 dataset.

In [7], the aim of the study was to create convolutional neural networks for detecting pneumothorax abnormalities in chest X-rays. They introduced a specialized neural network named CheXLocNet, crafted primarily for delineating pneumothorax lesions. Interestingly, CheXLocNet not only delivered precise classification results but also generated high-quality segmentation masks for pneumothorax lesions, making it a versatile tool for this medical imaging task. In the same year, another study documented in [8] presented an approach called "SIIM-ACR

Pneumothorax Segmentation." This approach focused on utilizing DICOM format images from the Kaggle dataset. They initiated the process by preprocessing these images and subsequently feeding them into the popular UNet architecture. Within this architecture, they employed the se_resnext50 32*4d configuration as the backbone of the network, enabling the detection of pneumothorax as a mask for semantic segmentation on X-ray images. Following the segmentation, they conducted postprocessing to eliminate noisy elements and pinpoint the mask associated with the pneumothorax region. Subsequently, they utilized the mask images to calculate the percentage of pneumothorax.

In Wang et al. [9], a pioneering learning framework named DeepSDM was introduced, strategically leveraging the abundant information embedded within the Signed Distance Map (SDM). This framework embraces a multitask paradigm, enabling concurrent acquisition of the binary segmentation mask and the SDM. Furthermore, in the SDM regression task, a boundary-centered weighting mechanism was implemented with the specific aim of prompting the model to assign greater significance to the precise delineation of the pneumothorax region and its intricate boundaries. To corroborate the efficacy of DeepSDM, a comprehensive array of experiments was conducted on two distinct datasets: the Kaggle SIIM-ACR Pneumothorax Segmentation dataset and the PTX-498 dataset.

In [10], an inquiry investigated the application of image search techniques based on deep features generated by DNN to distinguish pneumothorax in chest X-ray images. By utilizing image search as a classification method, all chest X-ray images were initially assigned feature vectors. When presented with a query chest X-ray image, the classifier employed majority voting among the top k retrieved chest X-ray images. Notably, they examined a database containing over 550,000 chest X-ray images for this study. This research elucidates the remarkable identification performance achievable through image search utilizing Auto Thorax-Net features, thereby paving the way for practical applications. They achieved an 82% AUC accuracy in a fully automated search across 551,383 images, covering normal cases, pneumothorax, and various other chest conditions. Additionally, 92% AUC accuracy was achieved for a semi-automated search in 194,608 images, specifically targeting pneumothorax and normal cases.

In a recent study, documented in [11], a novel approach to pneumothorax diagnosis was unveiled. This method hinged on the utilization of Bayesian pedagogy to instruct a computational model and embraced an integrated framework. Furthermore, it offered a formal rationale grounded in the domain of human learning cognitive science. The investigation revealed that medical professionals, when exposed to explanations generated through Bayesian pedagogy, exhibited the capability to proficiently anticipate the diagnostic determinations made by the AI system. Moreover, their proclivity to concur with these AI-based diagnostic verdicts was notably heightened. In empirical trials, experts, on average, accurately anticipated the AI system's decisions for six out of eight images.

For pneumothorax identification, a unique two-stage deep learning strategy based on residual neural networks is proposed by Tian et al. [12], where they employed local feature learning, succeeded by global multi-instance learning. To extract distinctive features, the methodology entailed the exclusion of the bulk of nonlesion areas within the images. The assessment of model effectiveness involved various measures, such as accuracy, precision, recall, specificity, F1 score, ROC, and AUC, with the objective of evaluating the proficiency of lesion identification. In the most recent study [13], a novel automated approach was employed to detect image segmentation techniques. The preprocessing tasks were managed using MATLAB 2020b software, incorporating image processing techniques. To distinguish between normal and abnormal lung chest X-rays, the study utilized SVM classification. Features were extracted from lung images using texture-based segmentation techniques, and rib boundaries were identified through Sober edge detection. Remarkably, the segmentation method based on gray-level co-occurrence matrices demonstrated a notable improvement in accuracy, achieving an AUC of 0.91.

In the current year, a new method presented in [14] extended the concept of transfer learning and revamped the existing UNet model to introduce a fresh architecture named PTXNet for region-of-interest segmentation. Within PTXNet, the traditional encoder underwent a transformation by integrating components from the EfficientNet, SE-ResNeXt50, and Xception convolutional neural network architectures. Additionally, the decoder phase was enhanced by introducing residual blocks, and a novel approach involved concatenating feature maps from previous decoder stages alongside the standard global skip connections. PTXNet underwent training using a significant dataset containing in excess of 15,000 chest radiography images, resulting in the attainment of a mean dice coefficient of 84.89%. Notably, the findings revealed that this proposed approach outperformed the UNet model, exhibiting an impressive 18.76% increase in the mean dice coefficient.

In their comprehensive analysis and meta-analysis, Takahiro et al. [15] scrutinized the body of literature regarding pneumothorax detection utilizing deep learning methods. They included 63 studies in their analysis and found that, despite the prevalence of significant bias in most of the studies, the diagnostic capabilities of deep learning models were found to be similar to those of physicians. Deng et al. [16] initially established a pneumothorax dataset by employing a transfer learning model, ResNeXt 101, on X-rays collected from Poh-Ai Hospital. The researchers achieved a maximum accuracy of 93.89%. Wang et al. [17] developed an optimal dual deep learning model by utilizing three models for localization and an additional three for the classification of pneumothorax.

Gourisaria et al. [18] introduced seven novel CNN models, out of which the PneuNetV1 model achieved the highest performance, reaching an accuracy of 91.23%. Furthermore, they compared their results with seven pretrained models. Agarwal and Choudhary [19] investigated a light-weight CNN with attention mechanism for parameter-efficient architecture for detecting pneumothorax. For channel-wise feature map, this model integrates stacked convolutional layers with an attention. This model achieved a sensitivity of 0.73 and an AUC of 0.79. In [20], Thian et al. investigated the influence of dataset size on the accuracy of CNN-based pneumothorax classification. They employed a combined dataset comprising ChestXray 14 and CheXpert for this analysis. ResNet-50 achieved an AUC score of 0.86 on the combined dataset.

Recently, Malik et al. [21] developed an automated classifier capable of classifying not only pneumothorax as well as additional respiratory conditions like COVID-19, lung carcinoma (LC), tuberculosis (TB), and pneumonia, all from a single chest X-ray. To achieve this objective, they designed an integrated architecture called CDC-net based on a residual network with dilation convolution. Additionally, statistical analysis was implemented on the proposed model to demonstrate its significance.

4.3 Materials and Methods

In the medical imaging domain, a computer-assisted evaluation is enabling higher image interpretations that have always been challenging. In the image understanding arena, recent breakthroughs in machine learning, especially deep learning, have made significant progress in assisting with identifying, classifying, and quantifying structures in medical data. The advances are supported by hierarchical visual features learnt directly from data rather than handwritten attributes developed and supported by subject experience. As a result, deep learning is quickly establishing itself as a cutting-edge platform, delivering better leads for a variety of medical applications. In this section, we will discuss the approaches, strategies, and tools that were selected for the work, including concepts and algorithms that are essential for carrying out this study and have also been examined on the datasets to get effective results.

4.3.1 Dataset

The data used to validate our proposed model in the context of medical diagnosis and on which the experiment is conducted is publicly available on Kaggle [22]. The total data size is 812.32 MB. The data is in the form of a CSV file, named train_data.csv, and has three columns containing the image id, image name (file name), and the target values "0" and "1," where "0" denotes the absence of pneumothorax and "1" denotes the presence of pneumothorax in respective images. The chest X-ray images are present in a folder named small_train_data_set and several images 2027 are in png file format with the size 1024*1024. The training dataset consists of 1824 total images, 1437 images with the presence of pneumothorax, and 387 images with the absence of pneumothorax. The validation and testing dataset comprised 101 images in each set, and the testing data contains 102 images in total, out of which 80 images are with the presence of pneumothorax and 22 images with no pneumothorax.

4.3.2 Convolution Neural Network (CNN)

CNN stands as a cutting-edge learning algorithm in the domain of computer vision, characterized by its fundamental utilization of mathematical convolution, a specialized variant of linear operations. This process of convolution merges two functions, producing a third function that vividly demonstrates how the shape of one function is modified by the other. Put differently, CNN operates by multiplying two matrices to produce an output that plays a crucial role in feature extraction from images. CNN serves two primary functions: first, "Feature Extraction" entails utilizing convolutional techniques to identify and extract distinctive characteristics within an image for subsequent analysis. Second, "Classification" involves employing a fully connected layer that utilizes the results of the convolutional process to assign the image to a specific class based on information gathered during an earlier phase. Different layers in CNN are

1. Convolutional Layer: In 1D, the convolution operation works with two signals, while in 2D, it works with two pictures. The basic goal of a convolutional layer is to detect edges, lines, color dips, and other visual elements from images.
2. Padding Layer: This concept pertains to the act of adding extra pixels to an image after it has undergone processing by a CNN's kernel. It is an effective method, especially when the convolutional kernel requires additional pixels to properly process the image's edges. In such cases, these extra pixels are appended to the image's outer edges, often by replicating the existing edge pixels. Alternatively, another common approach involves padding the image with zeros, commonly known as zero-padding, to ensure that the image receives a precise and accurate analysis during convolution.
3. Pooling Layer: Subsequent to a convolutional layer, it is conventional to employ a pooling layer. The principal function of this layer is to reduce the dimensions of the convolved feature map, consequently diminishing computational demands.

   Three distinct types of pooling operations are as follows:

   (a) Max pooling: This entails the selection of the highest pixel value within the batch.
   (b) Min pooling: In this case, the minimum pixel value within the batch is the chosen criterion.
   (c) Average pooling: This operation calculates the mean value derived from all the pixels in the batch.

4. Fully connected layer: The fully connected layer is composed of neurons, along with their associated weights and biases. Its purpose is to establish interconnections between neural connections spanning two layers. Typically, in a CNN architecture, the fully connected layer is situated toward the end, just prior to the output layer.
5. Activation function: The activation function plays a pivotal role within the CNN model. It serves as a crucial component for learning and approximating intricate relationships between network variables, especially when these associations are

continuous and complex in nature. In more straightforward terms, the activation function determines which pieces of model information should be activated in the forward propagation process and which should be disregarded as the network reaches its conclusion. In our proposed model, we employ "ReLU" and "Linear" activation functions.

4.3.3 Feature Extractor

In the past, feature extraction from image data involved the identification of the noteworthy elements within an image and converting them into a condensed feature vector. This task was accomplished using specialized algorithms for feature detection, extraction, and matching. However, the landscape of image and video analysis has undergone a significant transformation with the emergence of deep learning. Deep learning methodologies have gained prominence due to their innate capacity to directly process raw image data, eliminating the need for a separate feature extraction phase. Various computer vision applications, including tasks such as image alignment, object identification, categorization, and image retrieval based on content, all require the efficient representation of image attributes. This representation can be achieved either implicitly through the initial layers of a deep neural network or explicitly through the utilization of established techniques for image feature extraction.

4.3.3.1 Transfer Learning

In the realm of deep learning, transfer learning entails utilizing a model that has been previously trained for a different task to address a new task. This approach enables a machine to enhance its predictions for a novel task by building upon knowledge gained from prior work. Transfer learning essentially taps into the expertise of a deep learning model that has been previously trained and applies it to a different yet closely related problem. It is worth noting that popular models like VGG, Inception, and ResNet, renowned for their performance in tasks like image recognition on ImageNet, are readily accessible via Keras for utilization in transfer learning scenarios. ResNet-50 is a convolutional neural network comprising 50 layers. Within the domain of computer vision, ResNet (short for residual networks) is a widely used neural network architecture that serves as a fundamental building block for numerous applications. The key breakthrough introduced by ResNet was its ability to effectively train extremely deep neural networks, surpassing the challenge of handling networks with more than 150 layers.

ResNet-50 sits in a middle ground between very large neural networks, which can exceed 100 layers, and smaller ones with around 20 layers. The significant problem encountered by large neural networks is known as gradient vanishing, which leads to a halt in the learning process during training. The core function of

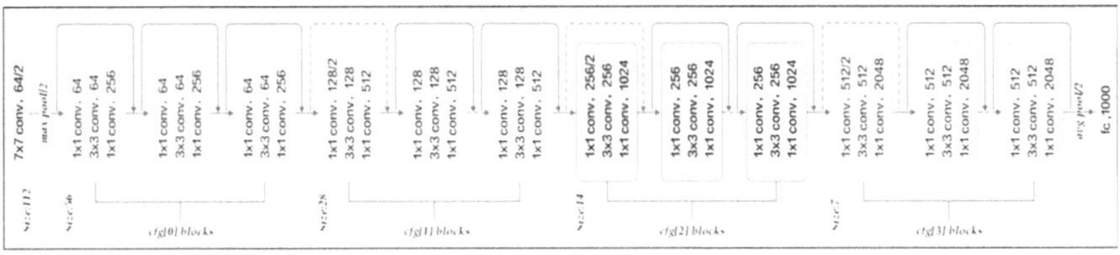

Fig. 4.2 ResNet-50 architecture

ResNet-50 is the incorporation of residual functions, primarily aimed at addressing the gradient vanishing problem. This problem of gradient vanishing is a critical challenge faced by deep, multilayered neural networks. You can observe the architectural layout of ResNet-50 in Fig. 4.2..

There are two kinds of blocks in ResNet-50:

(a) Identity block: If there is no change in the input or output dimensions, Identity Block is adopted.

  Input size == output size

(b) Convolutional block: Convolutional blocks are virtually identical to identity blocks, but a convolutional layer is used in a shortcut path to simply adjust the size so that the input and output sizes match.

  Input size! = output size The value of "x" is added to the output layer only if, Input size = = Output size, that is, the dimensions are the same
  There are two approaches to equalizing the input and output sizes: Performing 1*1 convolution and Padding the input volume.

The size of the output layer can be calculated by using the formula:

$$\left[\left\{\left(n+2^{*}p-f\right)/s\right\}+1\right]^{\wedge}2,$$

where

n = input size of image
p = padding,
s = stride
f = number of filters

4.3.4 *Classifier*

One of the most prominent applications of AI technology involves the utilization of machine learning classifiers to autonomously assess data, streamline procedures, and glean valuable insights. These algorithms, known as classifiers, automatically sort or categorize data into predefined "classes." A common example is email classification, where emails are examined and sorted as spam or legitimate. By leveraging machine learning algorithms, formerly manual tasks can be automated, resulting in significant cost savings, time efficiency, and increased productivity for businesses. A range of classifier categories, such as SVM, naïve Bayes, and decision trees, are deployed to fulfill these objectives.

Various classifier types, including decision trees, naïve Bayes, and SVM, are employed to accomplish these tasks. The focal point of the proposed study revolves around the binary categorization of chest X-ray images, and to tackle such tasks, the SVM is harnessed owing to its exceptional performance. SVM stands as a widely renowned supervised learning algorithm, celebrated for its proficiency in resolving both classification and regression quandaries. Nevertheless, its predominant utilization pertains to addressing classification challenges within the sphere of machine learning. The fundamental aim of the SVM algorithm is to ascertain the optimal hyperplane or decision boundary, effectively partitioning an n-dimensional space into discrete classes. This, in turn, facilitates the expeditious classification of new data points in subsequent applications.

In an n-dimensional space, there can exist multiple lines or decision boundaries capable of segregating different classes of data. However, the crucial task is to pinpoint the optimal decision boundary for effectively classifying the data points. This optimal boundary is referred to as the hyperplane in the context of SVM. The dimensions of the hyperplane are determined by the dataset's features. For instance, if there are only two features, the hyperplane will be a straight line. Conversely, if there are three features, the hyperplane will possess two dimensions. The hyperplane is always positioned to maximize the margin, which is essentially the space existing between the data points. Support vectors, on the other hand, refer to those specific data points or vectors that are nearest to the hyperplane and wield the most significant influence over its placement. These vectors earn their name because they provide crucial support to the positioning of the hyperplane.

4.3.5 Optimizers

Optimizers are methodologies or strategies used to modify the characteristics of the neural network, such as weights and learning rates, in order to minimize losses. Adaptive moment estimation (Adam) is a gradient descent optimization approach. The method is extremely productive and utilizes minimal memory when dealing with complicated situations requiring a large amount of data or components. It naturally mixes the "gradient descent with momentum" algorithm and the "RMSP" algorithm.

4.3.6 Regularization Technique

Overfitting is a typical problem that occurs while training not only deep learning models, but also machine learning models in general. When a model remembers examples and appears to perform effectively on the training data, it is termed overfitting. It is, however, impossible to apply what it has already learnt to previously encountered testing or validation data. Weight decay is a strategy that modifies the loss function to regulate the magnitudes of the weights and prevent the network from assigning exceptionally high values to individual weights. In practice, using this strategy to generate simpler models can assist to prevent overfitting. Weight decay is divided into two categories: L2 and L1 regularization.

4.4 Experiment and Results

In order to conduct the experiment, "Google Colaboratory" was used by utilizing GPU processor, TensorFlow API, and the Python programming language for code development. An outline of the model's implementation is visually represented in Fig. 4.3.

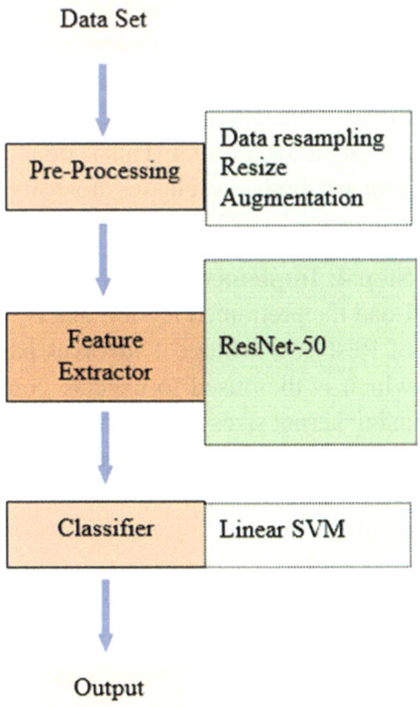

Fig. 4.3 Flow chart of the proposed work

Step 1: Import All the Libraries and Modules Required

Image processing tools are being used for a variety of machine learning and deep learning tasks. Python, on the other hand, is a very popular and trending programming language with a large number of Python libraries. As a result, it is important to use Python libraries for image processing to complete all machine and deep learning tasks. Different libraries used are Keras in TensorFlow, sequential, CV2, Scikit-image, PIL, NumPy, Matplotlib, Seaborn, OS module, Pandas, Train-test split, etc.

Step 2: Data Import and Data Splitting into Train, Validation, and Test Samples

Python has several modules that assist us in importing external data in different file formats into Python. With the train-test split function, the whole data is divided into train and validation parts, again validation data is further divided by the same function into validation and test parts. The training dataset comprises 1824 total images, 1437 images with the presence of pneumothorax, and 387 images with the absence of pneumothorax. The validation and testing dataset comprised 101 images in each set, and the testing data contains 102 images in total, out of which 80 images are with the presence of pneumothorax and 22 images with no pneumothorax.

Step 3: Preprocessing of Data

Now the training dataset for the images with the absence of pneumothorax is less in number, which gives rise to class imbalance. In order to overcome this, we created functions "trim" and "balance" to increase the number of images of this class by augmentation technique and balancing training data. A total of 387 images of class with no pneumothorax were available. The scarcity of data is handled by the augmentation technique by flipping them horizontally with zoom, height, and width range of 0.2 and 213 and new images were generated. Now 600 images of each class are taken for training. This is done in order to avoid biasness in training and to train data for both the classes honestly. Also, the image is resized from (1024,1024) dimension to (224,224) dimension.

Step 4: Implementation of the Proposed Models

Load the pretrained model, that is, ResNet-50. Four distinct stages are in the design of ResNet-50. Stage 1 involves passing a picture with dimensions of 224*224*3, which is then used to execute convolutions and max-pooling on the image array using kernel sizes of 7*7 and 3*3. In each phase of the network, there is a consistent pattern where three or more residual blocks are followed by three or more convolutional layers. Following these stages, the residual function comes into play within identity blocks after three consecutive convolutional layers. Notably, during the convolutional block, convolution is performed with a stride of two, effectively reducing the output size by half. These processes are consistently applied across all four stages of ResNet-50. Toward the culmination of the CNN architecture, a fully connected layer is incorporated. This layer's total number of outputs corresponds to the number of categories within ResNet-50's classification task.

4.5 Model Implementation

We initially have an image of size (1024,1024). However, for compatibility with the pretrained ResNet-50 model, which has specific input size requirements, we resize the image to (224, 224). This resized image is then fed into the ResNet-50 pretrained model. To employ the ResNet-50 pretrained model in Keras, we initialized it with the "include_top" parameter set to "False." This ensures exclusion of the fully connected layer responsible for classification in the original task. Additionally, we set the "weights" parameter to "imagenet" to leverage the pretrained weights from ImageNet, eliminating the need for random initialization. By utilizing ImageNet-trained weights, the model gains the advantage of calibrated parameters for making predictions.

In place of fully connected layer of the pretrained model, a global max pooling layer is incorporated. This pooling layer serves the dual purpose of feature extraction from the image and dimensionality reduction.

Following the pooling layer, a dense layer with 512 neurons is introduced, utilizing the "ReLU" activation function. Additionally, a dropout technique is applied with a dropout rate of 0.3, signifying that 30 percent of the activations from the preceding layer are randomly set to zero. To conclude the architecture, the output layer, structured as an SVM, is appended for data classification. SVM serves as a machine learning algorithm employed for classification, formally characterized by the allocation of a hyperplane. In binary classification, this hyperplane is represented as a line that partitions a plane into two segments, with each class positioned on opposite sides. In this context, a linear kernel SVM is utilized, which transforms the input data into a higher-dimensional space. This transformation is strategically designed to render the data linearly separable, facilitating the SVM's effectiveness in distinguishing between classes.

The optimization is handled by the Adam optimizer, set with a learning rate of 0.001. To enhance the performance of the SVM classifier, a "hinge loss" function is employed as the loss function. Additionally, L2 regularization is applied, and the activation function is designated as "Linear," given the binary nature of the classification task with just two classes.

After applying ResNet-50 for predicting the presence or absence of pneumothorax, the success rate of this method is presented in this section. Our model incorporates several Keras callbacks, including "Reduce Learning Rate on Plateau," "Early Stopping," and "Model Checkpoint." A noteworthy feature allows us to specify the number of epochs to train for before receiving a message that enables us to decide whether to stop training at the current epoch. We can either input "H" to halt immediately or an integer indicating how many more epochs should run before the message reappears. This functionality proves highly valuable during model training, enabling us to assess metric performance and determine when to conclude training. Moreover, the callbacks ensure that the model's weights are set to those of the epoch with the highest performance based on the tracked metric (AUC or validation AUC). Initially, the callbacks monitor the training AUC and adjust the learning rate until it reaches a predefined threshold. Once this training AUC level is attained, the callbacks shift their focus to monitoring validation loss and subsequently adjust the learning rate based on this metric. An overview of the model that is being suggested is illustrated in Fig. 4.4.

The training process employs a batch size of 20, meaning that each epoch involves training on a total of 1200 images distributed across 60 batches. The number of epochs set for training is 40. "Patience" is an integer value denoting the number of consecutive epochs that can transpire before an adjustment is made to the learning rate; for this specific model, it is configured as 3. The "float factor" parameter plays a part in determining the new learning rate, governed by the equation lr = lr * factor. In this study, the factor is set to 0.3. Moreover, a random state of 32 is established, and the training and validation datasets undergo shuffling during the training phase. Furthermore, there is the introduction of an integer parameter referred to as "stop patience," which denotes the consecutive epochs where

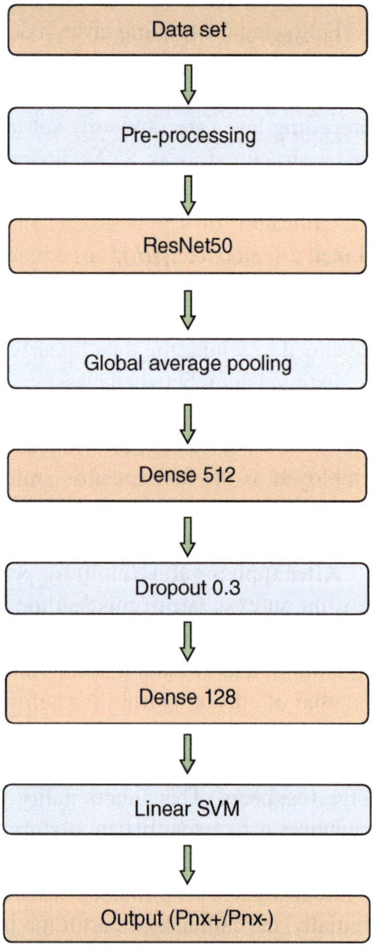

Fig. 4.4 A graphical representation of the model being proposed is depicted in the flow chart

adjustments have been made to the learning rate without any discernible enhancements in the monitored metric. It is instrumental in determining when to terminate training, and it aligns with the usage of Early Stopping callbacks.

As a consequence, after 40 epochs of training, the training set achieved an AUC of 92.91 with a training loss of 0.3418. Similarly, the validation AUC reached 88.31 with a validation loss of 0.4371, all under a learning rate of 5.9049e^09 as depicted in Fig. 4.5. Subsequent to the training phase, the model underwent testing on a set of test images, and the results were duly recorded in an array. The model exhibited an average accuracy rate of 79.261% in these assessments. It is worth noting that the initial results for the images, sourced from an Excel spreadsheet, were binary in nature, with "0" denoting the absence of pneumothorax in the X-ray and "1" signifying the presence of pneumothorax. Post-training, the obtained results were used to construct a confusion matrix, a task facilitated by the "sklearn" library. The values within this confusion matrix served as the foundation for computing a range of performance metrics and evaluating the classification model. While it is common in the literature for many authors to highlight the superiority of their proposed methods based on AUC, it is worth noting that only a limited number of researchers have determined additional performance metrics such as accuracy, sensitivity, precision, and specificity in their assessments. The formulas for the calculation of accuracy, recall/sensitivity, specificity, and precision are as written below.

$$\text{Recall or Sensitivity} = T_P / (T_N + F_N).$$

$$\text{Recision} = T_P / (T_P + F_P)$$

$$\text{F1 score} = 2\,(\text{Recall}^*\text{Precision}) / (\text{Recall} + \text{Precision})$$

$$\text{Specificity} = T_N / (T_N + F_N)$$

$$\text{Sorensen Dice Coefficient} = 2T_P / (2T_P + F_P + F_N)$$

In classification, a confusion matrix is a tabular representation employed to identify the locations where errors occurred in the model. Where the rows correspond to the actual classes for which the results should have been computed. The columns, on the other hand, indicate our predictions. This table clearly shows whose predictions are incorrect. The notations T_N (true negative), F_N (false negative), T_P (true positive), and F_P (false positive) pertain to distinct categories within the realm of classification assessment. For the proposed model, the required confusion matrix is

```
Epoch 38: ReduceLROnPlateau reducing learning rate to 5.904900035602622e-09.
60/60 [==============================] - 25s 408ms/step - loss: 0.3527 - auc: 0.9238 - val_loss: 0.4375 - val_auc: 0.8831 - lr: 1.9683e-08
Epoch 39/40
60/60 [==============================] - ETA: 0s - loss: 0.3395 - auc: 0.9296
Epoch 39: val_loss did not improve from 0.38934
60/60 [==============================] - 25s 407ms/step - loss: 0.3395 - auc: 0.9296 - val_loss: 0.4375 - val_auc: 0.8831 - lr: 5.9049e-09
Epoch 40/40
60/60 [==============================] - ETA: 0s - loss: 0.3418 - auc: 0.9291
Epoch 40: val_loss did not improve from 0.38934
60/60 [==============================] - 25s 407ms/step - loss: 0.3418 - auc: 0.9291 - val_loss: 0.4375 - val_auc: 0.8831 - lr: 5.9049e-09
```

Fig. 4.5 Training and validation AUC and loss

shown below with the prediction and actual values. False denotes a mistake or wrong prediction, whereas True denotes that the values were accurately predicted. Now, a confusion matrix in Table 4.1 is constructed and computes various measures to assess the model's quality. The confusion matrix as well as classification report of the proposed model is shown in Tables 4.1 and 4.2, respectively.

The graphical depictions of the training and validation loss metrics, alongside the validation AUC and training AUC of the proposed model, are visualized in Fig. 4.6. The relatively diminished size of the validation dataset in contrast to the

Table 4.1 Confusion matrix

$N = 102$	Negative (normal)	Positive (pathology)
Negative (normal)	17	5
Positive (pathology)	15	65

Table 4.2 Performance metrics

Metrics	Performance
AUC	0 0.8831
Recall or sensitivity	0.812
Specificity	0.773
Precision	0.929
F1 score	0.867
Dice coefficient	0.866

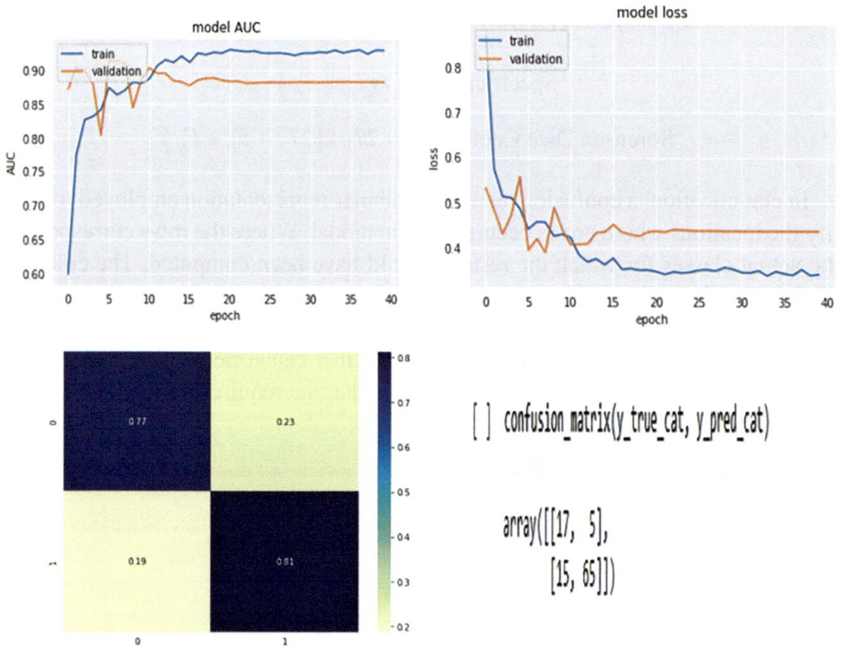

Fig. 4.6 AUC and loss graph with system-generated confusion matrices

training dataset elucidates the oscillations observed in the loss curves. If performance improvements on the validation data occurred after each epoch, the best weight configuration for each network was saved to storage media. We measured validation data performance using the AUC value, which is the area under the curve. As a result, we also included graphs displaying the AUC values of both the training and validation during training.

4.6 Discussion

Drawing inspiration from previously successful implementations, this study looks into ML and DL implementations and thus an integrated model of these two combines their benefits and provides a fresh look at the medical field. In this work, an integrated model is constructed using an SVM classifier and a ResNet-50 for deep feature extraction. An incomplete picture of a model's performance in the medical industry may be obtained from a single classification AUC result. Hence, an extensive evaluation encompassing accuracy, precision, specificity, sensitivity, and the F1 score was conducted. This holistic system exhibits a similarity to an automated mechanism that, with a single interaction, dissects and categorizes intricate information derived from chest images, discerning the presence or absence of pneumothorax. Furthermore, the amalgamation of DL and ML methodologies may be applied to domains characterized by limited data availability, even though they are optimally suited for classification tasks involving vast datasets. Despite the diminutive size of the dataset employed for feature extraction and classification, the capability to detect pneumothorax was significantly bolstered through the adept utilization of various potent techniques, encompassing drop-out, callbacks, data augmentation, global average pooling, and an optimizer.

DL methods have gained preference ML in the existing literature for pneumothorax detection because DL models are capable of making intelligent decisions without human intervention. A pivotal element influencing this inclination is the substantial influence of the size of the training dataset on the efficacy of deep learning models. It is noteworthy that the accessible chest datasets exhibit a scarcity of pneumothorax instances in comparison to the overall dataset magnitude. Consequently, there is a growing need for researchers to contribute to the collection of larger datasets that encompass a more substantial number of pneumothorax cases. It is a well-known challenge that most medical image datasets exhibit an inherent imbalance, typically containing fewer samples of pathological conditions than normal ones. This imbalance can potentially lead to biased results skewed toward the majority class samples. To mitigate this issue, many researchers have explored algorithmic approaches, such as introducing weighted loss functions that assign higher importance to the minority class samples, as evidenced in related studies.

The trend in medical image analysis has clearly shifted from traditional machine learning to more advanced deep learning techniques over the past decade. In the context of classification and localization, many proposed frameworks predominantly rely on pretrained CNN like DenseNet, VGG16, and ResNet. However, it is worth noting that there are ample opportunities for further exploration in this field. Although DenseNet and ResNet have been the favored choices for classification tasks, it could be intriguing to experiment with newer CNN models like EfficientNet in the context of automating the identification of thoracic pathologies, particularly pneumothorax

There are numerous methods for extracting features and machine learning techniques available. Pretrained models are deep feature extractors. Transfer learning offers a quick and effective technique to achieve this aim instead of having to build an ideal model from scratch for a particular task. It applies the weights learned from the pretrained model to our particular situation, leading to greater accuracy and shorter training times. This challenge is ideally poised for the application of deep learning, feature extraction techniques, and machine learning algorithms. Given that machine learning and deep learning methods in isolation yielded remarkable outcomes, the current study explored the prospect of further enhancing the performance of previously established machine learning and deep learning classifiers through aggregation.

From Table 4.3, we observe that in previous research [2] the data was balanced by weight cross-entropy loss function but still accuracy for feature extraction was so low that our model comparatively performed better. In [23], no proper explanation for class imbalance was provided, and in addition to this comparison with related works is missing while our study overcomes both its limitation by providing proper explanation for both. In [24], the problem of class imbalance was addressed by utilizing averaged binary cross-entropy during training. However, when training was conducted on a randomly split dataset, the model's performance indicated a need for improvement in this scenario.

In [25], the dataset taken for this research was balanced, but the criteria for selection were not mentioned and also a specific pathology should be the main focus. In [26], the research work was conducted without providing any proper explanation for class imbalance. The comparison with related works was unfair; also, the study lacks providing any experimental detail about data splitting, and performance in other metrics, for example, in terms of AUC, was missing. In [10], the dataset was balanced, but it was extremely small in addition to which performance also needs improvement as small training data is not adequate to train the model completely.

In [27], no proper explanation for class imbalance as well as data split is given, and also performance in terms of other metrics was missing. Issues such as unbalanced data, data scarcity, and feature redundancy contribute to lower accuracy rates. Optimal performance is achieved by deep and machine learning algorithms when there is a roughly equal distribution of samples across each class. The design of these algorithms focuses on maximizing overall accuracy and minimizing errors. In scenarios with imbalanced class data, a challenge arises as high accuracy can be achieved by simply predicting the majority class. However, this approach fails to adequately capture and address the minority class. Regrettably, prior studies have overlooked this significant concern regarding this dataset. To address this issue, the current research initially developed "trim" and "balance" functions. Additionally, data scarcity emerged as the second major challenge in this dataset. To overcome

Table 4.3 Comparison with previous related work

Work	Year	Technique	Methodology description	Class imbalance solution	Result (%)
[2]	2017	DL	For localization, DCNNs were used, and bounding boxes were created	Use loss function weighted cross-entropy	Classification: AUC:79.93
[23]	2018	DL	Utilize images with multiple resolutions and employ multi-instance learning techniques	Class imbalance not tackle	Classification: AUC:80.5
[24]	2019	DL	Pretrained ResNet-50 with and without fine tuning	Average binary cross-entropy calculated for a class	Random data ratio: AUC:87.0 Official data ratio: AUC: 84.0
[25]	2019	DL	Implement a generative adversarial network (GAN) for the purpose of discriminating between normal and abnormal X-ray images	Balanced dataset	AUC:84.1
[26]	2020	DL	VGG16, data augmentation	Class imbalance not tackle	Accuracy = 70.8 Val. Accuracy = 73
[10]	2021	DL	CheXNet-based transfer learning	A binary cross-entropy function with weighted components	AUC = 70.8
[27]	2021	DL	Create CNN architecture and pretrained VGG16	Class imbalance not tackle	Accuracy: 83.7
[28]	2022	DL	VGG19	Class imbalance not tackle	Accuracy: 89
Proposed work	2023	DL and ML	Pretrained ResNet-50 Along with support vector machine	Dataset is balanced by augmentation	AUC: 88.31 Recall: 0.812 Specificity: 0.773 Precision: 0.929 F1 score: 0.867 Dice coefficient: 0.866

this, augmentation techniques were employed to enhance the size of the balanced dataset.

The medical domain poses several challenges for machine learning and deep learning, despite their wide range of applications and promising potential. Architectures devised in existing literature are predominantly based on either of deep learning or machine learning algorithms. To harness the full benefits of these

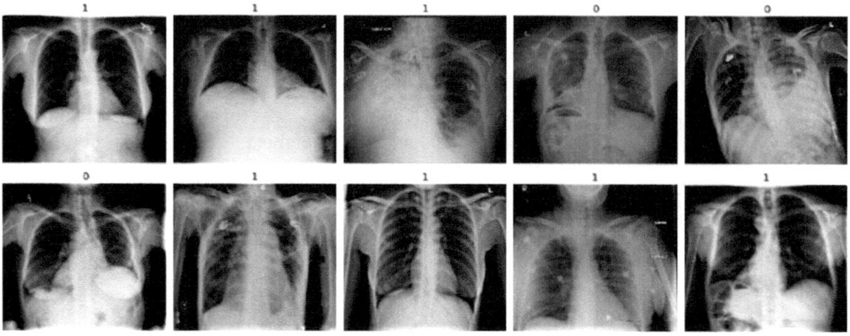

Fig. 4.7 Correctly classified X-ray images

techniques and address the challenges, our research explores combination of deep learning-based feature extractor and machine learning-based classifier. Ultimately, the proposed model's significance is assessed not only in terms of AUC but also through the measurement of recall, specificity, precision, F1 score, and dice coefficient.

In the proposed model, all the limitations of all the above models are properly explained along with class imbalance hurdle vanquished by augmenting technique. A comparison with the latest related works is done, and that too year wise and, most importantly, the performance of model matters. The proposed model outperforms in all the performance metrics.

Figure 4.7 shows the correctly classified X-ray with prediction value 0 for negative and 1 for positive result. Finally, we can conclude that our research outperforms previous works in the field of pneumothorax because by using publicly available datasets on Kaggle it allows other researchers to study, comprehend, and provide feedback. The results show that the proposed framework ResNet-50, a deep feature extractor, and support vector machine, a classifier that is aggregate deep transfer learning and machine learning, outperforms the existing approaches.

In conclusion, it can be affirmed that our research surpasses prior endeavors in the domain of pneumothorax, primarily attributed to the utilization of publicly accessible datasets from Kaggle. Our findings unequivocally demonstrate that the proposed framework, which combines ResNet-50 as a robust feature extractor and support vector machine as a classifier, representing an amalgamation of deep transfer learning and machine learning, excels in comparison to the preexisting methodologies. This endeavor will offer supplementary assistance to domain experts in making precise decisions. The practical implications of the proposed models in the medical domain in a clinical setting are far-reaching, encompassing improved diagnostic accuracy, early detection, enhanced efficiency, assisting diagnosis, risk reduction, alleviation of workload of domain expert, and the potential for transformative changes in patient care and medical research.

4.7 Conclusion

Cutting-edge AI methodologies, encompassing deep learning and transfer learning, occupy the leading position in the realm of prognosis and medical ailment diagnosis. In this investigation, a methodical assessment was carried out, culminating in the establishment of a fresh reference point for performance evaluation. The obtained outcomes mark the advent of a novel state-of-the-art standard, with an AUC reaching 0.8831 and a loss value of 0.4375, emblematic of a significant advancement within this field.

By obtaining these performances, we conclude that DL and ML are powerful tools as they give results of high accuracy to medical images. A comparison with various other models was done, and our model performs best among all of them. It is simple to understand and easy to implement over dataset. The presented model is efficient for achieving a relatively high diagnosis accuracy. This work will enable early treatment, increase the survival rate of pneumothorax patients, and give domain experts more assistance in making precise decisions. While the proposed model demonstrated superior outcomes in various aspects compared to prior research, there is still scope for improvement, particularly in terms of AUC. The model will be enhanced in the future utilizing approaches for selecting features across deep feature extractors and classifiers. Another future goal is to merge deep-radiomic features into a single model.

References

1. He K, Zhang X, Ren S, Sun J. Deep residual learning for image recognition. In: Proceedings of the IEEE conference on computer vision and pattern recognition; 2016. p. 770–8. https://doi.org/10.1109/10.1109/CVPR.2016.90.
2. Wang X, Peng Y, Lu L, et al. ChestX-ray8: hospital-scale chest X-ray database and benchmarks on weakly-supervised classification and localization of common thorax diseases. In: Proceedings of the IEEE conference on computer vision and pattern recognition; 2017. p. 3462–71. https://doi.org/10.1109/CVPR.2017.369.
3. Chan YH, Zeng YZ, Wu HC, et al. Effective pneumothorax detection for chest X-ray images using local binary pattern and support vector machine. J Healthc Eng. 2018;2018:1. https://doi.org/10.1155/2018/2908517.
4. Cheng Y, Feng J, Jia K. A lung disease classification based on feature fusion convolutional neural network with X-ray image enhancement. In: 2018 Asia-Pacific signal and information processing association annual summit and conference (APSIPA ASC); 2019. p. 2032–5. https://doi.org/10.23919/APSIPA.2018.8659700.
5. Gooßen A, Deshpande H, Harder T, Schwab E, Baltruschat I, Mabotuwana T, Cross N, Saalbach A. Deep learning for pneumothorax detection and localization in chest radiographs; 2019. http://arxiv.org/abs/1907.07324
6. Wang Y, Sun L, Jin Q. Enhanced diagnosis of pneumothorax with an improved real-time augmentation for imbalanced chest X-rays data based on DCNN. IEEE/ACM Trans Comput Biol Bioinforma. 2021;18:951–62. https://doi.org/10.1109/TCBB.2019.2911947.

7. Wang H, Gu H, Qin P, Wang J. CheXLocNet: automatic localization of pneumothorax in chest radiographs using deep convolutional neural networks. PLoS One. 2020;15:e0242013. https://doi.org/10.1371/JOURNAL.PONE.0242013.
8. Islam S, Rehana H, Asaduzzaman S, et al. Automated risk prediction by measuring pneumothorax size using deep learning. In: 2020 IEEE region 10 symposium (TENSYMP). IEEE; 2020. p. 1747–51.
9. Wang Y, Wang K, Peng X, et al. DeepSDM: boundary-aware pneumothorax segmentation in chest X-ray images. Neurocomputing. 2021;454:201–11. https://doi.org/10.1016/J.NEUCOM.2021.05.029.
10. Sze-To A, Riasatian A, Tizhoosh HR. Searching for pneumothorax in x-ray images using autoencoded deep features. Sci Rep. 2021;11:1–13. https://doi.org/10.1038/s41598-021-89194-4.
11. Folke T, Yang SC-H, Anderson S, Shafto P. Explainable AI for medical imaging: explaining pneumothorax diagnoses with Bayesian teaching. In: Artificial intelligence and machine learning for multi-domain operations applications III, vol. 11746. SPIE; 2021. p. 644–64. https://doi.org/10.1117/12.2585967.
12. Tian Y, Wang J, Yang W, et al. Deep multi-instance transfer learning for pneumothorax classification in chest X-ray images. Med Phys. 2022;49:231–43. https://doi.org/10.1002/MP.15328.
13. Mathumetha P, Rajagopal S, Vaidya S, Alhadidi B. Automated detection of pneumothorax using frontal chest X-rays. ECS Trans. 2022;107:861–72. https://doi.org/10.1149/10701.0861ECST/XML.
14. Patel A, Vidyarthi A. PTXNet: an extended UNet model based segmentation of pneumothorax from chest radiography images. Expert Syst. 2022;39:e12807. https://doi.org/10.1111/EXSY.12807.
15. Sugibayashi T, Walston SL, Matsumoto T, et al. Deep learning for pneumothorax diagnosis: a systematic review and meta-analysis. Eur Respir Rev. 2023;32:220259. https://doi.org/10.1183/16000617.0259-2022.
16. Deng LY, Lim X-Y, Luo T-Y, Lee M-H, Lin T-C. Application of deep learning techniques for detection of pneumothorax in chest radiographs. Sensors. 2023;23(17):7369. https://doi.org/10.3390/s23177369.
17. Wang CH, Lin T, Chen G, et al. Deep learning-based diagnosis and localization of pneumothorax on portable supine chest X-ray in intensive and emergency medicine: a retrospective study. J Med Syst. 2024;48:1. https://doi.org/10.1007/s10916-023-02023-1.
18. Gourisaria MK, Singh V, Chatterjee R, Panda SK, Pradhan MR, Acharya B. PneuNetV1: a deep neural network for classification of pneumothorax using CXR images. IEEE Access. 2023;11:65028–42. https://doi.org/10.1109/ACCESS.2023.3289842.
19. Agrawal T, Choudhary P. ALCNN: attention based lightweight convolutional neural network for pneumothorax detection in chest X-rays. Biomed Signal Proc Control. 2023;79:104126. https://doi.org/10.1016/j.bspc.2022.104126.
20. Thian YL, Ng DW, Hallinan JTPD, Jagmohan P, Sia SY, Mohamed JSA, Quek ST, Feng M. Effect of training data volume on performance of convolutional neural network pneumothorax classifiers. J Digit Imaging. 2022;35(4):881–92. https://doi.org/10.1007/s10278-022-00594-y.
21. Malik H, Anees T, Din M, et al. CDC_Net: multi-classification convolutional neural network model for detection of COVID-19, pneumothorax, pneumonia, lung cancer, and tuberculosis using chest X-rays. Multimed Tools Appl. 2023;82:13855–80. https://doi.org/10.1007/s11042-022-13843-7.
22. Gerry Pneumothorax Binary Classification Task. Available at: https://www.kaggle.com/code/gpiosenka/efficientnet-b2-f1-score-85/data; 2021. Accessed 3 Mar 2022
23. Yao L, Prosky J, Poblenz E et al. Weakly supervised medical diagnosis and localization from multiple resolutions; 2018. http://arxiv.org/abs/1803.07703
24. Baltruschat IM, Nickisch H, Grass M, et al. Comparison of deep learning approaches for multi-label chest X-ray classification. Sci Rep. 2019;9:1–10. https://doi.org/10.1038/s41598-019-42294-8.

25. Tang Y, Tang Y, Han M, Summers RM. Abnormal chest x-ray identification with generative adversarial one-class classifier. In: 2019 IEEE 16th international symposium on biomedical imaging (ISBI 2019). IEEE; 2019. p. 1358–61.
26. Bharati S, Podder P, Mondal MRH. Hybrid deep learning for detecting lung diseases from X-ray images. Inform Med Unlocked. 2020;20:100391. https://doi.org/10.1016/j.imu.2020.100391.
27. Choudhary P, Hazra A. Chest disease radiography in twofold: using convolutional neural networks and transfer learning. Evol Syst. 2021;12:567–79. https://doi.org/10.1007/s12530-019-09316-2.
28. Priya RK, Bimani AA, Bhupathyraaj M, Shabibi LA, Khusaibi MMMA. Multiple transfer learning-AI design for lung disease-pneumothorax and tension pneumothorax classification from chest-X ray. In: 2022 international conference on inventive computation technologies (ICICT); 2022. p. 1303–8. https://doi.org/10.1109/ICICT54344.2022.9850721.

Chapter 5
Prediction of Diseases Related to Heart by Using Different Techniques: Survey

Tanmay Kasbe and Sonal John

5.1 Introduction

In today's world, heart disease is the main cause of death. Due to cardiovascular diseases (CVDs), as per the latest report from the WHO, every year 17.9 million persons' lives are compromised. Heart disease has been brought by several risk factors, including alcohol and smoking, poor diet, and physical inactivity [1].

Heart attacks are considered more than four out of every five CVD deaths, and one in three deaths occurred before the age of 60 [2]. With the right diagnosis and treatment, many lives can be spared. A doctor identifies heart disease based on the patient's medical histories, physical examination results, and any additional symptoms if any. However, such types of methods of diagnosis do not precisely recognize patients with any type of heart disease. In addition, it is a very costly and challenging work to investigate computationally [3].

With all the above reasons, several computation techniques have been used to reduce the mortality rate. Identification at an early stage is the key point, and several techniques such as artificial intelligence, data mining, genetic algorithm, and hybrid are used by many researchers in terms of accuracy, Artificial intelligence and data mining techniques are the most popular; therefore, we will discuss these techniques in detail. In this chapter, we will also discuss the dataset and how without dataset any prediction model is not possible and for calculating the accuracy of any model, dataset is also very important [3].

T. Kasbe (✉)
Shri Vaishnav Vidyapeeth Vishwavidyalaya, Indore, Madhya Pradesh, India

S. John
Shri Vaishnav Vidyapeeth Vishwavidyalaya, Indore, Madhya Pradesh, India

Oriental University, Indore, India

T. Choudhury, A. Katal (eds.), *Machine Learning for Disease Detection, Prediction, and Diagnosis*, https://doi.org/10.1007/978-981-96-4241-0_5

5.2 Types of Heart Diseases

There are four main types of heart diseases:

- Coronary artery disease.
- Hypertensive heart disease.
- Rheumatic heart disease.
- Congenital heart disease.

5.2.1 *Coronary Artery Disease*

CAD ranks among the most common cardiac disorders. If any person has CAD, then that person's coronary arteries, which carry blood to the heart, may become blocked. As a result, less blood may flow to your heart muscle, supplying it with the oxygen it requires. Atherosclerosis, also known as artery hardening, is the condition that typically precedes the onset of the disease. Angina, also known as chest pain, is a symptom of coronary heart disease and can lead to a heart attack [4].

5.2.2 *Congenital Heart Disease*

CHD identifies when something is not correct with the heart of a developing baby while the organ remains in the mother's body. Septal anomalies are a type of congenital cardiac conditions. There are holes in the wall that separate the left and right sides of the heart. There is a procedure for closing the hole. Pulmonary stenosis is a distinct type of defect. A limited valve eliminates blood flow to the lungs. It can be opened or replaced with surgery [4].

5.2.3 *Hypertensive Heart Disease*

HHD is a combination of changes in the ventricle on the left side, atrium, and coronary arteries caused by a high blood pressure, which puts additional strain on the heart, because of which there are structural and functional changes in the heart. The changes include left ventricular hypertrophy (LVH), which can be one of the reasons for heart failure; patients with this condition have a significantly higher mortality rate. HHD refers to all of the short- and long-term effects of quite a high blood pressure, including diastole heart failure and conduction rhythm disturbances, particularly atrial fibrillation. This activity examines the evaluation of hypertensive heart disease and the role of the interprofessional team to handle it [5].

5.2.4 *Rheumatic Heart Disease*

RHD starts with the sore throat infection, which is caused by *Streptococcus pyogenes* belonging to group A, a bacterium that spreads easily from person to person, much like the upper respiratory tract infections. In some people, strep infections cause the immune system to attack body tissues, resulting in inflammation and scarring of the heart valves. RHD is caused by rheumatic fever-induced inflammation, and one of the main reasons is scarring of the heart valves [6].

5.3 Literature Survey

There are several machine learning algorithms introduced by researchers to predict the heart disease at early stages but only a few algorithms like SVM and KNN algorithms have the best accuracy, and in this chapter the authors used UCI datasets for training a model. The Catboost model introduced by the author has a 90.94% accuracy, and in this proposed model, the authors used the UCI dataset. The authors also mentioned and compared different models to predict heart diseases, but in terms of accuracy, their model performed better [1].

The authors proposed the K-model clustering techniques with 70,000 real-time datasets from Kaggle. In this proposed model, the authors achieved a higher accuracy of 87.28% and also compared many other algorithms like random forest (RM), decision tree (DT), and multilayer perceptron (MP) [7]. In terms of different models, machine learning models are best for diagnosing heart disease at the early stage, but the main challenge is real-time dataset because machine learning algorithms have the best accuracy in secondary datasets but still it is difficult to find how much high accuracy will achieve by researchers on primary datasets. This significant study was presented by the authors [8].

Many authors provide a comprehensive survey on heart disease diagnosis; here, the authors provide a feature selection technique on classification algorithms to predict heart diseases. The authors provide a detailed experimental evaluation of performance of different models of supervised learning algorithm [9]. A fuzzy rule-based system is also a very good model in terms of performance. The authors used the UCI dataset and MATLAB tool to develop an expert system that is based on the rule system [10].

5.4 Dataset Used for Prediction Models

As discussed in the introduction section, the accuracy of any model is very important for heart disease prediction and the actual patient dataset is very important regarding this. The UCL ML heart disease dataset, which consists of four independent datasets given by four independent medical institutions, is where the Cleveland dataset (UCI, 1990) that was used in most of the studies was obtained [9, 10].

There are 303 patient data instances in the Cleveland dataset, but 6 of them have missing values. The dataset that is used for heart disease prediction consists of 76 columns, but only 14 column or we can say attributes chosen by every study are the most common attributes for heart failure.

5.5 Common Architecture of Heart Disease Model (Fig. 5.1)

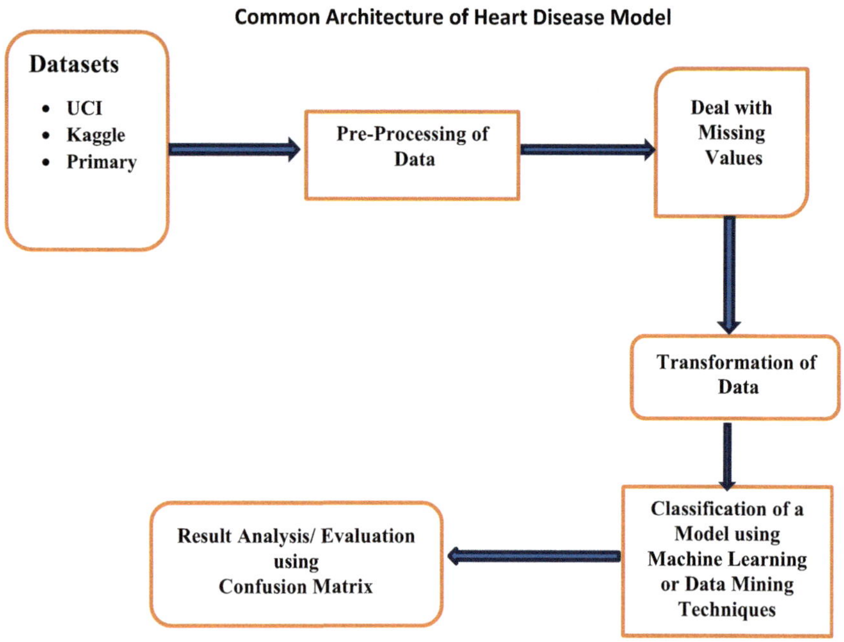

Fig. 5.1 Proposed methodology for the heart disease prediction (self-made)

5.6 Different Algorithms

In this section, we will explain all the algorithms that are best suited for heart disease detection at early stages. As there are many algorithms but in terms of accuracy and prediction level, we have considered only the mentioned algorithms.

5.6.1 Logistic Regression (Belongs to the Supervised ML Family)

LR is one of the main statistics-based techniques used to develop different ML models when the dependent variable is either divided or binary. Logistic regression is used to explain data and relationships between one dependent and more independent variables. Independent variables can be of nominal, ordinal, or interval types. The problems of classification and regression can be solved using logistic regression. Input values can be used gradually to predict the result using a sigma function or a logistic function and coefficients in order. It is one of the simplest models that is used to make proper predictions [8].

$$p = \frac{1}{1 + e - (a + bX)} \tag{5.1}$$

where in Eq. (5.1) a, b represent the model's parameters, X is a factor, and p is the probability.

5.6.2 Decision Tree (DT) (Nonparametric Supervised Learning)

DT algorithms are one of the most frequently supervised machine learning techniques, equivalent to a node-edge hierarchy ideal for classification problems. A node in the tree reflects the problem class, and each edge demonstrates the decision made in light of the results analysis. A classifier or classifiers are predictive machine learning models that demonstrate the relationship between features and dataset values. Every branch in the decision tree has its potential value indicated. It is widely accepted that the property with the highest entropy value is related to the root node. Classification and regression models (CART), iterative dichotomies 3 (ID3), and C4.5 are popular models of decision trees [1].

5.6.3 K-Nearest Neighbor (KNN) (Classification Algorithm)

It is also one of the types of supervised ML algorithm or we can also say a classification algorithm that may be used to solve the regression issues. In this algorithm, if an output value belonging to the target point is missing, then KN data points are found in the initial training sets, and the aggregate values of those points are supplied. In regression, the mean of the k labels is returned, and for the classification model, the mode of the k labels is assigned.

When prior knowledge of the data is not accessible, this fundamental algorithm is employed for categorization. The closest data points may be verified by using distance metrics like the Euclidean distance. Even with vast amounts of data, it can produce more accurate conclusions and forecasts [3].

5.7 Random Forest (RF) (Family of Classification and Regression)

It is also one of the supervised classification models made up of many decision trees interacting with one another. Regression as well as classification problems can be solved with random forest. The random forest, in which every tree predicts a class, has eliminated the drawbacks about the decision tree algorithm. Both accuracy and excess fitting of the dataset are reduced. Random forest is used to generate accurate predictions on large datasets, including high-dimensional data. Samples from the decision tree can be saved and applied to different kinds of data [7].

5.7.1 Support Vector Machine (SVM) (Family of Classification and Regression)

SVM algorithm is also one of the most popular ML algorithms that can be used in labeled data. This algorithm is perfectly suited to those models having small datasets. SVM can be used for the classification and prediction. Disease categories can be further divided into hyperplane in the context of heart prediction as having possibilities on one side of the margin and not having possibilities on another for developing cardiac disease. SVM uses the following mathematical formulas to divide data into hyperplanes:

$$\text{Positive hyperplane}: \boldsymbol{W}^{T}\boldsymbol{x}+\boldsymbol{b}=+\mathbf{1} \quad (5.2)$$

$$\text{Optimal hyperplane}: \boldsymbol{W}^{T}\boldsymbol{x}+\boldsymbol{b}=\mathbf{0} \quad (5.3)$$

$$\text{Negative hyperplane}: \boldsymbol{W}^{T}\boldsymbol{x}+\boldsymbol{b}=+\mathbf{1} \quad (5.4)$$

where x = vector point and w = weight vector.

The two types of SVM are linear and nonlinear, respectively. Nonlinear SVM is used when a line cannot be used to divide the data into discrete groups, whereas linear SVM can divide the data into discrete groups by using a line [9].

5.7.2 Naïve Bayes (Family of Classification Model)

Another supervised training approach for machine learning is naïve Bayes. It is typically applied to data-based Bayes theorem probabilistic classification. The naïve Bayes classifier assumes that features within a class are independent of one another and do not influence other features. No feature can influence the existence or non-existence of another feature quickly [1]. They contribute in some way to the total likelihood. The Bayes theorem looks like this:

$$\mathbf{P}(\mathbf{Q}|\mathbf{R}) = \frac{\mathbf{P}(\mathbf{R}|\mathbf{Q})\mathbf{P}(\mathbf{Q})}{\mathbf{P}(\mathbf{R})} \tag{5.5}$$

Equation (5.5) computes the probability (prob.) of Q given R; P(Q) is the independently calculated probability of Q; P(R) is the independently calculated probability of R; and P(R|Q) is the probability of R given Q. The naïve Bayes algorithm is useful for predicting heart disease and other diseases because it handles missing values effectively. It becomes even more advantageous when characteristics are given independence. It can operate on larger datasets and trains and classifies quickly [1].

5.7.3 Genetic Algorithms (Heuristic Search Algorithm)

An optimization method based on a natural, biologically favorable selection process is called a genetic algorithm. Iteratively, the genetic algorithm modifies a single population of solutions. Using a genetic algorithm, individuals from the current population are randomly selected as parent(s) to create offspring for the following generation. Over the next few generations, the population is "changing" toward the best possible outcome. The results of genetic algorithms show chromosomes. Genes that are individual components and represent the issue of make-up chromosomes. The population is called the collection of all chromosomes. The genetic algorithm uses at each step the following three main kinds of rules (operators) from the population:

(a) The choice of individuals is done in order to facilitate reproduction.
(b) Crossover is a technique used to blend the genes of two parents into the offspring.

(c) Mutation is used to alter new solutions in the quest for a better one. At a local minimum, mutation prevents GA from being trapped [1].

5.7.4 *Fuzzy Expert System*

Fuzzy expert system is one of the oldest but most accurate approach for prediction in any disease. The challenging task of fuzzy expert system is to create rule base system, which is the combination of "if..then" rule. These rules are the combination of input variable, or we can say combination of attributes, but these rules can be created under the supervision of experts in the field. This is one of the complex systems because if any modification is required in future, then it is very difficult to manage and sometimes again you need to create rules from the scratch [10].

5.8 Performance Evaluation/Result Analysis

Because disease prediction models deal with real-time situations, accuracy on the higher end of the spectrum is crucial. A similar reasoning applies to prediction models for heart disease: if a model's accuracy is high, it can be considered good; if not, it cannot be suggested. A number of assessment metrics—precision, accuracy, recall, and F1 score—are used to evaluate the accuracy of the input attribute selection techniques in this study. Researchers can determine the classification problem's performance rate based on four primary factors using a confusion matrix parameter, where true positive (TP), false positive (FP), true negative (TN), and false negative (FN) are the key parameters.

The following are the formulas for calculating accuracy, precision, and sensitivity.

$$\text{Accuracy} = \frac{\text{TP} + \text{TN}}{\text{TP} + \text{TN} + \text{FP} + \text{FN}} \tag{5.6}$$

$$\text{Precision} = \frac{\text{TP}}{\text{TP} + \text{FP}} \tag{5.7}$$

$$\text{Sensitivity} = \frac{\text{TP}}{\text{TP} + \text{FN}} \tag{5.8}$$

Irrespective of any techniques, for disease prediction above, the confusion matrix is always used and accuracy should be high for any successful prediction mode [7, 10].

Below, we explain how the accuracy of heart diseases will be calculated.

Let us say, we have the UCI dataset and we used the KNN algorithm for heart disease prediction. So, on 250 instances, the following are the values of confusion matrix. Considering the dataset values are

145 patients have heart disease, 105 patients are not having heart disease.

Now after applying the algorithm, the following are the values:

True positive = 139	True negative = 102
False positive = 6	False negative = 3

Now, for calculating accuracy, we are applying Eq. 5.6.

Accuracy = (139 + 102)/(139 + 6 + 102 + 3).
Accuracy = 241/250.

Accuracy = 0.964 in the term of percentage = 96.4%.

Now we can say, based on the above accuracy, our model is accurate.

In most of the heart disease diagnosis implementations, the mentioned algorithm is followed and the dataset is UCI, but sometimes few researchers work on the primary dataset, which is more reliable in terms of model building.

5.9 Case Study

5.9.1 Case Background/Presentation

A 65-year-old man with chronic systolic cardiac failure presents to the office with advancing heart failure symptoms (dyspnea, fatigue) and a 10-pound unintentional weight loss. He is also having more difficulty executing ADLs. He also suffers from nocturnal dyspnea. He complains of chest pain and has previously been diagnosed with 3-V CAD, which is not revascularizable. He had two ICD shocks a few weeks ago and a heart failure hospitalization a month before (advanced heart failure case study, ACC middle east conference 2016) (Fig. 5.2).

$$\text{Cr} = 1.4\,\text{mg}/\text{dL}$$

With the above case background, the following question is raised:

Question Which of the following is accurate about this patient?

Responses

A. This patient should be treated with more diuretics prior to deciding about candidacy for advanced heart failure therapy.
B. The degree of right heart failure precludes treatment with transplantation or VAD.

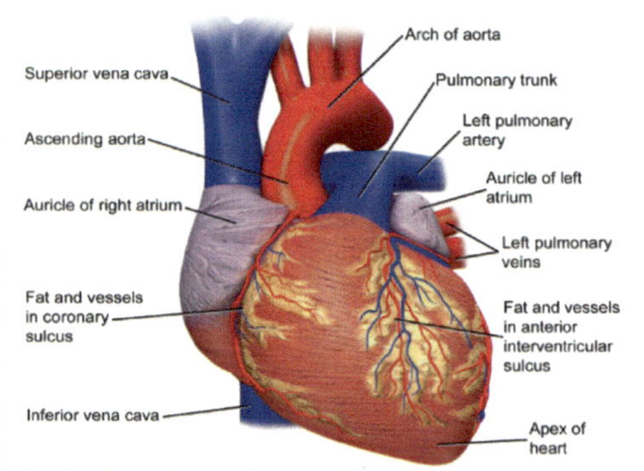

RHC:

RA=7

PA=58/25

PCWP=28

CI=1.8

Fig. 5.2 Front view of the heart (advanced heart failure case study, ACC middle east conference 2016)

C. Valsartan/sacubitril should be initiated based upon the results of the PARADIGM trial in this patient population.
D. This patient should be referred to a center that can provide VAD or cardiac transplantation.
E. This patient's age precludes successful treatment with either VAD or cardiac transplantation.

5.10 Justification and References

This patient has advanced heart failure and no obvious contraindications for transplant or VAD. Despite receiving inotropes and an IABP before VAD, the patient's symptoms did not improve. He was not stable enough to wait for transplants (Table 5.1).

While increasing diuretics may be a reasonable approach, the RA pressure remains constant.

Further diuresis could lead to hypotension and azotemia.

The issue was that we were unable to unload the LV further. Despite elevated PA pressures, the RA pressure is low, indicating adequate RV performance. Previous research has found that a CVP:PCWP ratio greater than 0.6 predicts RV failure after VAD.

The PARADIGM trial examined patients with class II–IV heart failure and found that valsartan sacubitril reduced hospitalizations and mortality compared to enalapril alone. However, less than 1% of patients in the trial experienced NYHA class IV symptoms (advanced heart failure case study, ACC middle east conference 2016).

Table 5.1 The UCI dataset attributes with definition (*UCI Machine Learning Repository*) [10, 11]

Attribute no.	Attribute name	Type	Definition
1	Age	Input–Integer	Age
2	Sex	Input–Categorical	Male or female
3	Cp	Input–Categorical	Four different types of chest pain: 1. Typical angina 2. Atypical angina 3. Nonanginal pain 4. Asymptotic
4	Trestbps	Input–Integer	Resting blood pressure/mm-gg
5	Chol	Input–Integer	Serum cholesterol/mm-gg
6	Fbs	Input–Categorical	Fasting blood sugar can be shown as FBS > 120 mg/dl then true
7	Restecg	Input–Categorical	It is known as resting electrocardiographic and the following are the values: 0 = normal 1 = ST-T wave abnormality 2 = left ventricular hypertrophy
8	Thalach	Input–Integer	Heart rate parameter
9	Exang	Input–Categorical	Angina related to exercise
10	Oldpeak	Input–Integer	ST depression parameter
11	Slop	Input–Categorical	Slop of the peek
12	Ca	Input–Integer	Values between 0 and 3
13	Thal	Input–Categorical	The following are the values of the parameter: 3 normal 6 fixed defects 7 reversible defects
14	Num	Output–Integer	This is the most important output parameter that provides the values between 0 and 4: 0 = absence 1, 2, 3, 4 = present

5.11 Conclusion

With different studies including research papers, book chapter, and survey papers, we can easily say that heart disease is one of the most deadly and risk-oriented diseases. In this chapter, we studied different parameters and algorithms that can be applied to the solution of heart disease diagnosis at early stages. This chapter also discussed several types of heart diseases that a person can have, with different or same symptoms. In this chapter, we also explained that the dataset is one of the key parameters for performance evaluation for any algorithm and for this purpose we have given details of the UCI dataset. Then, the most important part of any algorithm is accuracy and what are the procedures to calculate the accuracy, and for this objective, we have given details of confusion matrix with an example. We can conclude that all the mentioned predictive algorithms can reduce the mortality rate of heart disease.

References

1. Baghdadi NA, Farghaly Abdelaliem SM, Malki A, et al. Advanced machine learning techniques for cardiovascular disease early detection and diagnosis. J Big Data. 2023;10:144.
2. The World's First Market Research Aggregator. Global Information, Inc. (GII) – Premium market research reports; 2023. https://www.giiresearch.com/
3. Saboor A, Usman M, Ali S, Samad A, Abrar MF, Ullah N. A method for improving prediction of human heart disease using machine learning algorithms. Mob Inf Syst. 2022;2022(1):1410169.
4. Louridi N, Douzi S, El Ouahidi B. Machine learning-based identification of patients with a cardiovascular defect. J Big Data. 2021;8:133.
5. Tackling G, Borhade MB. Hypertensive heart disease. [Updated 2023 Jun 26]. In: StatPearls [Internet]. Treasure Island (FL): StatPearls Publishing; 2023.
6. World Health Organization. Rheumatic heart disease. World Health Organization; 2020. https://www.who.int/news-room/fact-sheets/detail/rheumatic-heart-disease
7. Bhatt CM, Patel P, Ghetia T, Mazzeo PL. Effective heart disease prediction using machine learning techniques. Algorithms. 2023;16(2):88.
8. Katarya R, Meena SK. Machine learning techniques for heart disease prediction: a comparative study and analysis. Health Technol. 2021;11:87–97.
9. Dissanayake K, Md Johar MG. Comparative study on heart disease prediction using feature selection techniques on classification algorithms. Appl Computat Intell Soft Comput. 2021;2021(1):5581806.
10. Tanmay TK. FRBF: a fuzzy rule based framework for heart disease diagnosis. Intel Artif. 2022;25(69):122–38.
11. Heart Disease. UCI Machine learning repository; 2023. Available at: https://archive.ics.uci.edu/dataset/45/heart+disease Accessed 19 Oct 2023.

Chapter 6
Automated Plant Disease Diagnosis with Machine Learning

Tanupriya Choudhury, Sumit Aich, Avita Katal, Madhur Aggarwal, Ayan Sar, and Subhangi Sati

6.1 Introduction

One of the major problems with globally sown crops is plant diseases, which can have far-reaching effects up to the level of hunger and poverty in certain regions caused by economic losses in the agricultural sector. Conventional disease diagnosis, using the visual inspection of the affected area with the use of manual controls, is very demanding and subjective, and the possibility of human error occurs often. Nevertheless, the emergence of artificial intelligence, machine learning, and other highly data-centric methods has disrupted the disease detection process in plant pathology, giving way to a more sensible and automatic machine-assisted approach. Deep learning network algorithms, along with machine and AI, show outstanding features in unveiling hidden patterns that lay within masses of data. These functions, collected from various sources such as images, sensor readings, and spectral signatures, are prerequisites for the training of ML models that can identify plant diseases at a highly accurate rate. While these modes provide the early resolve and the substitution with specific approaches and tell the crop, the ML model models these patterns and offers timely and necessary intervention, hence advancing the cause of

All authors are first authors for this chapter.

T. Choudhury (✉) · S. Aich · A. Katal · M. Aggarwal · A. Sar · S. Sati
School of Computer Science, University of Petroleum and Energy Studies (UPES), Dehradun, Uttarakhand, India
e-mail: tanupriya@ddn.upes.ac.in; akatal@ddn.upes.ac.in

T. Choudhury, A. Katal (eds.), *Machine Learning for Disease Detection, Prediction, and Diagnosis*, https://doi.org/10.1007/978-981-96-4241-0_6

crop sustainability and yield containment. The implementation of ML tech to agri-tech is not only ensuring better disease monitoring but also the accuracy and precision of disease control is done effectively. ML-based disease diagnosis systems offer several advantages over conventional methods, including ML-based disease diagnosis systems offer several advantages over conventional methods, including:

6.1.1 Automation

ML models accomplish the task of diagnosis of disease and, therefore, substitute the need for extensive labor-related interventions to get correct results and improve the degree of disease detection with speed.

6.1.2 Data-Centric Approach

Through machine learning methods, algorithms consider various massive amounts of data coming out of diverse resources. This assists in the comprehensive monitoring of plant health and disease presence. Such a data-oriented strategy for disease diagnosis increases the efficiency of accurate diagnosis.

6.1.3 Early Intervention

ML models serve to ensure all-encompassing disease detection, allowing farmers to react to the threats as soon as possible and block the dissemination of illnesses within the crop area, which, in the end, will minimize crop failure rates.

6.1.4 Customized Management

ML-driven disease detection systems enable a farmer to evaluate the outbreaks in a particular agricultural zone and determine the diseases that affect a specific crop in the particular surroundings.

6.1.5 Scalability

ML algorithms can be multiplied so that they are adaptable to a broad scope of agricultural terrain beyond the typical. Such scaled-up systems for large-scale production of automated disease diagnosis equipment would ease its popularity among farmers of all types.

Even though the above-said ML applications have their benefit, ML still faces issues in the development of plant disease diagnoses. This includes the need for large and labeled datasets, overcoming the problem of model explanation and sufficient generalization, and ensuring that the farmers, especially in areas of deficient resources, get easy access to this emerging technology. In conclusion, the AI and ML technology-based techniques in plant disease diagnostics have shown a remarkable step forward for current agriculture. Through the application of ML techniques, farmers, agricultural stakeholders, and the general public can realize crop management advancements, increase farmers' productivity, and ensure food security for a rising global population. A constant search and innovation in the field of concern still need to be carried on to deal with the remaining obstacles and to press for the optimal ones for the sake of agriculture sustainability and economic growth.

6.2 Literature Review

This is a growing area of application for understanding diseases of plants that exploit a broad range of machine learning algorithms, promising a great breakthrough in agricultural processes. Researchers utilize different approaches to deal with this problem., taking advantage of state-of-the-art machine learning engines, image processing techniques, and sensors. The reliance on manual inspection was labor-intensive and time-consuming, and consequently, the need for the introduction of automation was present. An important method implies the use of convolutional neural networks (CNNs) in image classification, which means deep learning models are trained in datasets comprehensively. This is done to recognize disease from the leaf's images with accuracy. An instance of this is the work of Liu et al. [1], in which the researchers employed a CNN-based network to precisely discriminate between healthy and diseased plant leaves. Although transfer learning has been an effective method in addressing the disadvantages of limited datasets, there are some remaining issues, such as the pretrained CNN models that often require fine-tuning to detect specific plant diseases [2]. However, the implementation of spectral imaging and hyperspectral imaging has further enhanced the possibilities of detecting various diseases, helping to identify the subtle physiological changes involved in these diseases [3]. Machine learning approaches with support vector machines and decision trees are another important tool for disease diagnosis whose effectiveness is seen in the nonsingular matrix and classifiers [4]. Moreover, including data from multiple sources, spectral ones, for example, as well as spatial and temporal, have been proven to enhance disease detection [5]. In the meantime, the wide application and good performance of deep learning algorithms, especially the brand-new techniques, such as recurrent neural networks (RNNs) and generative adversarial networks (GANs), have further expanded the scope of disease detection in plants by the brilliant analysis of temporal data and the synthetic images generation for data augmentation [6–8] mentioned its unique deep learning approach differently. They developed a framework that will ultimately be able to detect leaf disease in paddy crops. Sridevi et al. [9] did a novel work by the categorization and detection of tomato plant diseases, which are very frequent. This leveraged the power of convolutional neural networks to train and develop the methodology. Vamshi et al. [10] research work is the proposal of hybrid deep learning moderation for plant leaf disease detection, while [11] scientists with plantimage-specific deep learning model create a system "PlantLDNet" for disease detection in leaf images. Ramadan et al. [12] concentrate on enhancing the accuracy of wheat leaf pathogen detection by means of using synthetic-dependent and tiny datasets as well as the convolutional neural network architecture. Malhotra et al. [13] present agricultural AI as the new terminology for deep learning, with CNN at the core of it, which is also federated for the automated species identification of papaya leaf diseases. Patil et al. [14] come forward with a comprehensive concept that has various tendencies as well as technologies in trend. Mir et al. [15] advanced the multiclass classification of avocado leaf diseases by joining deep learning using CNN with random forests. Kumar et al. [16] excel in predicting citrus diseases by means of a hybrid CNN-SVM classifier. The publication of Suryavanshi et al. [17] reveals that federated learning CNN was instrumental in the cauliflower leaf disease diagnosis. Maitra and Damle [18] look at the impact of digital technological transformation on managing plants with new best practices. The two works of Rani et al. [19] and Banerjee et al. [20] are based on the usage of a convolutional neural network for the detection of plant diseases and the production of a hybrid model that includes CNN and SVM that classifies blackberry leaf disease accurately.

Irrespective of these breakthroughs, some complex challenges may still be prevalent among them, like access to adequate data, robustness to variations in the environment, and explanation of the model. These challenges, like adaptation of machine learning-based methods on a grand scale in plant pest detection, shall be indispensable to ensuring the widespread acceptance of these digital methods, thereby helping the world achieve the goal of sustainable agriculture and food security.

6.3 Methodology

There is a way by which plants are screened and diseases detected using machine learning approaches, which have several vital stages. First of all, the database is designed by integration of image or sensor data of healthy plants and their diseased counterparts obtained from different sources, such as agricultural research centers and plant pathology labs, along with the free resources available. The dataset goes into trouble-free treatment to improve its quality, and this is done with the use of data augmentation techniques such as rotation, flipping, and scaling, normalization to ensure consistency of image characters, noise removal to remove history artifacts, and image registration to have one plants of samples. Finally, we execute the feature extraction process that is used to capture especially relevant information from the data using the handcrafted features, deep learning-based feature extraction, and the transfer learning techniques that are applied. The features extracted are further used

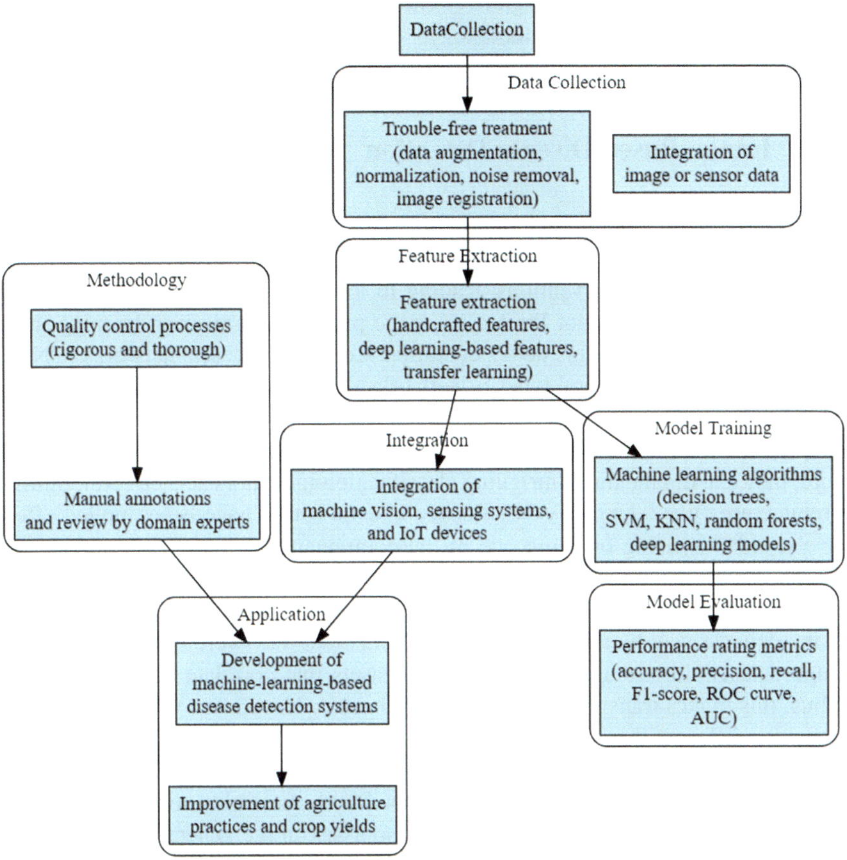

Fig. 6.1 Methodological diagram for disease detection in plants

as inputs to several machine learning algorithms, such as decision trees, support vector machines, for example, (SVM), K-nearest neighbors (KNN), random forests, and deep learning models, which are developed and trained on the labeled dataset to detect samples that are healthy and otherwise. Performance rating metrics like accuracy, precision, recall, F1 score, the receiver operating characteristic (ROC) curve, and the area under the curve(AUC) are among the tools used to measure the strength of the models that were trained. Regarding this, the integration of machine vision, sensing systems, and Internet of Things (IoT) devices makes the process more effective with real-time monitoring of plant health. It can be possible because of sensors' data, such as spectral reflectance and hyperspectral imaging used in recognizing the diseases at the earlier stage. Methodology refers to the quality control processes that are rigorous and thorough to guarantee the validity and authenticity of the dataset. In contrast, domain experts do manual annotations and review annotations periodically to ensure the dataset's wellness. This methodology will allow scientists to create computerized machine learning-based disease detection systems for plants and consequently develop more efficient agriculture practices and better yields of crops. A detailed diagram is cited in Fig. 6.1. from where the overview could be visualized for the methodology.

6.4 Image-Based Disease Detection

Especially today, where image-assisted disease detection in plants is an integral part of modern agriculture, early and accurate diagnosis of disease levels allows time for preventive interventions, leading to lower losses. The machine learning methods, including but not limited to image processing and deep learning, have played a leading part in this field's evolution. Convoluted neural networks (CNNs) are often the first of several types of networks that are used in most disease detection frameworks. Among various advantages of deep learning and convolutional, they are the masters of the picture features that appear at different levels; so, therefore, they can dig the most intriguing classification indicators at each level. Similar architectures like Alexnet, VGG, Resnet, and Inception have been applied effectively, either syncing or being one of the components of the more complex networks. Aspect of data scarcity, transfer learning has proved to be a powerful strategy in problem-solving, especially in cases when data is not enough. Models will obtain it the delicate way by retraining them, which is that they were initially trained on huge datasets like ImageNet on smaller datasets that correlate to plant diseases. Thus, this makes transfer learning very achievable. If we leave CNNs behind and consider different deep learning architectures such as recurrent neural networks (RNNs), generative adversarial networks (GANs), and Siamese networks, they all have also been explored. RNNs are equipped with the ability to model temporal dependencies over image sequences and also assist in the generation of synthetic images for training, just as GANs do. Finally, Siamese networks allow for classification according to the similarity of images. Ensemble methods compound predictions from various models, overcoming needing higher precision and endurance. Techniques like bagging, boosting, and combining involve the prediction's integration from various models, which exploit diverse data subsets or applications of different algorithms. A one-class classification is a go-to option when training on samples of only one category (usually when the training set consists of healthy plant data only) during the training phase. In this study, we are dealing with a lopsided dataset having quantities of diseased plants in fewer cases. Conventional machine learning methods based on architectures such as support vector machines (SVMs) and random forests integrate with deep learning techniques. These techniques use external properties of the images, such as color histograms, texture descriptors, and shape characters, especially when data is insufficient to do so. Active learning methodologies in neural networks have been shown to be more effective in the optimization of model performance as these methods iteratively select the most informative samples from the unlabeled data, essential for scenarios that have a shortage of labeled data. Overall, these machine learning techniques, along with preprocessing applications, data enhancement, and recursive iterations, make big progress in using images for identifying diseases of plants. Agriculture is benefiting more from those.

6.5 Sensor Data Fusion for Disease Diagnosis

Plant disease diagnostics using sensors with data fusion, consisting of sensor data integration, allow for more accurate and reliable disease detection. Varying machine learning techniques have been introduced together with the sensor fusion method for higher efficiency in detecting diseases in plant crops. Among the practical strategies one could employ is using the information acquired from various kinds of sensors, such as spectral sensors, thermal cameras, and hyperspectral imaging tools. Every sensor comes with different functions and captures different aspects of plant physiology at the same time. Machine learning algorithms, for instance, neural networks and support vector machines, can be trained on combined sensor datasets to discriminate between plants as healthy and diseased when different sensor modalities act associated with the features of the dataset. Another method of sensor data aggregation can be achieved by assessing the data of sensors located at different spatial points within a sowing field or greenhouse. Via the aggregation of data from different places, machine learning models are capable of better reflecting the spatial variability of the disease and, therefore, resulting in prospects for improving the robustness of the disease detection algorithms. This is an especially viable technique of large-scale agricultural monitoring and rapid diagnosis of disease hotspots that can be further managed to avert large-scale disease pandemics. Furthermore, a temporal combination of the sensor signals enables us to follow the changes in the plant's health over a certain period and detect a disease when it is still in the early stage. Through time-series data sensors, recurrent neural networks, and time-affected machine learning methods, which are capable of identifying temporal patterns that reveal disease development, could be used for the analysis. Incorporation of physiological data with disease symptom descriptions into the disease diagnosis models can help farmers make the best decisions as to when to implement the disease management strategies, and thereby, they can produce a better crop yield. Classification approaches, which include random forests and gradient boosting, are another way of achieving high-performing sensor data fusion systems by forwarding the observations from many machine learning programs, each modeled on a different subset of sensor data. The recognition of the inaccuracies and ambiguities in individual sensors' output using ensemble techniques leads to better knowing the real picture, which ensures more precise disease diagnosis outcomes. Hence, sensor data fusion, along with machine learning systems, can become an outstanding method for identifying diseases in plants using proper accuracy and effectiveness. Utilizing what is gathered from various sensors and applying machine learning technology that is on a sophisticated level, farmers can get the best of their crop health monitoring and increase agricultural productivity optimization.

6.6 Transfer Learning for Cross-Domain Disease Detection

Transfer learning has proved to be a valuable way of tackling the unique problem of small sample sizes for cross-domain plant disease identification. In this circumstance, while data with labels from different sources is readily accessible from the source domain, transfer learning is used to move knowledge from the source domain to the target domain; this accelerates the performance of the machine learning model. When it comes to fine-tuning, turning pretrained models of source domain data into target domain data is one of the useful approaches. Pretrained models, skilled at classes like ImageNet, contain general features that may be reused on other domains. With the aid of parameter fine-tuning, they can get accustomed to the target domain data, which can also allow them to adjust to the patterns of the specific domain of interest, such as different plant species or disease signs. With this approach, the model can learn from the source domain and make the adjustment to individual target domains through representation, leading to an enhanced detection beyond disease. Also, the domain adaptation algorithms are used to match the distribution of target and source domains to allow for benefit distribution. For instance, adversarial training uses a domain discriminator that can be trained together with the main model so that it is easier for the discriminator to learn the distinguishing features between the different domains. It guarantees that the model provides the features important for disease detection while filtering out variations specific to particular content. Keep away data that is not needed. Ensemble learning techniques feed into this by leveraging different source domains or transfer approaches and by combining predictions of multiple models obtained by such means of transfer. Through the integration of multiple types of knowledge, ensemble methods avoid tending to fit a single source domain and also enhance generalization gain to the target domain. Lastly, active learning methods can achieve information maximization by integrating only the most representative objects from the target domain into the annotation. This would eventually lead to a reduced reliance on the existing labeled data, thereby speeding up the process of model transfer. In particular, transfer learning stands out among all the machine learning techniques, which, when coupled with domain adaptation, ensemble learning, and the active learning approach, becomes a more effective approach than the rest that contributes significantly to cross-domain disease detection, using limited labeled data and improves the models' generalization performance. The transfer learning approach can be seen in Fig. 6.2.

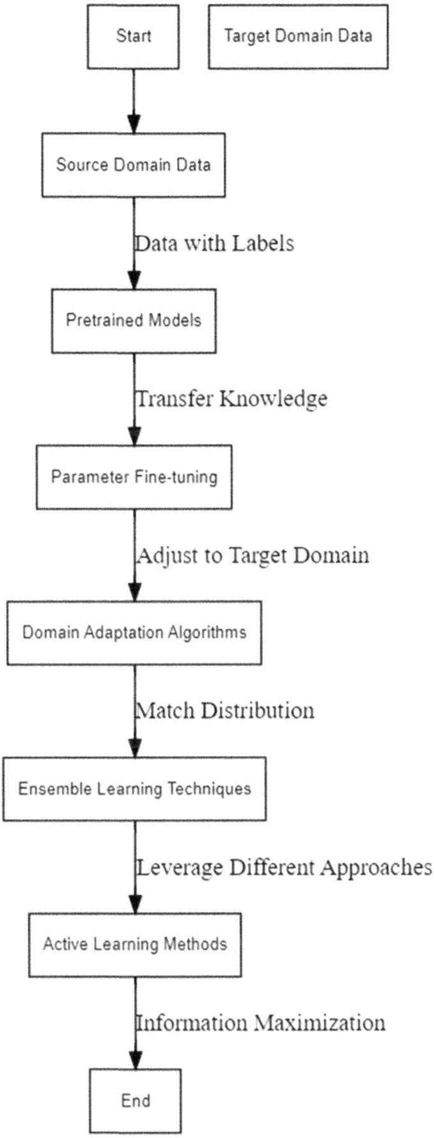

Fig. 6.2 Transfer learning approach for disease detection

6.7 Ensemble Learning Approaches

The integration of different machine learning methods by assembling learning strategies has demonstrated remarkable improvements in both the accuracy and robustness of the detection of diseases in plants by efficiently working together. Ensembles is an inclusive term for all methods, which include predictions from multiple basic learners to make a final prediction that outperforms individual models. As a form of plant disease detection, ensemble learning provides multiple benefits, such as better generalization, higher predictive power, and higher tolerance of the noise following uncertainty and imprecision of the data. Another approach of ensemble methods used is bagging (bootstrap aggregating). It consists of the training of a variety of base classifiers (learners) on different bootstrap samples of the dataset and averaging or voting of their predictions. Bagging is one of the techniques to improve the stability of learning algorithms; it accomplishes this by reducing noise in the training dataset and lessening the contribution of outliers. In plant disease recognition, wrapping can adapt to the different techniques of machine learning like decision trees, support vector machines, and artificial intelligence, among others. The boosting ensemble method is another popular method. This method trains the prediction of the base models one after the other, and the next model in the train aims to fix the errors made by the previous model. Enhancing algorithms, for example, AdaBoost and GBM, gradually increase the weights assigned to misclassified cases so that they are assigned higher weights, which in turn helps in identifying challenging scientific cases, thereby improving the overall performance. The new boosting techniques have been successfully employed in many machine learning algorithms for plant disease detection, having the ability to identify more accurate relationships in data indiscernible by the trained models. Error combination or voting is another powerful method in which the prediction output of the base learners is composed into a final prediction using a meta-reasoner who was trained for this task. Stacking is like a superimposition technique for combining different base learners, thus carrying also some complementary information, leading to overall improved performance. In plant disease detection, we can stack the multiple predictions by using different machine learning algorithms techniques such as CNNs, SVMs, and random forests to produce more accurate and reliable disease-identifying results. Briefly, ensemble learning techniques provide a framework for disease detection in plants with multiple machine learning models applied, which, after accumulating all their strengths into one, is much more powerful than each of them individually. Utilizing a variety of features like the ones mentioned above, boosting, bagging, and stacking allow these ensemble methods to improve the robustness and accuracy of plant disease detection systems, which consequently leads to better plant disease management systems and, as a result, improved yields.

6.8 Challenges and Solutions in Real-World Deployment

Implementation of machine learning for disease detection in plants exposes a set of challenges in practical cases of the farming sector. Still, it gives some prospects to get an insight into them.

6.8.1 Data Collection and Labeling

Reliable and appropriately labeled data to feed into computer learning models can present an obstacle owing to the diversity of symptoms associated with the diseases, the unavailability of experts, and the requirement for vast, large-volume, and heterogeneous datasets. Solutions are crowdsourcing labeling campaigns, transferring learning from related fields by leveraging them, and active learning policies to let a system know which part needs priority labeling to move forward.

6.8.2 Generalization to New Environments

Data that is carried out in one location or season may not generalize for new environments or times as environmental conditions may differ greatly from those of the data population or season. Resolving this challenge demands diverse feature selections, domain adaptation methods, and frequent model retraining using new data from respective deployment locations supplied within time intervals.

6.8.3 Resource Constraints

The outside world, which is deployed in resource-limited and constrained conditions with no computational power, memory, and energy availability, usually expects practical solutions. Efficiency problem stands as solutions that will be the optimization of model architectures as well as deployment of the models that can fit edge devices and, lastly, distributed computations.

6.8.4 Interpretability and Trust

It is extremely significant to understand and interpret machine learning model prediction, given that farming and other agronomic activities done by end users such as farmers and agronomists usually result in a loss of trust in these technologies. One of the key solutions is using explainable AI to gain insights into how model decisions are made, building UX interfaces for output visualization, as well as including stakeholders in the model maintenance and validation processes.

6.8.5 Data Privacy and Security

Sensitive data about agriculture, such as the images of the plants and the notes on diseases, poses great threats of theft and misuse and thus should be effectively protected from the perpetrators. The actions integrate secure data transmission basic elements, the privacy of the data requiring anonymization where possible, as well as compliance with applicable privacy regulations and standards.

6.8.6 Model Maintenance and Updates

Machine learning models require periodical maintenance and updates so they can cope with circumstances change, with the acquisition of new information including more functions, and also allow for performance improvement. In response, we propose deploying automated model monitoring systems, version control procedures for model updates, and retraining the model periodically using fresh data.

Therefore, introducing different scenarios in which machine-hunting techniques for disease detection in plants could be utilized appropriately is the way to make this approach more effective, scalable, and consistent. As a result, contexts such as agricultural productivity and food security could benefit.

6.9 Integration of Multimodel Data Sources

Data blending, especially for different data types, is a critical endeavor that leads to the precise representation of the chance and validity of plant disease detection by several machine learning techniques. The integration of data from multiple sources such as photos, spectrum data, environmental parameters, and genetic features leads to a more deep-rooted conclusion of the primary health status of plants, which in turn guides proper diagnosis and suitable treatment strategy. One way of incorporating multimodel data sources into models is through feature fusion, where features that manifest from different data dimensions are blended to create a uniform feature representation. For example, the features extracted using convolutional neural networks (CNNs) can be brought out together with spectral data acquired from hyperspectral apparatus and environmental factors such as temperature or humidity. The techniques of early fusion, late fusion, and hybrid fusion can be used to amalgamate features across the whole AI framework at different levels. A yet nontraditional way of approach may be the construction of multimodal machine learning models that may an information from different sources directly. Due to some problems (e.g., colors), which can be easily distorted in some images, ensemble learning methods such as multimodal ensembles join the results of the individual models trained on different data modalities to produce accurate output. An example of this would be a multilevel ensemble that involves CNNS, which is trained using visual data; SVM, which is trained using spectral data; and RNN, which is trained using rotating environmental data. Also, deep learning architectures, including multimodal neural networks and graph neural networks, have been tailor-made to deal with multidata sources as well as to harness those complex relationships among the data sources. In addition, these AS models can optimize overall representations from many data modalities, which in turn contributes to more efficient usage of multimodal datasets. Drawbacks of combining multimode data sources are the fact that the data have various heterogeneity, mismatch between the features, and scalability problems. The solution to these problems needs to be based upon a correct choice of preprocessing techniques, alignment methods for the features as well as a machine learning approach that is efficient enough to handle both a variety of and large datasets. Overall, the combining of varied data sources with multiple machine learning tools demonstrates higher chances of success in the development of the approach to plant disease detection. Using synergistic information from a variety of data models, the multimodel techniques can increase the reliability, tolerance, and generalization capacity of disease diagnosis systems, which help the agricultural industry to go toward a more stable and sustainable direction.

6.10 Real-Time Case Studies

In such a short period, crop production has encountered many innovations where machine learning techniques have been integrated into the early detection and control of plant diseases. An exemplary case has tomato crop disease detection machine learning technique application in its core as the major component. The situation on the fungal infection front being early blight and bacterial diseases as obvious bacterial spots, the farmers have had to deal with huge obstacles in addressing their crop health and producing yield. Consequently, scientists resorted to the development of a holistic framework based on the use of machine learning algorithms. Firstly, convolutional neural networks (CNNs) have been used for image analysis of plant leaves, and they have indeed succeeded in distinguishing between both healthy and diseased plants by recognizing their print and color patterns. The second type of implication is classification trees and the random forest algorithm, which are used to classify data based on the environment variables, including humidity, temperature, soil characteristics, and leaf characteristics. On the other hand, SVMs have benefited from this feature of handling nonlinear data and diagnosing diseased plants into various categories effectually. The creation of integrated systems that have enabled real-time plant health monitoring, at the same time, has also been able to empower early intervention and targeted treatment. Through machine learning adoption power, smallholder farmers around the globe have been able to curb yield loss, cut down on pesticide use, and boost resource utilization efficiency. This case study symbolizes the next step into the future, where machine learning is changing the way we fight diseases in crops and, as a result, moving agroecosystems into sustainable and reliable production systems.

6.10.1 PlantVillage

The PlantVillage is an app (application) designed by researchers at Penn State University and based on machine learning (computer learning) technology to diagnose plant diseases. The app renders a user interface where users can capture a picture of a plant with any disease and submit it to the website by simply uploading the images. The machine learning algorithms then scan the uploaded images to detect any diseases. The system deals with all value-chain crops and diseases used, helping farmers to effectively monitor crops and control diseases.

6.10.2 Deep Learning for Citrus Disease Detection

The following example comes from an article published in the journal *Plant Methods*, where researchers detected diseases in citrus plants using deep learning approaches. To accomplish this, they fed CNNs with citrus leaf images from the dataset and showed how to tell when plants are healthy and when they are not. The well-trained model rendered high accuracy in detecting citrus diseases like citrus canker and citrus greening, giving an insight into the possibility of deep learning in the identification of plant diseases.

6.10.3 Fusarium Head Blight Detection in Wheat

Fusarium head blight (FHB) is a serious disease embedded in such crops as wheat and is proliferating all over the world. A team at the University of Minnesota used machine learning to extract FHB from wheat fields when the drones flew by using the aerial images captured. The algorithms looked for the same symptoms and found the areas that proved to be associated with FHB infection, which enabled farmers to apply place-specific measures that led to lower yield losses.

6.10.4 Early Detection of Potato Late Blight

While the first occasion was a devastating potato late blight, which decimated crops and plunged Europe into starvation, the second event is the ubiquitous sprouting of spuds all over our surroundings. The collaboration between the researchers of

Cornell University and the International Potato Center (CIP) resulted in the development of models based on machine learning that were able to forecast the time of nature-driven late blight infections by looking into environmental features and the distribution of the disease. The models alerted farmers heedily before the onset of pests and diseases, thus making farmers take timely preventive actions to stop crop damage.

6.11 Conclusion

Finally, machine learning techniques have been formulated as a highly effective diagnostic instrument for identifying plant diseases and, as such, hold the potential of a new agritech approach that would change agriculture as we know it and improve food security. By way of employing various techniques, including convolutional neural networks (CNNs), support vector machines (SVMs), ensemble methods, and transfer learning, remarkable progress has been made in effectively discriminating and diagnosing certain plant diseases based on their images, quantitative or qualitative spectral data, and environmental parameters. These computational approaches have shown remarkable accuracy, sensitivity, and efficiency in building the systems for disease recognition, which, in some cases, even outperforms the best conventional methods and human experts as well. It can be achieved by utilizing big and comprehensive datasets. As a result, machine learning models can discover complicated patterns and relationships that are characteristic of plant diseases, thus providing enhanced accuracy in case of early disease diagnosis, which, in turn, can be applied to minimize crop losses, on the one hand, and to increase agricultural productivity, on the other. Nevertheless, obstacles remain, such as the demand for large annotated datasets that hinder the model generalization to different environments as well as new plants and the interpretability of the output data, which is rather complicated due to machine learning models that are often too advanced. It will be necessary for the challenges to be tackled by multidisciplinary collaboration, new research, standard datasets, evaluation criteria, and also open-source software. Moving forward, future research will aim at investigating more advanced machine learning techniques, measuring model interpretability, broadening model generalization for different contexts and applications, and creating an integration platform that creates a smooth transfer of disease detection systems from the research environment to real agricultural settings. On top of this, it is essential that efforts to tackle ethical, privacy, and security issues that appear with the exploitation of agricultural data would be crucial in the deployment of machine learning technologies for agriculture. What is more, it appears that machine learning is a key technology that helps to reach significant results in disease detection in plants. Therefore, research activity in this area will be necessary to build a sustainable agricultural industry to meet the growing demand of the population for the production of food.

Acknowledgments We want to acknowledge the PlantVillage portal [21] for providing us with a case study, which is beneficial to this chapter.

References

1. Liu X, Li Y, Fei S, Wang Z. A survey of image classification methods and techniques for improving classification performance. Int J Remote Sens. 2017;38(9–10):2630–57.
2. Fuentes A, Yoon S, Kim SC, Park DS. A robust deep-learning-based detector for real-time tomato plant diseases and pests recognition. Sensors. 2017;17(9):2022.
3. Mahlein A-K, Oerke E-C, Steiner U, Dehne H-W. Recent advances in sensing plant diseases for precision crop protection. Eur J Plant Pathol. 2012;133(1):197–209.
4. Mohanty SP, Hughes DP, Salathé M. Using deep learning for image-based plant disease detection. Front Plant Sci. 2016;7:1419.
5. Barbedo JGA. Factors influencing the use of deep learning for plant disease recognition. Biosyst Eng. 2019;180:80–91.
6. Toda Y, Matsuoka M, Kiyoshi T, Komatsu T. Plant disease recognition using deep learning with residual networks. In: Proceedings of the IEEE conference on computer vision and pattern recognition workshops; 2019. p. 2201–7.
7. Xie C, Li Z, He H, Zhao J, Fu K. Deep learning-based methods for plant disease recognition. Comput Electron Agric. 2020;170:105264.
8. Bhaskar N, Tupe-Waghmare P, Shetty P, Shetty SS, Rai T. A deep learning hybrid approach for automated leaf disease identification in Paddy crops. In: International conference on distributed computing and optimization techniques (ICDCOT), vol. 2024. Bengaluru; 2024. p. 1–5. https://doi.org/10.1109/ICDCOT61034.2024.10515926.
9. Sridevi S, Famila S, Mariammal G, Hemalatha K, Havish MS. Detection and categorization of tomato plant diseases using a convolutional neural network. In: 2024 international conference on distributed computing and optimization techniques (ICDCOT). Bengaluru; 2024. p. 1–7. https://doi.org/10.1109/ICDCOT61034.2024.10515432.
10. Vamshi J, Tanwar S, Thatipudi JG, Mukherjee S, Mandhare JB. Novel plant leaf disease detection approach using hybrid deep learning strategy. In: 2024 5th international conference on intelligent communication technologies and virtual Mobile networks (ICICV). Tirunelveli; 2024. p. 266–73. https://doi.org/10.1109/ICICV62344.2024.00047.
11. Nikhileswar GS, Pampana SS, Bhavana CL, Nanditha RL, Krishna KC. PlantLDNet: disease detection in leaf images using deep learning models. In: 2024 5th international conference on intelligent communication technologies and virtual mobile networks (ICICV). Tirunelveli; 2024. p. 190–8. https://doi.org/10.1109/ICICV62344.2024.00036.
12. Ramadan STY, et al. Improving wheat leaf disease classification: evaluating augmentation strategies and CNN-based models with limited dataset. IEEE Access. 2024;12:69853. https://doi.org/10.1109/ACCESS.2024.3397570.
13. Malhotra S, Garg A, Kukreja V, Mehta S. Agricultural AI: integrating federated learning and CNN in papaya leaf disease detection. In: 2024 3rd international conference for innovation in technology (INOCON). Bangalore; 2024. p. 1–6. https://doi.org/10.1109/INOCON60754.2024.10512288.
14. Patil SA, Bhosale A, Patil A. Advancements in plant leaf disease detection: a comprehensive review of latest trends and technologies. In: 2024 3rd international conference for innovation in technology (INOCON). Bangalore; 2024. p. 1–5. https://doi.org/10.1109/INOCON60754.2024.10511970.
15. Mir TA, Gupta S, Chauhan R, Singh M, Banerjee D, Kumar BV. Enhanced multiclassification of avocado leaf diseases: CNN and random forest integration. In: 2024 3rd international conference for innovation in technology (INOCON). Bangalore; 2024. p. 1–6. https://doi.org/10.1109/INOCON60754.2024.10512211.

16. Kumar V, Banerjee D, Chauhan R, Kukreti S, Singh Gill K. Optimizing citrus disease prediction: a hybrid CNN-SVM approach for enhanced accuracy. In: 2024 3rd international conference for innovation in technology (INOCON). Bangalore; 2024. p. 1–6. https://doi.org/10.1109/INOCON60754.2024.10511309.
17. Suryavanshi A, Kukreja V, Malhotra S, Mehta S, Choudhary A. Transforming crop diagnosis: the role of federated learning CNN in cauliflower leaf disease detection. In: 2024 3rd international conference for innovation in technology (INOCON). Bangalore; 2024. p. 1–6. https://doi.org/10.1109/INOCON60754.2024.10512092.
18. Maitra A, Damle M. Revolutionizing plant health management with technological digital transformation to enhance disease control & fortifying plant resilience. In: 2024 3rd international conference for innovation in technology (INOCON). Bangalore; 2024. p. 1–8. https://doi.org/10.1109/INOCON60754.2024.10511728.
19. Rani KP, Lavanya G, Madhuri KS, Vamsinath J, Sabitha C, Rao KS. Plant disease detection using convolution neural network. In: 2024 international conference on advancements in smart, secure and intelligent computing (ASSIC). Bhubaneswar; 2024. p. 1–4. https://doi.org/10.1109/ASSIC60049.2024.10507921.
20. Banerjee D, Kukreja V, et al. CNN and SVM-based model for effective blackberry leaf disease classification. In: 2024 international conference on advancements in smart, secure and intelligent computing (ASSIC). Bhubaneswar; 2024. p. 1–7. https://doi.org/10.1109/ASSIC60049.2024.10507945.
21. Plantvillage. https://plantvillage.psu.edu/

Chapter 7
Exploring Disease Prediction Techniques Through Data Mining: A Comprehensive Overview

Mitali Chugh

7.1 Introduction

New technologies, a quadrillion-fold increase in the amount of data collected, and the shift of the focus to the prevention of diseases led to a paradigm shift in the healthcare industry. This chapter covers the relevance of data mining in disease forecasting and control in the present context of a constantly evolving healthcare system. It can be noted that the healthcare industry is one area where significant development as well as innovation has taken place in the recent past. For instance, complexities related to diagnosis and treatment have opened new opportunities after the discoveries in genomics and precision medicine. Systems like telemedicine and wearable health devices have enhanced the actual delivery of health and are also personalized [1, 2]. Its swift growth is illustrated by the present value of the global healthcare market expected to accrue to USD 85.95 billion by 2027 [3].

Healthcare data mining is of utmost significance. The pool of healthcare data generated, including electronic health records, medical imaging, and patient-generated data, is enormous [4]. We use data mining tools to obtain useful insights from this large amount of information. Data mining enables early disease prediction, risk assessment, and individualized therapy by revealing hidden patterns [5].

Moreover, data mining aids the discovery of fresh biomarkers and novel previously unrecognized risk factors. These findings can alter our understanding of diseases while at the same time improving prevention strategies as well as treatment approaches.

Healthcare is one of the best examples where data mining can be used to identify diseases at an early stage and prevent their occurrence. By looking into previous medical records, data mining techniques can detect individuals who have higher chances of developing a disease, thus allowing for timely interventions [6]. In this

M. Chugh (✉)
School of Computer Science, UPES, Dehradun, Uttarakhand, India

T. Choudhury, A. Katal (eds.), *Machine Learning for Disease Detection, Prediction, and Diagnosis*, https://doi.org/10.1007/978-981-96-4241-0_7

case, instead of needing quick solutions, healthcare authorities would need to respond earlier by making more considerations in terms of preventing diseases than treating them.

Data mining has played a major role in public health aspects of epidemic identification and tracking. It helps to detect diseases early by examining epidemiological data and social media trends, thus allowing for prompt response and control [7].

The goal is to make sure the readers are well-versed in how data mining can help with disease prediction and prevention in healthcare. The chapter deals with the following: Sect. 7.3 reveals the basics of data preprocessing, feature selection, and dimensionality reduction in healthcare. These are building blocks for accurate disease prediction models. In Sect. 7.4, there is a comprehensive study of machine learning approaches for disease prediction with their respective merits, applicability, and disadvantages. These have real-world case examples that prove their efficiency.

Section 7.5 describes how deep learning can revolutionize healthcare through the analysis of sophisticated medical information and images. In Sect. 7.6, the moral challenges and dilemmas that supplement data mining for disease prediction are discussed, and this calls for ethical and transparent types of processes. Section 7.7 studies the state of development of the sector in the present day and potential new directions of its evolution, as well as ways of integrating such technologies and certain recent studies in clinical areas. Consequently, by the end of this chapter, the researchers, physicians, and data scientists will be informed about the proactive and crucial facts on the significance of data mining in creating the setting for the identification and prevention of diseases in healthcare.

7.2 Basics and Background Work

To extract knowledge from large data repositories, a healthcare system must include highly elaborate and complex mechanisms for managing, storing, processing, and analyzing data and information [8]. Data mining as pointed out by Dash et al. [9] provides the necessary tools and techniques that enable the conversion of big data to valuable insights that are useful in decision-making [9].

Classification is thus a procedure of model construction to predict target characteristics or class labels [10]. This method is used appropriately when data has to be classified into predetermined categories related to goal attributes when making predictions of the label outcomes from the data [11]. Integrated models of classification have been utilized in many healthcare applications. For instance [12], the classification method was applied to enhance the identification of dementia among patients given the outcome of neuropsychological tests. In this, [13] pointed out relationships between colonic fauna and the WITH or WITHOUT diabetes employing the methods of SVM and ANN to assist in predicting metabolic diseases like diabetes. Besides, to predict seminal quality, [14] utilized particle swarm optimization integrated with the SVM approach. Also to predict seminal quality, [14] employed particle swarm optimization along with the SVM technique. Thus, the KStar algorithm was used to predict the result of the single sperm implant in intracytoplasmic in humans [15] to enhance the implantation rate of the intracytoplasmic implant. Moreover, classifier algorithms were employed in all the listed publications in the detailed data mining applications in cancer prognosis and prediction as provided by [16]. Still, Delen et al. [17] on breast cancer employed decision trees for the prognosis of survival rates, whereas Exarchos et al. employed a Bayesian network, to predict oral cancer recurrence based on multiple data sources that include clinical imaging and genomic data derived from both tissue and blood [18].

Clustering, which is insistently used in the connotative domains, involves the process of partitioning a set of data objects or interpretations into subgroups or clusters [10].This occurs if there is limited knowledge of the several data objects that make up a population; hence, a need to apply this approach. Clustering is another type of machine learning approach belonging to the unsupervised learning type that looks at a group of similar objects of data without regard to the existence of specific labels [11]. The clustering techniques are mostly applied during the descriptive analysis stage and are commonly used to categorize the data before the classification stages. Such an example can be seen in Sharma, Singh, and Khatri [19] where research thereof based on classification and clustering techniques analyzed medical literature [19].

Also, other methods such as clustering techniques, for instance, the K-means algorithm were used for colon malignancies' classification [20] and for diagnosing heart disease patients [21]. Moreover, clustering algorithms have also been employed to find disease synoptic groups in medical image data [22], the likelihood approximation of the occurrence of a disease [23], and the functional grouping of proteins [10].

In association analysis, rules for associations between attributes are obtained especially when links in a dataset are being investigated [11]. The identification of relations between different characteristics is possible with the help of this approach. For example, the association analysis can help determine the relationship between high blood pressure and salt intake patterns that may facilitate deriving rules of association. For instance, [24] found connections in clinical data from diabetic patients by using the well-known APRIORI association rule mining technique. In the same direction, other studies suggested using association rule mining algorithms to identify relationships between time, place, and patient infections in clinical data and treatments [25], healthcare data, and disorders related to rhinitis [26], and patient statistics for the diagnosis of coronary heart disease [27], and public health surveillance data [28].

Finding characteristics or results that demonstrate anomalous behavior or set off from the norm is known as outlier detection. Studies like those of V [29, 30]. demonstrate how often this technique is applied to discover inconsistencies in data with the goal of either dataset cleaning or identifying anomalous values within medical databases.

By investigating further, it has been verified that the different data mining techniques that have been found are used in many kinds of healthcare research projects.

7.3 Data Mining Principles in Healthcare

7.3.1 Data Preprocessing

Something as basic as data preparation can be considered one of the most crucial processes when it comes to using healthcare data to forecast diseases and prevent them from happening. To get accurate forecasts that will prove to be useful, it is necessary to work with high-quality, nonambiguous, and given data. In this part, the various activities of preprocessing data in the healthcare industry are reviewed, which includes data cleansing and quality assurance, data acquisition and transformation, and missing data handling.

7.3.2 Data Cleaning and Quality Assurance

The accuracy of data mining models can highly depend on the errors, inconsistencies, and outliers that are often inherent to the data originating from the healthcare domain [31]. The data-cleaning techniques include

1. Outlier detection: to find the outliers from the datasets and respective techniques that may address issues that are raised due to their existence.
2. Elimination of duplicates: preprocessing the data by scrubbing the datasets and by dealing with high dimensionality by removing the irrelevant data.
3. Noise reduction methods: processes to reduce or remove unwanted or unnecessary variability in data.
4. Checking consistency: the process of confirming that the data is within defined rules or limits.

Besides, to ensure that the data being used are clean and accurate, other quality assurance measures like data audits, and validation tests are also important [32]; data quality therefore is a key to reliable disease prediction models.

7.3.3 Data Integration and Transformation

Data in healthcare is collected from multiple sources, providing information from wearable devices, tests, and electronic health records. However, to allow a detailed investigation, data integration entails combining the above sources of data into a single format [4].

The integrated data may be transformed using the following methods to make it appropriate for analysis: The integrated data may be transformed using the following methods to make it appropriate for analysis:

- Normalization: standardization of the data to a common range to avoid any sort of overshadowing in the entire analysis by a particular variable.
- Aggregation: aggregating the data at different levels of detail, for example, compiling and averaging daily patients' notes monthly.
- Feature engineering: creating new attributes based on the already-existing ones to obtain the required info [5] enhanced and more accurate.

Hence, through data integration and transformation, it would be possible to come up with a consolidated database that may be useful in disease prediction.

7.3.4 Handling Missing Data

Missing data is one of the difficulties that are typical for healthcare data: to avoid shifting the direction of error and bias, such data needs to be addressed [33]. When dealing with missing data, several methods can be employed:

- *Imputation*: taking alternate means of the missing data in terms of the data that is remaining.
- *Deletion*: thus, deleting records or characteristics that use missing data, although could be done, would mean losing some data.
- *Predictive modeling* [6]: utilizing the methods of ML for estimating missing values with the help of other characteristics of the dataset.

Missing data can be a critical aspect of data analysis since the accuracy of the disease prediction models and the quality of the dataset considered are often highly affected by how the missing information is dealt with. Therefore, this study establishes that data preparation is a crucial step in the data mining process employed in the framework of disease prediction in healthcare. The datasets used by the researchers and various practitioners can be checked and made ready for modeling through integration, transformation, and imputations of missing values.

7.4 Feature Selection

Feature selection is considered one of the important steps in data preprocessing when applying the methods of data mining for disease prognosis. Hence, feature selection can be defined as the ability to select the most relevant and significant attributes from databases belonging to the health sector. The subsequent segment presents an explanation of why feature selection should be implemented in disease prediction.

7.4.1 Importance of Feature Selection in Disease Prediction

Feature selection is a crucial stage in creating powerful disease prediction models that may assist in early detection and intervention.

It is therefore important that the right features are selected for disease prediction, especially when the causes are many. Often, the data on healthcare contains hundreds of predictors, of which a part is potentially irrelevant, duplicate, or otherwise ineffective. Nonrelevant features may worsen overfitting, increase costs when solving problems on a computer, and hinder model interpretation [34].

The following list of factors is important for feature selection: the following list of factors is important for feature selection:

- Better model performance: evaluations of models have suggested that models apply more brilliance and reliability when the most significant features are incorporated [5].
- Reduced dimensionality: in Feature selection, the curse of dimensionality is avoided, which in turn helps to ease the handling of the data and the need for processing it [35].

Enhanced interpretability: the models with lesser numbers and closer numbers to the real characteristics are less complex to understand, which is essential to making decisions for the patients in the field of medical CG [36].

Thus, feature selection is the significant step in developing effective disease prediction models necessary for early disease diagnosis and treatment.

7.4.2 Techniques for Feature Selection

Concerning feature selection in healthcare data mining, it is possible to work with different approaches. Filter methods, wrapper methods, and embedding methods are three major categories for these techniques:

Filter methods: Filter approaches assess features: Independent of the predictive model. To assess the relevance of every characteristic they often use statistical requisites or scoring functions [34]. Some of the representative methods of filtering techniques are the chi-squared test, mutual information, and feature selection based on correlation measures.

Wrapper methods: Wrapper methods are used for training and testing the prediction models that are used to evaluate the feature subsets. Based on the performance of the model, these variables include cross-validation accuracy or the area under the ROC curve-AUC [37]. Recursive feature elimination (RFE) and forward selection are examples of wrapper approaches.

Embedded methods: In this sense, the selection of features does not occur apart from the model-building process but is indeed a part of the model-building process and is immanent to the given model used. Various methods used in linear models now include, for instance, LASSO (least absolute shrinkage and selection operator). In line with this, in the tree-based approaches, decision trees popularly known as random forests fall under the frequently used embedded methods [38].

The nature of dataset attributes and the kind of predictive model to be adopted should determine the technique to be used in feature selection. This is because the attempt of the various methods in practice may be beneficial to know that is best for the healthcare utility.

To summarize, in healthcare's data mining process for disease prediction, feature selection is a crucial step. It decreases dimensionality, boosts interpretability, and improves the model's functioning. Researchers and practitioners must meticulously select and execute appropriate feature selection methods to effectively use healthcare data for early disease prediction.

7.4.3 *Dimensionality Reduction*

Usually, healthcare data mining requires dimensionality reduction for characteristics of high dimensionality. This section describes difficulties with high-dimensional healthcare data and the possibility of decreasing its dimensionality. As a result, datasets in healthcare have more features than samples present, hence the large size. This poses challenges:

More computational power needed: While handling high-dimensional data, an analyst/technician must spend more time analyzing data and training models [39].

- Occurrence of overfitting: Not all the features are beneficial, models with lots of features might end up learning noise rather than real patterns. This means that during mapping poor predictions of novel data are made [35].
- Challenging interpretation: [40] The authors opined that high dimensions hinder the ability to visualize data and hence interpret them. This restricts knowledge for physicians and policy implementers.
- Data sparsity: In a high dimensionality, the data points are usually isolated and sometimes it is challenging to pinpoint distinctive patterns [39].

For high-dimensionality in healthcare data and the usefulness of such data for disease prediction, the issues stated are critical.

7.4.3.1 Dimensionality Reduction Methods

The methods of dimensionality reduction are designed in a way that decreases the number of features but retains the most necessary information. The following are typical dimensionality reduction methods used in healthcare data mining: the following are typical dimensionality reduction methods used in healthcare data mining.

- *Principal component analysis (PCA)*: In the opinion of Jolliffe and Cadima, PCA detects the new orthogonal linear feature combination or principal components that jointly account for most of the variability in the data. This allows for a reduction of dimensions while at the same time maintaining as much variation as is desirable [41].
- *t-distributed stochastic neighbor embedding (t-SNE)*: t-SNE is a nonlinear dimensionality reduction technique that is exceptionally fit for visualization as affirmed in Maaten and Hinton [40]. Specifically, it is intended to preserve the pairwise similarities of high-dimensional data in a lower-dimensional space.
- *Linear discriminant analysis (LDA)*: LDA can be therefore looked at as a supervised nonlinear technique of dimensionality reduction, which aims at maximizing the distance between the classes in the created low-dimensional space [42]. Hence, it is useful for classification in healthcare data analysis.
- Feature selection-based techniques: Thus, selection methods like the recursive feature elimination (RFE) can also be used as a feature dimensionality reduction tool because they find the most informative features [34].
- Autoencoders: Autoencoder based on deep learning is a technique where data gets encoded into a compact feature space through a process of unsupervised learning [43]. They are especially useful in applications that are related to healthcare for feature learning and for reducing the dimensionality of data.

Depending on the objectives of the research and the characteristics of the given dataset, the type of the required dimensionality reduction should be selected. It is very crucial to balance between keeping important information and minimizing the dimensionality for disease prediction.

In conclusion, dimensionality reduction is an essential first step in handling high-dimensional healthcare data. It reduces issues with computing complexity, overfitting, and data interpretability. Researchers and practitioners need to select the most effective dimensionality reduction approaches to fully realize the potential of healthcare data for precise disease prediction (refer to Fig. 7.1).

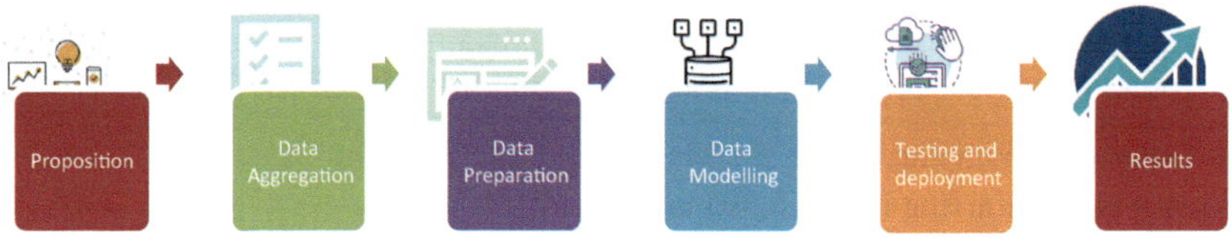

Fig. 7.1 Data mining in healthcare

7.5 Machine Learning Algorithms for Disease Prediction

7.5.1 *Supervised Learning*

One of the most used methods of healthcare data mining for disease prediction is known as supervised learning. This technique entails applying the labeled data on forecasting models on which the labels are used in place of the target qualities, for instance, the prevalence or absence of a certain disease. In this section, various aspects of applied supervised learning methods random forests, decision trees, and support vector machines are discussed.

- Decision Trees
  Decision trees are scalable and logical algorithms generally used in healthcare to forecast sickness [44]. They are useful for clinical interpretation as they relay a semblance of unbiased decision-making. Decision trees' key characteristics include
  - Decision trees are easily understandable by physicians regarding why, where, and when a particular decision has been made [45].
  - Managing nonlinearity: it can capture complex and nonlinear patterns of the observed data.
  - Feature importance: one of the prominent strategies of decision trees is the ability to order the features regarding their significance for the target variable that proves to be very useful in feature selection [44].

Decision trees are error-prone models, and the reason often lies in growing the tree deeper than is useful. Many machine learning algorithms that are currently available can perform more tasks than just providing prognosis, so finding the perfect balance between interpretability and prognosis accuracy is very important. Thus, there is a need to achieve the right balance and find the desired level of interpretability and predictive capacity.

7.5.1.1 Random Forests

Random forests are one type of ensemble learning that combines the forecasts made by several decision trees [46]. Random forests have several benefits when it comes to disease prediction: random forests have several benefits when it comes to disease prediction:

Enhanced accuracy: By aggregating predictions of several trees, random forests can have higher predictive accuracy than individual trees but at the same time reduce the issue of overfitting.

Robustness: They are less inclined to the out-of-bag and noisy values compared to single decision trees. To facilitate the task of feature selection, random forests can generate the score of feature importance (Leo [46]). Random forests can be used for disease prediction as they are devoid of some of the problems of using decision trees yet include beneficial factors.

7.5.1.2 Support Vector Machines

Support vector machines, as described by [47], are a subtype of supervised learning machines that strive to maximize class separability through the determination of the best hyperplane. Due to its subsequent advantages, SVMs are very useful in healthcare for disease prediction.

- *Effective in high-dimensional spaces*: SVMs are good in higher-dimensional spaces having many features.
- *Robust to overfitting*: notably, the SVMs are less sensitive to overfitting, even when operating with less data hence, not easily overfit.
- *Versatile kernel functions*: it should be noted that dealing with nonlinear data in SVMs may use different kernel functions [48].

On the other hand, SVMs may be troubled by the hyperparameters and more so the kernel function to fine-tune.

7.5.2 Unsupervised Learning

In the context of healthcare data mining, unsupervised learning is an essential branch of ML, particularly when working with datasets that lack outcomes that are labeled. This section summarizes two important unsupervised learning methods: principal component analysis (PCA) and clustering algorithms (such as hierarchical clustering and K-means). These techniques are essential for spotting patterns, clusters, and latent structures in healthcare data.

7.5.2.1 Clustering Algorithms (e.g., Hierarchical Clustering, K-Means)

Unsupervised learning is a significant part of the ML that is used intensely in data mining of healthcare data and might deal with the data that do not include outcomes that are labeled. This section summarizes two important unsupervised learning methods: Some of the most common measurements that are employed in the study of customers' data are principal component analysis (PCA) and clustering algorithms like hierarchical clustering and K-means clustering. These techniques are crucial in the identification of patterns and clusters, as well as the underlying structures in the healthcare datasets.

In clustering methods of unsupervised learning, the data points are categorized according to the similarity or dissimilarity of the data points. Some of the uses of these algorithms include disease subtyping, patient stratification, and anomaly detection. Important components to emphasize include

- *K-means clustering*: Clustering is done, and the data is partitioned into the "k" number of clusters using each of the "k" means that are generated using the K-means algorithm. Due to the separation between the given data points and receiving cluster centroids, it categorizes data points with clusters [49].
  Based on Breunig et al. [50], it is established that K-means is useful for cohorting as well as segmenting patient data.
- *Hierarchical clustering*: Hierarchical clustering is a process of clustering that results in the tree structure of clusters. This setup demonstrates how data are related at different levels of aggregation. It offers a glance at how diseases or patient groups are affiliated with each other. The method constructs a tree-like structure with all relationships shown in different levels, explaining how info points are connected. This speaks about the stratified nature of disease or patients' clusters where this approach helps in mapping and analyzing [51].

Specific to the cure and epidemiological works, the clustering techniques for generating tailor-made schedules are the identification of the right clusters that are nested in the datasets and can be found using the algorithm [39].

7.5.2.2 Principal Component Analysis (PCA)

For the dimensionality reduction of high-dimensional data, principal component analysis (PCA) is best suited, and thus, it is employed in healthcare [41] for the dimensionality reduction of voluminous data, assisting improvised visualization and extracting the latent variables.

Key elements of PCA include

- *Dimensionality reduction*: PCA helps in decreasing the dimension of the data and preserving the most important features that assist in feature selection and improving visualization of data; hence, it is deemed important for the process [41].
- *Noise reduction*: this way, the noise can be minimized and better identification of signal in the context of healthcare data can be made because PCA only highlights the most variance of the dataset required [52].
- *Feature engineering*: according to [53], PCA can reveal latent variables and connections between features, enabling more reliable modeling.

In healthcare, PCA enables us to discover unseen patterns in complex datasets such as medical imaging and EHRs. Also, in machine learning PCA is used to reduce features, label biomarkers, and categorization of disease.

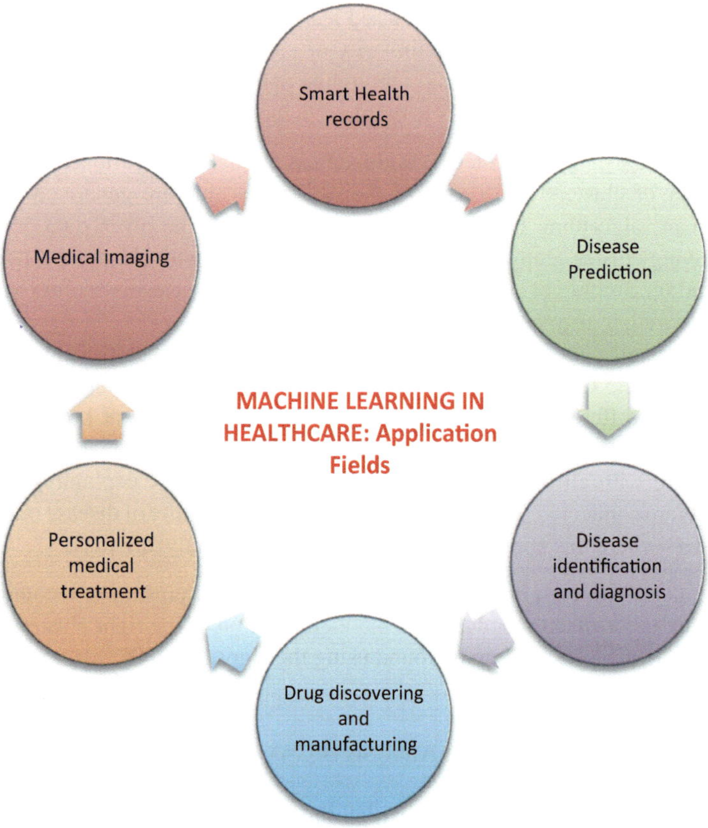

Fig. 7.2 Machine learning application fields in healthcare

All in all, existing techniques such as extracted PCA and clustering algorithms can be applied in healthcare data mining. They assist clinicians as well as researchers in uncovering latent relationships, casualties, and the nature of the information that is vital in the improvement of the quality of patients' care as well as advancing medical research (refer to Fig. 7.2).

7.6 Case Studies

Case studies show dedicated descriptions of how individuals can predict and control diseases, which focuses on how useful ML is in the health sector. This section points out three important healthcare case studies: ML for the prediction of heart disease and diagnosis of cancer and employing predictive analytics for diabetes treatment.

7.6.1 Application of Machine Learning in Predicting Cardiovascular Disease

Cardiovascular diseases (CVDs) are the main reason for ill health and mortality all over the world. Machine learning is very important in making better predictions and preventing CVDs. Researchers and healthcare professionals use huge sets of patient data, health records, and test results to build models that predict diseases.

Here, the case studies mainly include the following:

- Risk stratification: According to D'Agostino et al. [54], based on factors such as gender, genetics, age, cholesterol level, and selection of lifestyle, it is predicted how likely it is that an individual might suffer a CVD as heart attack, etc. [54].
- Detecting early: To diagnose CVDs quicker doctors, apply ML for analysis of ECG signals, medical images, and wearable devices. As stated by Ahmed et al. [55], Rajpurkar et al. [56]. Recent research has shown this.
- Personalized treatment: It involves the analysis of patient-level characteristics including genetic information, previous medication prescription and use, and patients' habits [57].

The above cases show how machine learning is useful in cardiovascular disease since it assists in recommending treatment, evaluating the risk for the patients, and diagnosing CVD, hence improving the lifespan of the patients.

7.6.1.1 Using ML for Cancer Diagnosis

Leveraging the potential of ML cancer detection and categorization is enhanced. In this context, case studies present the areas where ML is used in cancer diagnosis. These case studies' main features include

- *Pathology and radiology*: to help pathologists and radiologists analyze medical images including X-rays, MRIs, and histopathological slides, machine learning models are being used in these fields [58, 59].
- *Early detection*: developing methods for identifying signs of cancers at one's initial stage during screenings so that appropriate procedures can be initiated, and patients' well-being enhanced [60].
- *Subtypes and prognosis*: cancer subtypes and the use of machine learning to predict the outcome of patients determine the treatment and the subsequent survival rate [61].

These case studies show that machine learning can make the diagnosis of cancer as well as the subsequent treatment much more efficient, ultimately resulting in a better strategy for patients.

7.6.1.2 Predictive Analytics in Diabetes Management

From these case studies, it is evident that predictive analytics could enable patients to lead healthier lives while reducing the risk of diabetes complications, hence enhancing the management of diabetes. Predictive analytics has made it possible to encourage the true prospect that covers patients' outcomes and therapeutic tactics for diabetes a significant portion of the healthcare system. Diabetes management case studies frequently involve

- Usage of machine learning algorithms for blood glucose prediction that help diabetics to better manage insulin intake and meals.
- Analyzing the data of patients to predict the probability of the acute manifestation of diabetes-related complications, for example, neuropathy or retinopathy [62].
- Patient engagement: to promote patient engagement and self-management personal suggestions and feedback based on the predictions can be given.

As evident from the foregoing case studies, the application of predictive analytics in the management of diabetes has the potential of letting patients live healthier lives while also easing the management of complications arising from the disease.

7.7 Deep Learning for Disease Prediction

Deep learning has proven to be a valuable weapon in the armory of healthcare when it comes to disease prediction and identification. This section introduces deep learning approaches, particularly adopting sequential approaches contained in the RNN for treating healthcare data and the CNN for image interpretation.

The application of deep learning has significantly impacted healthcare by utilizing current techniques to enhance patients' and diagnose illnesses. CNN and RNN are two of the most important branches of deep learning, together with others that are less relevant for this development. This is imperative because CNNs are more confident at deciphering images, hence capable of identifying complex patterns as well as structures in an image such as an X-ray or a magnetic resonance imaging (MRI) scan. This is critical in diagnosing early-onset diseases such as cancer and diabetes, among others.

Complex paths embedded within medical images such as X-rays and MRI scans can be assessed automatically due to strong image interpretation abilities through CNNs. It depends upon this capability that diseases such as cancer or diabetic retinopathy may be diagnosed early enough.

Because CNNs have stronger capabilities for image interpretation, they can automatically identify complex patterns, structures, or features that are found in medical images, particularly CT scans ruling out the need for human participants. Such functionality is crucial since it helps doctors diagnose such early-stage disorders as cancer, retinopathy, etc.

RNNs can be used to reliably forecast the course of diseases, the consequences of strokes, and dependencies in sequential data as it is designed to handle sequential data. These are best suited in instances such as analyzing EHRs, time-series data, and patient histories Better patient outcomes could result from the revolutionary potential of CNNs for visual interpretation and RNNs for sequential data processing to transform disease prediction and personalized medical care.

CNNs have completely changed the way that images are interpreted in healthcare. Both customized CNN architectures and pretrained, fine-tuned CNNs are being used for medical imaging tasks. CNNs could be utilized for many different purposes, such as automating radiological diagnostics, classifying diseases, and identifying cancers. These could include applications such as screening for early signs of diseases and improving patients' care.

They consist of complicated patterns and structures on the body images extracted from the MRIs, X-rays, and histopathological slides. Convolutional layers apply filters on images in a spatial pyramid way and at the same time maintain the spatial hierarchies of the layers. This is one of CNN's USPs since people can now get updated on important events on the go and at their convenience. Although fully connected layers decrease computational complexity and increase the model's stability, layer pooling down samples the feature maps to combine features for classification or prediction.

Records from electronic health as well as patient monitoring in the healthcare sector are often offered in sets of sequences known as time-series data. So, the handling of such sequential data can be effectively addressed using RNNs. In forecasting disease progression, prediction, or assessing the clinical note, RNNs apply recurrent layers that allow information to flow across the time steps. Some of the architecture for RNN is GRU and LSTM to solve and get rid of the vanishing gradient problem to capture long-term dependency for sequences. Also, the application of RNNs is for the disease progression modeling, patient outcome prediction, and decision support analysis of electronic health information. A combination of CNN and RNN can help revolutionize healthcare since the prediction of illnesses is improved and healthcare plans are adapted to be precise alongside early identification of illnesses.

7.8 Deep Learning Applications in Healthcare: A Closer Look Through Case Studies

7.8.1 Early Detection of Retinal Diseases Using CNNs

Hence, identifying the diseases in their early stages is very important to avoid vision impairments. This paper concentrates its case studies on the application of CNNs in the early detection of diseases such as diabetic retinopathy and age-related macular degeneration from retinal images. These case studies depict how CNN can be applied in the identification of numerous retinal diseases such as in the early stage, which can enhance the treatment of patients and clinical benefits.

7.8.1.1 Detection of Diabetic Retinopathy with CNNs [63]

CNN was trained in detecting DR in retinal images. These findings implied that because of its high accuracy and sensitivity fused with low specificity, CNN provided evidence for early detection and continuous tracking of preliminary symptoms of the disease. Hence, the impact of the DL in improving the efficacy of the testing for diabetic retinopathy was discovered in this study.

7.8.1.2 Detection of Age-Related Macular Degeneration Using CNN [64]

Through this study, the participant's retinal images were decoded by the CNNs for early-detection AMD before reaching the sufficiently treatable stage. This was also going to help to show that CNN was very specific and sensitive to identifying AMD-related traits. The study's outcomes also suggest that deep learning applies to the early detection and evaluation of age-related retinal diseases.

7.8.1.3 Detection of Glaucoma Using CNNs [65]

The focal area of this research is the detection of glaucoma at the initial stage by using CNNs. The network CNN was trained with glaucomatous features that were used to learn features unique to glaucoma in the retinal images. The analyzed results revealed that glaucoma diagnosis using the CNNs is possible through early diagnosis and treatment.

7.8.1.4 Retinopathy Classification Using CNN [66]

In the present case, the architecture of the CNN was used, just diagnosing the retinal images concerning the grades of retinopathy. The accuracy and efficiency of demarcation between the diseased and normal areas of the retina enabled the use of the CNN model in the therapy of retinopathy.

7.8.1.5 Automated Retinal Disease Detection System [67]

To perform the classification in the given case study, CNNs were employed for grouping the retinal images based on the severity of the retinopathy. The CNN model aids in the treatment and early detection of retinopathy since it proved efficient in differentiating between the patients' retinal diseases and a normal retina.

7.8.2 Predicting Neurological Disorders with RNNs

These case studies raise questions about recurrent neural networks (RNNs), and their ability to forecast and treat different neurological diseases like multiple sclerosis, epilepsy, stroke, Parkinson's disease, and Alzheimer's disease, among others. RNNs can potentially improve the patient's experiences, early detection of diseases, and an optimal treatment regimen in neurology.

7.8.2.1 Predicting Alzheimer's Disease Progression with RNNs [68]

The study employs RNNs to predict Alzheimer's disease drawing inferences from the patient's clinical dataset. The model training involved the usage of data extracted from various sources that included images and clinical data for the simulation of disease progression.

The work illustrated how RNNs can be used to enhance early disease diagnosis and intervention, and such a development would be helpful for Alzheimer's patients.

7.8.2.2 RNN-Based Prediction of Epileptic Seizures [69]

This paper discussed the use of RNNs on the prognosis of epileptic episodes utilizing electroencephalogram (EEG) input data. To predict seizures at the initial stages, preictal patterns of the EEG signals were used to train the RNN model. The research presented the role of RNN in getting timely notifications for epilepsy patients.

7.8.2.3 Stroke Outcome Prediction with RNN [70]

In the present study, an effort was also made to briefly discuss the possibility of using RNNs in the prediction of stroke patients. Sequential healthcare data involved in the model entails the health indicators and clinical evaluations. The study also presented research that provides an understanding of how severe and probable prognosis of strokes can be forecasted by employing RNNs to boost the intervention.

7.8.2.4 Prediction of Parkinson's Disease Based on Deep Long Short-Term Memory Networks [71]

This case study focused on the following objectives: RNNs: construction and utilization in the context of predicting the progression of Parkinson's disease. Based on the sequential motor performance data, the model aimed at providing a prognosis of the disease's evolution. This line of work showed the functions of RNNs as the manner of designing distinct rehabilitative approaches and the determination of the onset of Parkinson's illness.

7.8.2.5 RNN-Based Prediction of Multiple Sclerosis Progression [72]

In this case study, the chosen approach involved the use of RNNs to forecast the progression of the diseases of multiple sclerosis (MS) patients. RNNs were trained to learn how sequences of those clinical data and imaging biomarkers could be applied to forecast the development of the disease. The paper depicted how RNNs could help medical practitioners in managing and forecasting the development of MS.

7.9 Ethical Considerations in Data Mining for Disease Prediction

Ethics are fundamental when using data mining to predict illnesses. This part presents key ethical matters in healthcare data mining, which are privacy and data security, bias and fairness, informed consent and data ownership, transparency and interpretability, and regulatory compliance as sensitive healthcare data usage, as well as sophisticated algorithms, may evolve.

7.9.1 Privacy and Data Security

In predicting diseases using patient data, ensuring privacy and data security is essential. There is a need for confidentiality to be maintained since medical information usually contains very personal details. Normative safeguards include.

- Data anonymization: to protect patient identity by removing PII from datasets [73].
- Encryption: ensures that such information stays protected while in use or at rest [74].
- Access controls: implementing measures to make sure only authorized persons can access the information [75].

Putting effective privacy and security measures into practice is necessary to gain the trust of both the patient side as well as other stakeholders.

7.9.2 Bias and Fairness

Patients' distinguished classifications could receive unequal or unfair care due to biased healthcare data analysis. Issues of bias in data and algorithms must be considered and reduced to the possible minimum.

- Bias detection: one way is to further augment bias-detection tools that are used to discover and fix biases in the data [76].
- Fair algorithms: developing procedures for making predictions that are fair for all subgroups of a population [77].
- Consistent audits: it was recommended that such evaluation checks should be performed regularly to periodically detect and reverse prejudice in the predictive models [78].

In evaluating healthcare data mining, it is therefore required that it must be ethical and meet the requirement of being fair in its allocation.

7.9.3 *Informed Consent and Data Ownership*

Basic principles are preserving the necessary ethics of voluntary and informed consent, as well as the patient's self-determination. The ownership of generated data remains ambiguous; thus, the patients should be granted some sort of control over the use of their data.

- Informed consent: obtaining the patient's clear and voluntary consent before using or disclosing the information [79].
- Data ownership: outlining who owns the data and the patient's role and right in the utilization of their data [80].
- Data access control: enabling the patient to control who gets to view their record [80].

Therefore, allowing ethical healthcare data mining is the prior respect, and recognition of the patient's autonomy and the ownership of their data.

7.9.4 *Transparency and Interpretability*

- Another issue pertains to the ethical repercussions of the "black-box" system of operation in certain machine learning models. To understand how predictions are done, there is require transparency and interpretability. Ethics-related factors include:
  - Interpretability techniques: these include the use of models or methodologies that can explain the predictions made in a model such as interpretable models [81].
  - Recording the steps to develop the model and the rationale for choosing the assumptions [45].
  - Explanations: enabling patients and healthcare professionals to get a brief rationale of the forecast [82].

Another important element that healthcare data mining should address is accountability and confidence in the results obtained and that would be achieved through the interpretability of the process.

7.10 Regulatory Compliance in Healthcare Data Mining

The legal requirements and morals need to be addressed so regulatory compliance is needed in healthcare data mining. Observing the Health Insurance Portability and Accountability Act.

- (HIPAA) in the United States and comparable laws in other nations are ethical issues [83].
- GDPR compliance [84]: adhering to the General Data Protection Regulation (GDPR) for data processing in the European Union.
- Ethical surveillance: founding ethics commissions and strategies for healthcare data mining in healthcare corporations [85].

To sum up, when using patient datasets in data mining for disease prediction, categorization, and developing customized treatment plans ethical issues must be taken into consideration to ensure privacy, fairness, and transparency.

7.11 Challenges and Future Directions

To fully realize the potential of this game-changing strategy, several hurdles and future directions must be addressed as the healthcare sector embraces data mining for disease prediction and prevention. The main issues are covered in this part, along with some promising future directions.

7.11.1 Data Accessibility and Integration

Perhaps, the most significant issue in healthcare data mining is to ensure that numerous sorts of healthcare data are available and compatible. They are complex to develop because it entails creating intelligible and cohesive data that may be tucked away in miscellaneous systems and a variety of formats within the healthcare industry. Thus, it will require collective action to design data-sharing compatibility, and standard security systems to address this problem [86]. If new developments in healthcare data integration can integrate a broad spectrum of clinical, genetic, environmental, and lifestyle information, this can contribute to a better understanding of the patient's state of health as well as enhance disease risk estimations.

The last bullet is closely connected to the previous one as both the interpretability and explainability of models are crucial in many applications to ensure end users understand how predictions are made.

7.11.2 Interpretability and Explainability of Models

Interpretability and explainability in finding the real cause of a particular result or conclusion is crucial in many fields, a huge problem in the healthcare sector mainly due to the black-box nature of many machine learning algorithms. The outcome decisions are supposed to be made by predictive models, which are also comprehensible and believable to clinicians and patients. To better achieve this goal, the construction of models that are accurate but also comprehensible has to be the priority of future trends in healthcare data mining [45]. Making it possible to explain how such predictions are made will entail the use of methods such as Explainable Artificial Intelligence (XAI) and models with an explanatory structure [82]. Incorporation of interpretability into the healthcare model will go a long way in enhancing the confidence of the healthcare givers while the use of data-based predictions in the handling of healthcare givers will also be assured to be ethical.

7.11.3 Integration of Clinical Expertise and Machine Learning

The main issue that lies in developing applications for virtual care is in the integration of clinical knowledge with machine learning. The authors of the study [78] claimed that healthcare data mining should be done in conjunction with data scientists, physicians, and domain specialists to ensure that the models created are suitable to meet the patient's needs as well as align with clinical practice [78]. Clinicians must engage themselves in the model development process to validate predictions and give additional information that may accompany the data but can hardly be observed in it. The development of effective methods of forming interdisciplinary teams and the construction of tools that would facilitate the work of such teams should be the major trends to follow concerning the development of models that would be true to the nature of therapeutic processes.

7.11.4 *Leveraging Emerging Technologies (e.g., Explainable AI, Federated Learning)*

Thus, to consider data mining as predictive for diseases and as a tool to bring revolution in the healthcare industry it largely depends on the future technology. A huge enhancement of model transparency is anticipated to be accomplished by explainable AI (XAI) in the future [87]. Moreover, federated learning offers a practicable approach to protecting the privacy of patients' information and facilitating positive model training cooperation between distantly located healthcare entities [88]. With that, the former also helps in paving the way for new approaches and solutions to more responsible, privacy-preserving, and understandable healthcare data mining, which can fit present problems.

7.12 Future Possibilities for Disease Prediction and Healthcare Transformation

Data mining can open great opportunities relating to the prediction of future diseases and further development of the healthcare business. Thus, one may expect better disease predictions that are personalized and more accurate due to advancements in machine learning, integrated databases, and interdisciplinary cooperation. This could radically transform early disease diagnosis, intervention, and proof by patient orientation. Further, patients and providers can benefit from the ethical and responsible use of these predictive models if healthcare organizations adopt the practices of responsible data mining that respect patient's privacy and are fair. These advancements can result in patient health statuses and outcomes that are better and more knowledgeable, thus transforming the face of health.

The latter suggests the necessity of keeping an eye on problems and adopting innovative technology in the rapidly growing field of healthcare and data mining so that the future of disease prediction will be effective yet equally ethical.

7.13 Conclusion

To sum up this chapter, the revolutionary potential of data mining in changing the face of healthcare through disease prediction and prevention has been described. The excellent tools for data mining are helping the booming healthcare sector by detecting diseases at an early stage and providing preventive medicine. When summarizing the key points of this chapter, we stressed the importance of ethical aspects again, as the approaches to patients' privacy violation, prejudice, and fairness, as well as the matters of transparency and compliance with the requirements, were described in detail.

These ethical principles are vital to preserving the patients' confidence and guarantee that the private medical data will be employed ethically. Presenting the examples of the diseases' prediction in this chapter, the practical usage of data mining in treatment is shown as its capability to enhance patients, refine disease diagnosis, and improve curative measures. The chapter can be very useful to researchers, clinicians, and data scientists because it equips them with the knowledge, they need to transform disease prediction through data mining, to enhance patient care and the progress of healthcare. Ethical challenges must remain central to an optimistic process of real and radical improvement in healthcare while progressing this important field of data mining in the healthcare sector.

References

1. Kvedar JC, Fogel AL, Elenko E, Zohar D. Digital medicine's march on chronic disease. Nat Biotechnol. 2016;34(3):239–46.
2. Steinhubl SR, Muse ED, Topol EJ. The emerging field of mobile health. Sci Transl Med. 2015;7(283):1–12. https://doi.org/10.1126/scitranslmed.aaa3487.
3. Statista. Health care – worldwide; 2023. https://www.statista.com/outlook/emo/otc-pharmaceuticals/worldwide?currency=USD
4. Wang Y, Kung L, Byrd TA. Big data analytics: understanding its capabilities and potential benefits for healthcare organizations. Technol Forecast Soc Chang. 2018;126:3–13.
5. Bellazzi R, Zupan B. Predictive data mining in clinical medicine: current issues and guidelines. Int J Med Inform. 2008;77(2):81–97. https://doi.org/10.1016/j.ijmedinf.2006.11.006.
6. Rajkomar A, Oren E, Chen K, Dai AM, Hajaj N, Hardt M, Liu PJ, et al. Scalable and accurate deep learning with electronic health records. NPJ Digit Med. 2018;1(1):1–10. https://doi.org/10.1038/s41746-018-0029-1.
7. Chew C, Eysenbach G. Pandemics in the age of twitter: content analysis of tweets during the 2009 H1N1 outbreak. PLoS One. 2010;5(11):1–13. https://doi.org/10.1371/journal.pone.0014118.
8. Pramanik MI, Lau RY, Azad MAK, Hossain MS, Chowdhury MKH, Karmaker BK. Healthcare informatics and analytics in big data. Expert Syst Appl. 2020;152:113388.
9. Dash S, Shakyawar SK, Sharma M, Kaushik S. Big data in healthcare: management, analysis and future prospects. J Big Data. 2019;6(1) https://doi.org/10.1186/s40537-019-0217-0.
10. Han J, Kamber M, Pei J. Data mining concepts and techniques. J Chem Inf Model. 2012;53
11. Tekieh MH, Raahemi B. Importance of data mining in healthcare. In: Proceedings of the 2015 IEEE/ACM international conference on advances in social networks analysis and mining 2015; 2015. p. 1057–62.
12. Maroco J, Silva D, Rodrigues A, Guerreiro M, Santana I, de Mendonca A. Data mining methods in the prediction of dementia. BMC Res Notes. 2011;4:1–14.
13. Elhoseny M, Abdelaziz A, Salama AS, Riad AM, Muhammad K, Sangaiah AK. A hybrid model of internet of things and cloud computing to manage big data in health services applications. Futur Gener Comput Syst. 2018;86:1383–94. https://doi.org/10.1016/j.future.2018.03.005.
14. Sahoo AJ, Kumar Y. Seminal quality prediction using data mining methods. Technology Healthcare. 2014;22(4):531–45.
15. Mirroshandel SA, Ghasemian F, Monji-Azad S. Applying data mining techniques for increasing implantation rate by selecting best sperms for intra-cytoplasmic sperm injection treatment. Comput Methods Prog Biomed. 2016;137:215–29. https://doi.org/10.1016/j.cmpb.2016.09.013.

16. Kourou K, Exarchos TP, Exarchos KP, Karamouzis MV, Fotiadis DI. Machine learning applications in cancer prognosis and prediction. Comput Struct Biotechnol J. 2015;13:8–17. https://doi.org/10.1016/j.csbj.2014.11.005.
17. Delen D, Walker G, Kadam A. Predicting breast cancer survivability: a comparison of three data mining methods. Artif Intell Med. 2005;34(2):113–27. https://doi.org/10.1016/j.artmed.2004.07.002.
18. Exarchos KP, Goletsis Y, Fotiadis DI. Multiparametric decision support system for the prediction of Oral cancer reoccurrence. IEEE Trans Inf Technol Biomed. 2012;16(6):1127–34. https://doi.org/10.1109/TITB.2011.2165076.
19. Sharma R, Singh SN, Khatri S. Medical data mining using different classification and clustering techniques: a critical survey. In: 2016 2nd international conference on computational intelligence and communication technology, CICT 2016. IEEE; 2016. p. 687–91. https://doi.org/10.1109/CICT.2016.142.
20. Kumar P, Wasan SK. Analysis of X-means and global k-means using tumor classification. In: 2010 the 2nd international conference on computer and automation engineering, ICCAE 2010, vol. 5. IEEE; 2010. p. 832–5. https://doi.org/10.1109/ICCAE.2010.5451883.
21. Shouman M, Turner T, Stocker R. Integrating decision tree and K-means clustering with different initial centroid selection methods in the diagnosis of heart disease patients. In: Proceedings of the international conference on data science (ICDATA); 2012. p. 1–7.
22. Bruse JL, Zuluaga MA, Khushnood A, McLeod K, Ntsinjana HN, Hsia TY, Sermesant M, Pennec X, Taylor AM, Schievano S. Detecting clinically meaningful shape clusters in medical image data: metrics analysis for hierarchical clustering applied to healthy and pathological aortic arches. IEEE Trans Biomed Eng. 2017;64(10):2373–83. https://doi.org/10.1109/TBME.2017.2655364.
23. Paul R, Hoque ASML. Clustering medical data to predict the likelihood of diseases. In: 2010 5th international conference on digital information management, ICDIM 2010. IEEE; 2010. p. 44–9. https://doi.org/10.1109/ICDIM.2010.5664638.
24. Stilou S, Bamidis PD, Maglaveras N, Pappas C. Mining association rules from clinical databases: an intelligent diagnostic process in healthcare. Stud Health Technol Inform. 2001;84(February 2001):1399–403. https://doi.org/10.3233/978-1-60750-928-8-1399.
25. Ting SL, Wang WM, Kwok SK, Tsang AHC, Lee WB. RACER: rule-associated CasE-based reasoning for supporting general practitioners in prescription making. Expert Syst Appl. 2010;37(12):8079–89. https://doi.org/10.1016/j.eswa.2010.05.080.
26. Yang L, Li Z, Luo G. MH-ARM: a multi-mode and high-value association rule mining technique for healthcare data analysis. In: 2016 international conference on computational science and computational intelligence, CSCI 2016. IEEE; 2017. p. 122–7. https://doi.org/10.1109/CSCI.2016.0030.
27. Orphanou K, Dagliati A, Sacchi L, Stassopoulou A, Keravnou E, Bellazzi R. Combining naive Bayes classifiers with temporal association rules for coronary heart disease diagnosis. In: 2016 IEEE international conference on healthcare informatics, ICHI 2016. IEEE; 2016. p. 81–92. https://doi.org/10.1109/ICHI.2016.15.
28. Brossette SE, Sprague AP, Michael Hardin J, Waites KB, Jones WT, Moser SA. Association rules and data mining in hospital infection control and public health surveillance. J Am Med Inform Assoc. 1998;5(4):373–81. https://doi.org/10.1136/jamia.1998.0050373.
29. Bellaachia A, Bari A. A flocking based data mining algorithm for detecting outliers in cancer gene expression microarray data. In: 2012 international conference on information retrieval and knowledge management, CAMP'12. IEEE; 2012. p. 305–11. https://doi.org/10.1109/InfRKM.2012.6204996.
30. Kumar V. Outlier mining in medical databases: an application of data mining in health care management to detect abnormal values presented in medical databases. J Comput Sci. 2008;8(8):272–7.

31. Hamdi A, Shaban K, Erradi A, Mohamed A, Rumi SK, Salim FD. Spatiotemporal data mining: a survey on challenges and open problems. Artif Intell Rev. 2022;55(2):1441–88. https://doi.org/10.1007/s10462-021-09994-y.
32. Sarkar IN, Chouchane L, Bouchard K, Ayers GD. Data quality assessment and control in health sciences data. Int J Comput Biol Drug Des. 2015;8(4):347–59.
33. Little RJA, Rubin DB. Statistical analysis with missing data. Wiley; 2002.
34. Guyon I, Elisseeff A. An introduction to variable and feature selection Isabelle. J Mach Learn Res. 2003;3:1157–82. https://doi.org/10.1016/j.aca.2011.07.027.
35. Dy JG, Brodley CE. Feature selection for unsupervised learning. J Mach Learn Res. 2004;5:845–89. https://doi.org/10.1007/978-3-642-34487-9_47.
36. Safavian SR, Landgrebe D. A survey of decision tree classifier methodology. IEEE Trans Syst Man Cybern. 1991;21(3):660–74. https://doi.org/10.1109/21.97458.
37. Kohavi R, John GH. Wrappers for feature subset selection Ron. Artif Intell. 1997;97:273–324. https://doi.org/10.1007/978-3-642-39038-8_27.
38. Chen T, Guestrin C. XGBoost: a scalable tree boosting system. In: Proceedings of the ACM SIGKDD international conference on knowledge discovery and data mining; 2016. p. 785–94. https://doi.org/10.1145/2939672.2939785.
39. Hastie T, Tibsirani R, Friedman J. The elements of statistical learning data mining, inference and prediction. Springer Ser Statist. 2009;32:441. https://doi.org/10.1111/j.1532-5415.1984.tb02220.x.
40. van der Maaten L, Hinton G. Visualizing data using T-SNE laurens. J Mach Learn Res. 2008;9:2579–605. https://doi.org/10.1007/s10479-011-0841-3.
41. Jollife IT, Cadima J. Principal component analysis: a review and recent developments. Philos Trans R Soc A Math Phys Eng Sci. 2016;374(2065):20150202. https://doi.org/10.1098/rsta.2015.0202.
42. Duda TF, Palumbi SR. Gene expression and feeding ecology: evolution of Piscivory in the venomous gastropod genus conus. Proc R Soc B Biol Sci. 2004;271(1544):1165–74. https://doi.org/10.1098/rspb.2004.2708.
43. Hinton GE, Salakhutdinov RR. Reducing the dimensionality of data with neural networks. Science. 2006;313(5786):504–7. https://doi.org/10.1126/science.1127647.
44. Breiman L, Ihaka R. Nonlinear discriminant analysis via scaling and ACE; 1984. Technical Report No. 40.
45. Caruana R, Lou Y, Gehrke J, Koch P, Sturm M, Elhadad N. Intelligible models for healthcare: predicting pneumonia risk and hospital 30-day readmission. In: Proceedings of the ACM SIGKDD international conference on knowledge discovery and data mining; 2015. p. 1721–30. https://doi.org/10.1145/2783258.2788613.
46. Breiman L. Random forests. Mach Learn. 2001;45(5) https://doi.org/10.1007/978-3-030-62008-0_35.
47. Cortes C, Vapnik V. Support vector networks. Mach Learn. 1995;20(1):273–97. https://doi.org/10.1088/1742-6596/628/1/012073.
48. Hearst MA, Dumais ST, Osuna E, Platt J, Scholkopf B. Support vector machines. IEEE Intelligent Systems and their applications. 1998;13(4):18–28.
49. Lloyd SP. Least squares quantization in PCM. IEEE Trans Inf Theory. 1982;28(2):129–37. https://doi.org/10.1109/TIT.1982.1056489.
50. Breunig MM, Kriegel HP, Ng RT, Sander J. LOF: Identifying density-based local outliers. SIGMOD Record 2000;29(2):93–104.
51. Everitt BS, Landau S, Leese M, Stahl D. Hierarchical clustering. Cluster Analysis. 2011;5:71–110.
52. Hotelling H. Analysis of a complex of statistical variables into principal components. J Educ Psychol. 1933;24(6):417–41. https://doi.org/10.1037/h0071325.
53. Abdi H, Williams LJ. Principal component analysis. WIREs Comp Stats. 2010;2(4):433–59.

54. D'Agostino RB, Vasan RS, Pencina MJ, Wolf PA, Cobain M, Massaro JM, Kannel WB. General cardiovascular risk profile for use in primary care: the Framingham heart study. Circulation. 2008;117(6):743–53. https://doi.org/10.1161/CIRCULATIONAHA.107.699579.
55. Ahmed AF, Attia FA, Liu Z, Li C, Wei J, Kang W. Antioxidant activity and total phenolic content of essential oils and extracts of sweet basil (Ocimum basilicum L.) plants. Food Sci Human Wellness. 2019;8(3):299–305.
56. Rajpurkar P, Hannun AY, Haghpanahi M, Bourn C, Ng AY. Cardiologist-level arrhythmia detection with convolutional neural networks; 2017. http://arxiv.org/abs/1707.01836
57. Khera AV, Chaffin M, Wade KH, Zahid S, Brancale J, Xia R, Distefano M, et al. Polygenic prediction of weight and obesity trajectories from birth to adulthood. Cell. 2019;177(3):587–96. https://doi.org/10.1016/j.cell.2019.03.028.
58. Esteva A, Kuprel B, Novoa RA, Ko J, Swetter SM, Blau HM, Thrun S. Dermatologist-level classification of skin cancer with deep neural networks. Nature. 2017;542(7639):115–8. https://doi.org/10.1038/nature21056.
59. Litjens G, Kooi T, Bejnordi BE, Setio AAA, Ciompi F, Ghafoorian M, et al. A survey on deep learning in medical image analysis. Med Image Anal. 2017;42:60. https://doi.org/10.1016/j.media.2017.07.005.
60. Nahon P, Najean M, Layese R, Zarca K, Segar LB, Cagnot C, Ganne-Carrié N, et al. Early hepatocellular carcinoma detection using magnetic resonance imaging is cost-effective in high-risk patients with cirrhosis. JHEP Reports. 2022;4(1):100390. https://doi.org/10.1016/j.jhepr.2021.100390.
61. Giordano TJ. The cancer genome atlas research network: a sight to behold. Endocr Pathol. 2014;25(4):362–5. https://doi.org/10.1007/s12022-014-9345-4.
62. Vieira MN, Lima-Filho RA, De Felice FG. Connecting Alzheimer's disease to diabetes: underlying mechanisms and potential therapeutic targets. Neuropharmacology. 2017;136:160–71.
63. Gulshan V, Peng L, Coram M, Stumpe MC, Derek W, Narayanaswamy A, Venugopalan S, et al. Development and validation of a deep learning algorithm for detection of diabetic retinopathy in retinal fundus photographs. JAMA. 2016;316(22):2402–10. https://doi.org/10.1001/jama.2016.17216.
64. Dvornek N, Ma J, Gao L, Ou Y. Convolutional neural networks for age-related macular degeneration recognition based on fundus images. In: 2019 IEEE 16th international symposium on biomedical imaging (ISBI 2019); 2019. p. 505–9.
65. Rama M, Mookiah K, Hogg S, Macgillivray T, Trucco E. On the quantitative effects of compression of retinal fundus images on morphometric vascular measurements in VAMPIRE. Comput Methods Prog Biomed. 2021;202:105969. https://doi.org/10.1016/j.cmpb.2021.105969.
66. Chen W, Yang B, Li J, Wang J. An approach to detecting diabetic retinopathy based on integrated shallow convolutional neural networks. IEEE Access. 2020;8:178552–62. https://doi.org/10.1109/ACCESS.2020.3027794.
67. Tayal A, Gupta J, Solanki A, Bisht K, Nayyar A, Masud M. DL-CNN-based approach with image processing techniques for diagnosis of retinal diseases. Multimedia Systems. 2022;28(4):1417–38. https://doi.org/10.1007/s00530-021-00769-7.
68. Tabarestani S, Aghili M, Shojaie M, Freytes C, Cabrerizo M, Barreto A, Adjouadi M. Longitudinal prediction modeling of Alzheimer disease using recurrent neural networks. In: 2019 IEEE EMBS international conference on biomedical & health informatics (BHI); 2019. p. 1–4.
69. Aslam MH, Usman SM, Khalid S, Anwar A, Alroobaea R, Hussain S, Almotiri J, Ullah SS, Yasin A. Classification of EEG signals for prediction of epileptic seizures. Applied Sciences (Switzerland). 2022;12(14) https://doi.org/10.3390/app12147251.
70. Chantamit-o-Pas P, Goyal M. Long short-term memory recurrent neural network for stroke prediction. In: Lecture notes in computer science (including subseries lecture notes in artificial intelligence and lecture notes in bioinformatics); 2018. p. 312–23. https://doi.org/10.1007/978-3-319-96136-1_25.

71. Che C, Xiao C, Liang J, Jin B, Zhou J, Wang F. An RNN architecture with dynamic temporal matching for personalized predictions of Parkinson's disease. In: Proceedings of the 17th SIAM international conference on data mining, SDM 2017; 2017. p. 198–206. https://doi.org/10.1137/1.9781611974973.23.
72. Zhang X, Mehta D, Zhu C, Merlo D, Hu Y, Gresle M, Ge Z. Deep survival analysis in multiple sclerosis. In: International workshop on PRedictive intelligence in MEdicine. Cham: Springer Nature; 2023. p. 108–19.
73. Ohm P. Broken promises of privacy: responding to the surprising failure of anonymization. UCLA Law Rev. 2010;57(6):1701–77.
74. Cavoukian A, Jonas J. Privacy by design in the age of big data information & privacy commissioner Ontario, Canada; 2012. https://jeffjonas.typepad.com/Privacy-by-Design-in-the-Era-of-Big-Data.pdf, p. 1–17
75. Fung BC. Privacy preserving data portals. In: Encyclopedia of portal technologies and applications. IGI Global Scientific Publishing; 2007. p. 842–7. https://doi.org/10.1561/9781601983251.
76. Hardt M, Price E, Srebro N. Equality of opportunity in supervised learning. Adv Neural Inf Proces Syst. 2016:3323–31.
77. Chouldechova A. Fair prediction with disparate impact: a study of bias in recidivism prediction instruments. Big Data. 2017;5(2):153–63. https://doi.org/10.1089/big.2016.0047.
78. Obermeyer Z, Powers B, Vogeli C, Mullainathan S. Dissecting racial bias in an algorithm used to manage the health of populations. Science. 2019;366(6464):447–53. https://doi.org/10.1126/science.aax2342.
79. Haw R, Murdoch TB, Lin Z, Wang Y, Liu Z, Hong P, Chen J. Privacy preserving prognostic models for hospital readmission. J Am Med Inform Assoc. 2019;26(4):305–13.
80. Mittelstadt BD, Allo P, Taddeo M, Wachter S, Floridi L. The ethics of algorithms: mapping the debate. Big Data Soc. 2016;3(2):1–21. https://doi.org/10.1177/2053951716679679.
81. Ribeiro MT, Singh S, Guestrin C. 'Why should I trust you?': Explaining the predictions of any classifier. In: Proceedings of the 22nd ACM SIGKDD international conference on knowledge discovery and data mining (KDD'16); 2016.
82. Chen J, Song L, Wainwright M, Jordan M. Learning to explain: an information-theoretic perspective on model interpretation. In: Proceedings of the 35th international conference on machine learning, PMLR 80; 2018. p. 883–92.
83. U.S. Department of Health & Human Services. Summary of the HIPAA privacy rule; 2020. https://www.hhs.gov/hipaa/for-professionals/privacy/index.html
84. Hoofnagle CJ, Van Der Sloot B, Borgesius FZ. The European Union general data protection regulation: what it is and what it means. Inform Commun Technol Law. 2019;28(1):65–98.
85. Mittelstadt B, Russell C, Wachter S. Explaining explanations in AI. In: FAT* 2019 – proceedings of the 2019 conference on fairness, accountability, and transparency; 2019. p. 279–88. https://doi.org/10.1145/3287560.3287574.
86. Hersh W. A stimulus to define informatics and health information technology. BMC Med Inform Decis Mak. 2009;9(1):1–6. https://doi.org/10.1186/1472-6947-9-24.
87. Adadi A, Berrada M. Peeking inside the black-box: a survey on explainable artificial intelligence (XAI). IEEE Access. 2018;6:52138–60. https://doi.org/10.1109/ACCESS.2018.2870052.
88. Kairouz P, Brendan McMahan H, Avent B, Bellet A, Bennis M, Bhagoji AN, Bonawitz K, et al. Advances and open problems in federated learning. Foundat Trends® Mach Learn. 2021;14(1–2):1–210.

Chapter 8
Detection of Parkinson's Disease Using Different Machine Learning Techniques: A Comparative Analysis

Prachi Mall and Deepika Raina

8.1 Introduction

Parkinson's disease (PD) is defined as a progressive neurodegenerative disorder that affects the neurons producing the dopamine in the brain area called substantia nigra [1]. Symptoms usually grow constantly over the years. Due to the diversity of the disease, symptom progression is often somewhat different from one individual to another [2]. There are two categories for Parkinson's disease symptoms: motor and nonmotor. Tremor, bradykinesia (slowness of movement), shivering, rapid eye movement dysfunction, gait problems (locomotory actions), and unstable posture are some of the motor symptoms. Hypotension, exhaustion, weight loss, perspiration, constipation, urinary issues, sleep disturbances, and depression are examples of nonmotor symptoms. However, bradykinesia, dysphagia, loss of reflexive movements, walking difficulties, changes in writing, speech, and voice issues, and slowness of movement are the most prevalent early indications of this illness. There is currently no recognized cure for PD, but a variety of pharmaceutical treatments can greatly lessen the symptoms, especially in the early stages [3, 4]. In the United States, Parkinson's disease complications are counted as the 14th cause of death by the Centers for Disease Control and Prevention (CDC) classification. Treatment approaches included anticholinergics, until levodopa (dopamine agonist) was introduced, which is the most widely used drug nowadays. Dopamine agonists are widely preferred than used alone. Although monoamine oxidase B inhibitors enhance early and advanced motor function, research has shown that people with Parkinson's disease experience symptoms later during the disease because a major amount of substantia nigra neurons have vanished or damaged. Patients having PD show the presence of Lewy bodies in the substantia nigra region of the brain. Lewy bodies are

P. Mall · D. Raina (✉)
School of Pharmacy, Graphic Era Hill University, Dehradun, Uttarakhand, India
e-mail: deepikaraina@gehu.ac.in

T. Choudhury, A. Katal (eds.), *Machine Learning for Disease Detection, Prediction, and Diagnosis*, https://doi.org/10.1007/978-981-96-4241-0_8

abnormal protein aggregates of alpha-synuclein protein that are associated with cognitive and motor functions. Researchers are scrutinizing methods to recognize Parkinson's disease biomarkers that may lead to earlier diagnosis and more tailored medication to decrease the severity of symptoms of the diseases. All therapies used for Parkinson's disease are currently improving symptoms without progression of the disease. It is recommended that regular exercise (aerobic) may decline the cause and effect of Parkinson's disease (PD) [5]. More people are dying and becoming disabled globally from Parkinson's disease (PD) than from any other neurological condition. In the last 25 years, the prevalence of Parkinson's disease has doubled. Over 8.5 million people were estimated to have Parkinson's disease (PD) in 2019. Based on available statistics, there has been an estimated 81% increase since 2000, resulting in 3,29,000 deaths—a rise of more than 100% since 2000 [6].

8.2 Pathophysiology

The primary cause of Parkinson's disease is the degeneration of dopaminergic neurons in the substantia nigra region of the brain, which produces dopamine [7]. Although intensive research is been carried out, yet it is considered idiopathic (of unknown cause). It probably includes host sensitivity, environmental and genetic variables interacting [8]. It has been determined that a number of genetic mutations are risk factors for Parkinson's disease. These mutations increase the likelihood of developing the disease. Some of these include PINK1, PARK2, LRRK2, and SNCA [9]. Physiologically, Parkinson's disease-related symptoms lead from the reduction of several neurotransmitters, particularly dopamine. In PD, dopaminergic connection is lost with the neostriatum (putamen). This results in less inhibition and enhanced excitation from subthalamic nucleus to globuluspalidus internal segment and due to which inhibitory output to thalamus and brainstem is increased [10].

PD is a progressive disease, the symptoms of which progress with age and no complete cure of the disease has been known to date, but the treatment of PD is symptomatic. Levodopa has been used from years to treat the clinical signs and symptoms of PD. However, the first-line treatment options have been available in the past decade with new compounds including two MAO-B inhibitors: safinamide and zonisamide, a catechol O-methyltransferase inhibitor: opicapone and one amantadine extended-release formulation. Inclusive of the oral levodopa dispersible tablets and apomorphine injection, new and more innovative drug therapies such as levodopa inhalations and sublingual apomorphine have shown tremendous efficacy (Fig. 8.1). In cases when treatment of PD becomes difficult with the available therapies, surgeries, deep brain stimulation, or device-aided therapies with subcutaneous apomorphine or subcutaneous or intestinal levodopa are considered [11]. Some innovative advanced therapies related to different genetic forms of PD include lithium, prasinezumab, deferiprone, and cinpanemab [12].

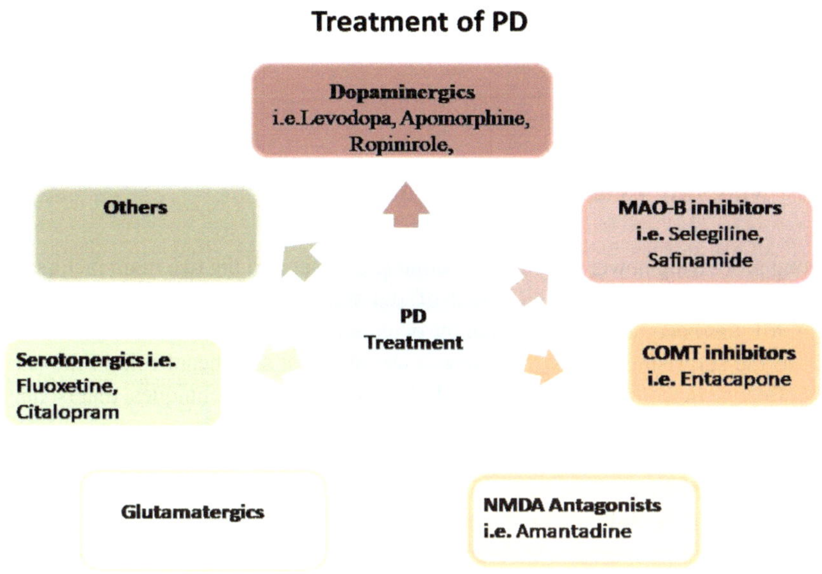

Fig. 8.1 Treatment approaches of Parkinson's disease

8.3 Role of Artificial Intelligence in the Diagnosis of PD

PD is a progressive neurodegenerative disease, the misdiagnosis rate of which is significant due to the absence of a definitive test. The physician's main goal is to identify Parkinson's disease (PD) and determine its severity before treating the patient since the disease's severity affects both motor and nonmotor symptoms. In such cases, it will be helpful to aid the doctor by providing an ML model. ML has helped in the development of the predictive models [13].

In the past years, it was difficult to diagnose PD by just counting on an instinctive assessment scale that are the characteristics of the disease, that is, limb's rigidity, speech loss, and control on motor activities, which can be used to detect the disease. As the research progresses, it was discovered to have several clinical signs and symptoms. Besides, the motor symptoms commonly display nonmotor symptoms such as cognitive and neurocognitive issues, sleep disturbances, and postural instability that can later progress to depression and dementia [14].

AI is a scientific approach that enables the machine to work based on its intellect or think without human intervention [15, 16]. AI is a broad area that deals with the creation of intelligent machines that can carry out any work without the need of human guidance and instructions, such as decision-making, learning, and problem-solving. Recent advances have emphasized on AI-based ML techniques for early diagnosis of PD using motor symptoms monitoring, brain imaging, speech monitoring, handwriting analysis, face video analysis, and using multimedia approaches [13]. ML is a subset of AI that uses statistical techniques to search and analyze data in order to train digital computers to perform tasks without obvious instructions by identifying patterns and insights in the data. Hence, we can call ML an integral part of AI [17].

8.4 Machine Learning Techniques

Strong processing power and readily available data sets are the two main factors that have contributed to the development of statistical data-driven machine learning [17]. ML has become unavoidable in the detection and diagnosis of several diseases due to the surprising results; similarly, it has played a significant role in the early and effective detection of PD along with the neuroimaging techniques, that is, magnetic resonance imaging (MRI), functional magnetic resonance imaging (FMRI), positron emission tomography (PET), and single-photon emission computed tomography (SPECT) [18].

Several ML techniques have been introduced to classify the patients as PD patients or healthy individuals based on voice/speech analysis, handwriting analysis, and gait patterns, and the performance can be measured based on accuracy, sensitivity, specificity, and F1 score (Fig. 8.2). Accuracy generally measures the performance of the method that depends on the availability of the features, selection of the most appropriate subset to improvise the classification, and accuracy of the

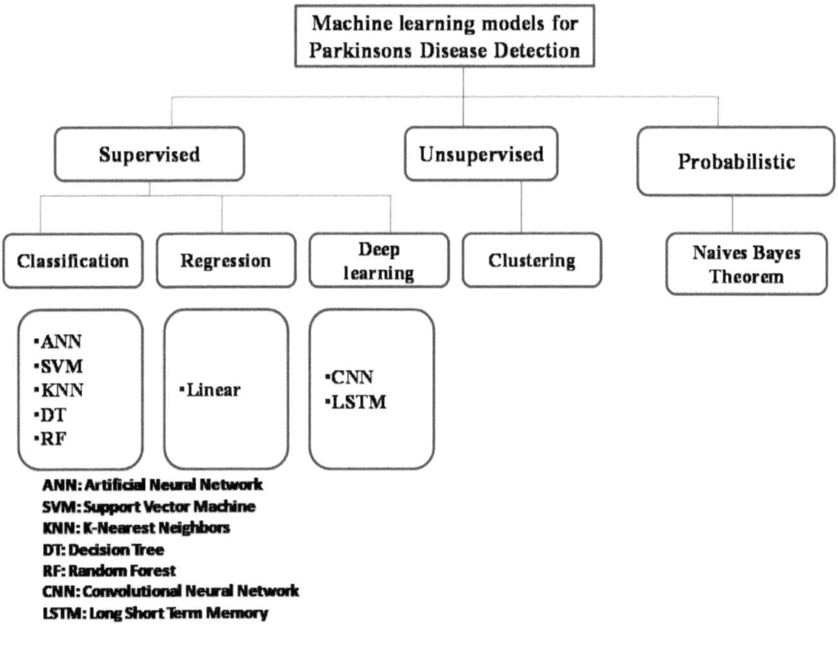

Fig. 8.2 Different ML models for PD detection

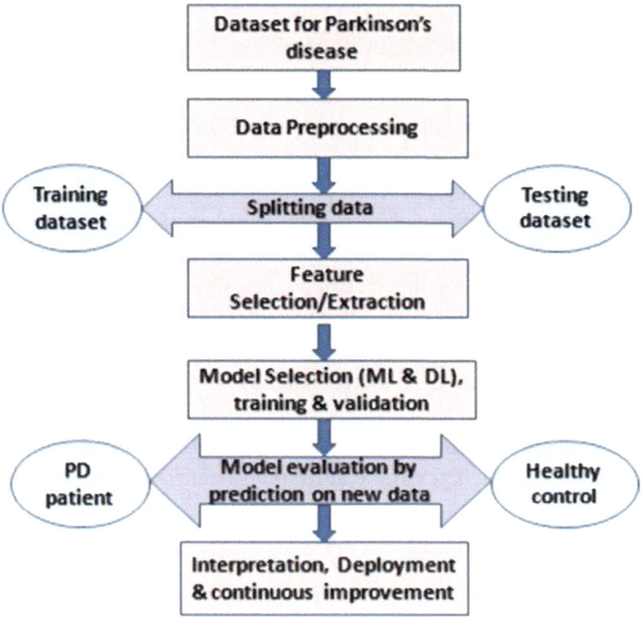

Fig. 8.3 Classification pattern using ML techniques

applied ML technique [3]. It is therefore required to create a feature subset selection (FSS) method that optimizes only the required number of features to maximize the efficacy and minimizes the confusion among the different available features.

A feature subset selection method basically comprises four steps:

1. Generation of the subset.
2. Assessment of the subset.
3. Stopping norms.
4. Validation of the results [19]

The detection and diagnosis of any neurodegenerative disease usually consist of the following steps as shown in Fig. 8.3.

8.5 ML Models for the Detection of Parkinson's Disease

8.5.1 K-Nearest Neighbors' Classifier

The K-nearest neighbor (K NN) classification is a simple nonlinear comparative learning classification method. Out of the entire dataset, 70% is considered training subset and 30% is considered testing subset.

The method basically employs two steps:

- Determination of the K-closest neighbors.
- Classification based on the K-closest neighbors.

Using the approach, the patient is categorized based on the K closest sample classes, that is, by comparing the test sample to the training sample that are the closest or similar. The feature space is checked for the largest K-nearest neighbors and based on K-value, assigning the class to the testing subset. Through the cross-validation method, the method determines the number of neighbors, k [20].

Several distance metrics are available to measure the distance between the training and test sample, that is, Euclidean, Chebychev, Cosine, Mahalanobis, Hamming, and Spearman metrics but Euclidean distance metric is highly considered [21]. To determine the K values, usually cross-validation approach is applied, that is, by experimenting with a range of K values, such as 1–10, and then optimizing the K value, the one that gives the least classification error [22]. It is the method of choice when information on the given data is not sufficient. However, the classification method can show poor accuracy if the attribute is irrelevant or noisy.

8.5.2 Support Vector Machines

The support vector machine (SVM) model is established on the basis of new statistical learning concepts used for both classification and regression tasks. It converts the original data to higher dimensional feature space. The SVM method involves defining a separator, often called hyperplane that efficiently differentiates the two classes by augmenting the margin between the two classes with the aid of support vectors [23]. Support vectors are the data points that lie closest to the hyperplane. It is called a good separation when the distance between the nearest training data point (support vectors) of any class is the largest. Hence, the best hyperplane is the one that divides the dataset into distinct classes [24]. The class of the new data point is assigned according to the side of the plane. The fruitfulness of an SVM technique for PD detection lies on the quantity and quality of data, also including the careful selection of features and proper model tuning [25].

8.5.3 Decision Tree

Decision tree classifier (DTC) is a supervised learning algorithm that is basically used for classification. Decision tree can transform the most complex decision into several simpler decisions and make problems easy, thus providing an easy and comprehensive desired solution [25]. In context to Parkinson's detection, DTC splits the dataset into smaller subsets called branches and nodes and thus resembles the structure of a tree. DTC can also determine the stage/severity of the disease where initially training model is created to predict the variables by learning the decision rules, concluded from the dataset [26]. The classification is usually done in the following pattern:

- One root node, consisting of the class labels.
- Internal nodes that generally have a single incoming and outgoing.
- Terminal or leaf nodes are nodes with no descendent edges [27].

DTC can be constructed simply with the use of "IF-THEN" rules, which will serve as a tool to conclude the results of the dataset based on the gained information and insights. The major drawback of DTC is the overlapping of subsets when there exist a large number of classes [27, 28].

8.5.4 Naïve Bayesian Classifier

The naïve Bayesian classifier is a supervised learning algorithm that is based on Bayes theorem to make strong independent assumptions. It is well known for its simplicity and speed, making it a good option for disease detection. This classifier is based on classifying the test sample based on the probabilities/experiences. It predicts strong assumptions about the current condition based on the past experience [29].

The major advantages of the classifier are.

- It requires only a small amount of training data to calculate parameter estimation.
- It is particularly used for large datasets and does not involve the estimation of the parameters.
- It is considered "naïve" as the occurrences of the features are independent of one another [30, 31].

The common variants of naïve Bayes algorithm include multinomial naïve Bayes and Gaussian naïve Bayes [30, 31].

8.5.5 Random Forest

Random forest (RF) classifier is a hierarchical group of decision trees. It is an extensive classifier where decision trees are implemented for all the classification-related activities. It is applicable to the classification and regression problem [32]. In this method, decision tees are used for collection of data, training, planning, and every single tree/decision is given equal consideration to predict the output [28]. When the dataset contains a large number of attributes, the RF algorithm uses simple probability to select the most relevant attributes to construct multiple decision trees and thus predict the results [33].

The algorithm is basically divided into two steps:

1. RF tree is constructed that involves the selection of features, determination of nodes, further distribution into internal nodes, terminal nodes, and repeating the same procedure till the desired number of trees is obtained.
2. Classification of the data based on the developed RF tree, using voting results of each decision tree, the results are concluded with the highest predicted value.

However, the accuracy of the results is directly proportional to the number of trees in the forest, that is, with an increase in the number of trees, accuracy increases and the probability of overfitting decreases [34, 35].

8.5.6 Logistic Regression

A logistic regression model is a predictive regression analysis that provides a simple yet interpretable way to model the relationship between features and the probability of having a particular condition. This method is best suited when the relationship is to be established between the outcome variable and one or more predictor variables [36, 37]. Logistic regression predicts an S-shaped sigmoid curve that is appropriate for PD classification as here the data is not linearly co-related [32, 38]. In cases where the dependent and independent variables do not have a linear relationship, logistic regression algorithm is applied [39]. It is particularly used for binary classification, making it suitable for tasks like PD detection. Based on the calculated probabilities, allocate a threshold above which a person is classified as having Parkinson's and below which they are classified as healthy [39].

8.5.7 Boosting

Boosting is a supervised ML ensemble meta-algorithm in which a set of weak learners are converted to a strong one that primarily reduces the chances of bias and variance [40].The concept behind boosting method is to combine the number of simple learners/weak learners to obtain a strong complex learner with a very small error probability. To train a complex classifier is a more difficult job than to train several smaller classifiers and ultimately summing up the outputs [41]. The final classifier provided by this method has a majority votes of the individual classifiers. The process involves adjusting the weights of miscategorized samples to emphasize on difficult cases, leading to improved accuracy and efficiency in recognizing PD patients [42]. AdaBoost is thc first practical model of this method. Several new variants of AdaBoost have been launched showing improvement in performance. AdaBoost and Gradient Boost are used to differentiate between healthy individuals and patients with PD [43, 44].

8.5.8 *Bagging*

Bagging is an ML ensemble algorithm. It is also commonly referred to as Bootstrap aggregating. This algorithm is devised to improve the accuracy and stability of ML algorithms used in statistical regression and classification [45]. Similar to boosting, it also uses the approach of aggregation/averaging, that is, predictions from multiple decision trees through a majority voting, then the prediction with the highest majority votes is taken into consideration. In Bagging, a subset of feature is picked up randomly to develop a forest [46]. However, it can be used with almost every method but it is usually deployed in combination with decision tree. Bagging is a learning method that reduces errors due to variance, noise, and bias. Errors by variance generally occur due to sampling, noise, and bias [47].

8.5.8.1 Deep Learning (DL) Models for the Detection of Parkinson's Disease

Deep learning methods are developed and utilized extensively for the clinical diagnosis and detection of multiple diseases due to their high sensitivity and accuracy. Deep learning models have shown excellence in the detection of PD as these models automatically learn complicated patterns from complex data. These models have been used in adjunct with neuroimaging techniques for classification of samples. Deep learning models have the advantage over ML models that the feature selection, extraction tools are not required [48]. Also, these models have the ability to learn complex high-dimensional data and mimic the neuron's activity [49]. The model consists of several layers that are interconnected with the nodes/artificial neurons [50, 51].

8.5.8.2 Artificial Neural Networks

Artificial neural network (ANN) models consist of three main layers that are all connected with each other via nodes/artificial neurons.

- Input layer.
- Hidden layer.
- Output layer.

The weight of the connected neurons increases as the neural network learns from the learning algorithm. Each neuron itself acts as a classifier, classifying the output signal by determining the weights from its previous connections [52]. The model requires a large amount of datasets for training purpose and is the main disadvantage of the method. The number of hidden layers differs in different models of neural networks [53].

8.5.8.3 Deep Neural Networks

When an ANN is constructed in such a way that the number of hidden layers is greater than one, it is known as a deep neural network. The DNN can learn more complex data than the ANN [54].

8.5.8.4 Convolutional Neural Network

A CNN model can be called a subset of ANN that consists of multiple convolutional and pooling layers. The initial convolutional layer will help to extract features and provide feature map from the feed-in image that itself will act as an input to the subsequent convolutional layers. The subsequent convolutional and pooling further improve the features and reduce the chances of overfitting [55]. The feature maps are converted into a one-dimensional vector at the flattened layer once they have reached the final pooling layer. The CNN models are commonly used for image processing and medical imaging. This model is also used for Parkinson's detection and diagnosis with a high accuracy and sensitivity [56].

8.5.8.5 Recurrent Neural Network

RNN remembers the earlier information to compute the current output. RNNs are useful for sequential data such as voice recordings or time-series data. However, the major drawback associated with the RNN is the limited long-term memory and vanishing gradient [54].

8.5.8.6 Long Short-Term Memory (LSTM)

The LSTM copies the brain's memory storage and predictions. The model can be applied to time-series data where the relationship between past and current observations is important, like voice recordings over time. The model serves for both the drawbacks of RNN, that is, vanishing gradient and limited long-term memory. The model uses a unique gate structure, with the input gate, output gates, and forget gate to eliminate vanishing gradient problems. To avoid inconsistencies, the forget gate chooses which data from the earlier states to remove. LSTM classifiers are particularly effective for tasks involving sequential patterns to memorize patterns over time, making it suitable for speech and handwriting recognition in PD [57] (Table 8.1).

Table 8.1 Principle of classification of different ML and DL models

Classifier	Assessment mode	Principle of classification
K-nearest neighbor	Speech and clinical features (like age, gender, and motor UPDRS (unified Parkinson's Disease rating scale) scores	It compares the features of a new data to those of known cases, assigning the patient the most frequent Parkinson's status among its closest neighbors
Support vector machine	Speech, clinical features, and movement patterns	It creates a hyper-plane that distinguishes patients with Parkinson's from healthy individuals by finding the most informative features to distinguish between the two classes
Decision trees	Speech and handwriting, clinical features, and movement patterns	The dataset is divided based on the features to construct a tree-like structure, allowing classification of patients based on the features decision
Naïve Bayesian classifier	Speech and handwriting patterns, clinical features	It applies probabilistic reasoning, assuming independence between features given the disease label, to calculate the likelihood of PD based on the observed features
Random forest	Speech and biomedical data	Random forest creates multiple decision trees making use of different subsets of data and features, combining their results to enhance robustness and accuracy in classifying patients with Parkinson's disease
Logistic regression	Speech and clinical features	It uses a linear combination of features, converted through a logistic function, allowing binary classification based on the calculated probability threshold
Boosting	Speech, clinical features, and movement patterns	It assembles multiple weak classifiers into a strong ensemble, where each subsequent classifier emphasizes more on correcting the mistakes of antecedent, improving overall accuracy of classification
Bagging	Speech, clinical features, and movement patterns	It trains several instances of the same model on bootstrapped subsets of data and pooling their predictions improves classification accuracy
Artificial neural network/ deep neural network	Speech signals, handwriting patterns, movement trajectories and neuroimages/brain scans.	It learns the complex relationships within data by imitating interconnected neurons, allowing automatic feature extraction and classification to distinguish Parkinson's from healthy individuals

(continued)

Table 8.1 (continued)

Classifier	Assessment mode	Principle of classification
Convolutional neural network	Speech signals, handwriting patterns, movement trajectories, and neuro	It masters in capturing spatial patterns from medical images like brain scans by convolving filters over image regions to automatically learn features that helps in the classification
Recurrent neural network	Speech signals, handwriting patterns, movement trajectories, and neuro	It works on sequential or time-series data, considers past patterns to identify patterns and temporal relationships indicative of the disease
Long short-term memory	Speech signals, handwriting patterns, movement trajectories, and neuro	It specializes in modeling time-series data with long-range dependencies for accurate disease identification

8.5.8.7 Literature Review

Deep learning model was developed by Wang et al. who compared the developed model with other machine learning models based on the premotor features for the classification of healthy individuals and patients suffering from Parkinson's. The deep learning model, especially the ensemble method, showed the highest accuracy for the early detection of PD [50]. Three machine learning classifiers were employed for the detection of PD and compared by Pahuja et al., that is, support vector machine, K-nearest neighbor, and artificial neural network (multilayer perceptron) based on voice dataset. ANN with Levenberg–Marquardt algorithm has provided the highest classification accuracy of 95.89% over other classifiers [19]. A CNN model was developed by Taleb et al. based on the online handwriting for the early detection of Parkinson's disease. The CNN models outperformed the SVM but require huge data for training the model [53]. Four machine learning classifiers were compared for the detection of PD by Govindu et al., that is, support vector machine, logistic regression, K-nearest neighbor, and random forest model. The ensemble model, random forest, provided the highest accuracy for the detection of PD based on voice modulation [32]. Gradient boosting model was proposed by Varalakshmi et al. to predict PD with the highest accuracy for the detection of PD based on the speech [57]. A deep neural network (DNN) and a long short-term memory (LSTM) network-based model were developed by Rizvi et al. for the prediction of PD using voice samples. They also compared the developed models with the conventional machine learning models. Both the developed methods proved efficient in PD prediction and better than the other machine learning model. The LSTM model showed the highest accuracy [54]. The artificial neural network suggested by Singh et al. provided the most precise and accurate results among the different machine learning models based on the voice dataset [58].

The literature mentioned above demonstrates the efficacy of machine learning models in predicting Parkinson's disease. However, compared to the conventional machine learning models, the ensemble classifiers and neural network models produced superior outcomes (Table 8.2).

Table 8.2 Comparison of the different ML and DL models based on the literature

S. no.	Title	Dataset	Outcome	References
1.	Early detection of Parkinson's disease using deep learning and ML	Premotor features (i.e., rapid eye movement, sleep behavior disorder, and olfactory loss)	Proposed the deep neural network provides better efficacy over other ML methods	[50]
2.	A comparative study of existing ML approaches for Parkinson's disease detection	Voice input dataset using UCI ML repository	Suggested that artificial neural networks with Levenberg–Marquardt algorithm give the highest classification accuracy of among SVM and KNN for voice dataset	[19]
3.	Detection of Parkinson's disease from handwriting using deep learning: a comparative study	Handwriting dataset	Suggests that the deep learning method (CNN) outperforms over SVM	[53]
4.	Early detection of Parkinson's disease using ML	Audio dataset (vowel phonation data)	Suggested using the random forest model rather than SVM and KNN to categorize the disease's progression	[32]
5.	Parkinson disease detection based on speech using various ML models and deep learning models	Speech analysis	Suggested the boosting methods shows the highest accuracy over other ML methods	[57]
6.	An LSTM-based deep learning model for voice-based detection of Parkinson's disease	Voice dataset	Suggested that the DNN and LSTM performed better than the existing ML models	[54]
7.	Voice-based detection of Parkinson's disease through ensemble ML approach: a performance study	Voice dataset	Suggested that the boosting ensemble method outperformed all other models (logistic regression, naive Bayes, KNN, random forest, decision tree, SVM, MLP) in terms of accuracy, precision, and F1 score	[59]
8.	An improved approach for prediction of Parkinson's disease using ML techniques	Rapid eye movement (REM) sleep behavior disorder (RBD), olfactory loss, and the important biomarkers.	Demonstrated that the boosted logistic regression method proved superior over other ML models, hence the use boosting ensemble method was encouraged	[36]

(continued)

Table 8.2 (continued)

S. no.	Title	Dataset	Outcome	References
9.	Parkinson disease prediction using ML algorithm	Voice dataset	ANN combined with KNN models yields better results, hence multi-classifier or ensemble methods provides better and early disease detection	[60]
10	A comparative study of different ML algorithms in detection of Parkinson's disease	Voice dataset	ANN gives the most accurate and precise results in comparison to other ML models	[58]
11.	Comprehensive and comparative analysis of different types ML algorithm for detecting Parkinson's disease	Voice dataset	XGBoost (boosting ensemble method provides better accuracy, sensitivity and specificity than SVM, KNN, logistic regression, AdaBoost)	[61]
12.	A novel sample and feature dependent ensemble approach for Parkinson's disease detection	Voice dataset	Proposed that the feature selection integration with DNN improved PD detection accuracy than other ML methods and conventional neural networks	[62]

The choice of classifier to be used relies on the type of the dataset and the features considered. However, researchers/clinicians can experiment with multiple classifiers/hybrid models to perform rigorous evaluation to identify the best-performing one for Parkinson's disease detection and progression [62].

8.5.8.8 Case Study

Case 1 The real-time case study proposed by freezing of gait (FOG) detection technique via wearable sensor. The study used data from 21 healthy senior adults and 118 Parkinson's disease patients. The data was acquired during simulated daily living activities using a single inertial sensor that was worn around the waist. By analyzing inertial data at various spatial resolutions, the multi-head convolutional neural network that was constructed was able to reduce computational complexity and optimize performance. The classification performance was good to exceptional, with over 50% of FOG episodes predicted on average 3.1 s prior to commencement, according to the results. Additionally, the system showed above 88% specificity on the main dataset and 100% specificity on aged, healthy participants. With implications for better patient outcome and lowering fall risks, the study's methodology provides a workable alternative for real-time FOG detection, potentially improving the monitoring and management of Parkinson's disease symptoms. Subsequent efforts may concentrate on enhancing the algorithm's validation and improvement for broader application in home and healthcare environments [63].

Case 2 A web-based framework was developed with the aim of providing Parkinson's disease (PD) screening in the remote areas. Data was gathered from 726 individuals, both with and without Parkinson's disease from different demographic and geographical backgrounds. In a simple speech exercise, participants were asked to speak a common phrase, which was subsequently subjected to several machine learning algorithms for analysis. According to the results, the model was able to detect PD with a promising accuracy [64].

Cases have been discussed for insight into the early detection of Parkinson's disease; however, the major challenges associated with PD detection are accurately tracking motor and nonmotor symptoms in real-life situations. The evaluation and titration of therapy can be enhanced by the nonintrusive recording of symptoms and their realistic changes within or between days [57, 65].

8.5.8.9 Discussion on Future Work

There are numerous arenas for future work related to the detection of Parkinson's disease (PD) with the use of various machine learning techniques. The ensemble learning strategies like random forest, AdaBoost, and Gradient Boosting may increase the classification efficiency for PD detection. Ensemble approaches have the potential to improve the performance of PD detection systems by combining numerous models to yield better results than any single model could [66–68]. Further research can be conducted on feature selection and engineering techniques tailored specifically for PD detection. This may involve exploring advanced feature extraction methods from biomedical signals and investigating automated feature selection algorithms to identify the most appropriate and discriminative features for classification [69, 70]. Deep learning achitecture is sophisticated neural network structures used in AI to process information and make predictions, including convolutional neural networks, recurrent neural networks, or their combinations (e.g., convolutional recurrent networks). The ability of deep learning models to automatically generate hierarchical representations from raw data has shown promise in a number of medical applications, which could lead to an improvement in the classification accuracy of Parkinson's disease [71, 72]. Exploring and integrating multiple data modalities (e.g., clinical assessments, imaging data, genetic data, voice recordings, wearable sensor data) for the detection of PD using fusion techniques. Combining information from different sources may provide complementary insights into the underlying disease mechanisms and improve the overall classification performance [73]. To promote the use of machine learning models for Parkinson's disease (PD) detection in clinical settings, concentrate on improving their interpretability and explainability. Provide methods for producing comprehensible feature representations, illustrating model predictions, and offering justifications for categorization choices so that medical professionals may comprehend and have faith in automated diagnostic systems [74].To assess the generalization performance of the created models, carry out external validation investigations using separate datasets gathered from various populations and clinical contexts [75]. Examine the practical difficulties and factors that need to be taken into account when implementing PD detection systems based on machine learning in actual clinical settings. These factors include data security, legal compliance, and system integration [76]. By exploring these different avenues for future work, researchers can further advance the field of PD detection using machine learning techniques and contribute to the development of more accurate, robust, and clinically relevant diagnostic tools for this debilitating neurological disorder [77].

8.6 Conclusion

The KNN and SVM models are extensively used for Parkinson's detection due to the simplicity of ML models, but on the basis of the various studies conducted, it has been determined that deep learning models are superior to ML models in terms of achieving the highest accuracy in Parkinson's disease detection, even when combined with neuroimaging techniques, providing more concrete evidence of the disease. However, these models are known as black-box algorithms because it is difficult to interpret the trained neural networks and comprehend how deep learning achieves good results. However, ensemble networks and hybrid models have also demonstrated outstanding disease detection and progression accuracy.

Acknowledgments We are extremely thankful to the Graphic Era Hill University for providing us the platform for the research work.

References

1. Ahmad MH, Rizvi MA, Ali M, Mondal AC. Neurobiology of depression in Parkinson's disease: insights into epidemiology, molecular mechanisms and treatment strategies. Ageing Res Rev. 2023;101840
2. Marti-Masso JF, et al. Drugs inducing or aggravating parkinsonism: a review. Therapie. 1996;51(5):568–77.
3. Rana A, Dumka A, Singh R, Panda MK, Priyadarshi N, Twala B. Imperative role of ML algorithm for detection of Parkinson's disease: review, challenges and recommendations. Diagnostics. 2022;12(8):2003.
4. Krasko MN, Rudisch DM, Burdick RJ, Schaen-Heacock NE, Broadfoot CK, Nisbet AF, et al. Dysphagia in Parkinson disease: part II – current treatment options and insights from animal research. Curr Phys Med Rehabil Rep. 2023;11:1–11.
5. de Souza Silva MA, Mattern C, Häcker R, Tomaz C, Huston JP, Schwarting RK. Increased neostriatal dopamine activity after intraperitoneal or intranasal administration of L-DOPA: on the role of benserazide pretreatment. Synapse. 1997;27(4):294–302.

6. WHO. https://www.who.int/news-room/fact-sheets/detail/parkinson-disease#:~:text=The%20prevalence%20of%20PD%20has,of%20over%20100%25%20since%202000
7. Ross CA, Smith WW. Gene–environment interactions in Parkinson's disease. Parkinsonism Relat Disord. 2007;13:S309–15. https://doi.org/10.1016/S1353-8020(08)70022-1.
8. Weisskopf MG, et al. Plasma urate and risk of Parkinson's disease. Am J Epidemiol. 2007;166(5):561–7. https://doi.org/10.1093/aje/kwm127.
9. Vázquez-Vélez GE, Zoghbi HY. Parkinson's disease genetics and pathophysiology. Annu Rev Neurosci. 2021;44:87–108.
10. Schwarzschild MA, et al. Parkinson study group PRECEPT investigators. Serum urate as a predictor of clinical and radiographic progression in Parkinson disease. Arch Neurol. 2008;65(6):716–23. https://doi.org/10.1001/archneur.2008.65.6.nct70003.
11. Fabbri M, Barbosa R, Rascol O. Off-time treatment options for Parkinson's disease. Neurol Ther. 2023;12(2):391–424.
12. Cavallieri F, Cury RG, Guimarães T, Fioravanti V, Grisanti S, Rossi J, et al. Recent advances in the treatment of genetic forms of Parkinson's disease: hype or Hope? Cells. 2023;12(5):764.
13. Dixit S, Bohre K, Singh Y, Himeur Y, Mansoor W, Atalla S, Srinivasan K. A comprehensive review on AI-enabled models for Parkinson's disease diagnosis. Electronics. 2023;12(4):783.
14. Bohnen NI, Yarnall AJ, Weil RS, Moro E, Moehle MS, Borghammer P, et al. Cholinergic system changes in Parkinson's disease: emerging therapeutic approaches. Lancet Neurol. 2022;21(4):381–92.
15. Su J, Ng DTK, Chu SKW. Artificial intelligence (AI) literacy in early childhood education: the challenges and opportunities. Comput Educ Artif Intell. 2023;4:100124.
16. Sheth A, Roy K, Gaur M (2023) Neurosymbolic ai-why, what, and how. arXiv preprint arXiv:2305.00813
17. Holzinger A, Keiblinger K, Holub P, Zatloukal K, Müller H. AI for life: trends in artificial intelligence for biotechnology. New Biotechnol. 2023;74:16–24.
18. Khanna K, Gambhir S, Gambhir M. Comparative analysis of ML techniques for Parkinson's detection: a review. Multimed Tools Appl. 2023:1–27.
19. Pahuja G, Nagabhushan TN. A comparative study of existing ML approaches for Parkinson's disease detection. IETE J Res. 2021;67(1):4–14.
20. Shirvan RA, Tahami E. Voice analysis for detecting Parkinson's disease using genetic algorithm and KNN classification method. In: 2011 18th Iranian conference of biomedical engineering (ICBME). IEEE; 2011. p. 278–83.
21. Ouhmida A, Raihani A, Cherradi B, Terrada O. A novel approach for Parkinson's disease detection based on voice classification and features selection techniques. Int J Onl Eng. 2021;17(10):111.
22. Bourouhou A, Jilbab A, Nacir C, Hammouch A. Comparison of classification methods to detect the Parkinson disease. In: 2016 international conference on electrical and information technologies (ICEIT). IEEE; 2016, May. p. 421–4.
23. Bind S, Tiwari AK, Sahani AK, Koulibaly P, Nobili F, Pagani M, et al. A survey of ML based approaches for Parkinson disease prediction. Int J Comput Sci Inf Technol. 2015;6(2):1648–55.
24. Nissar I, Mir WA, Shaikh TA. ML approaches for detection and diagnosis of Parkinson's disease-a review. In: 2021 7th international conference on advanced computing and communication systems (ICACCS), vol. 1. IEEE; 2021, March. p. 898–905.
25. Pardoel S, Kofman J, Nantel J, Lemaire ED. Wearable-sensor-based detection and prediction of freezing of gait in Parkinson's disease: a review. Sensors. 2019;19(23):5141.
26. Balaji E, Brindha D, Balakrishnan R. Supervised ML based gait classification system for early detection and stage classification of Parkinson's disease. Appl Soft Comput. 2020;94:106494.
27. Safavian SR, Landgrebe D. A survey of decision tree classifier methodology. IEEE Trans Syst Man Cybern. 1991;21(3):660–74.
28. Yadav S, Singh MK, Pal S. Artificial intelligence model for Parkinson disease detection using ML algorithms. Biomed Mater Devices. 2023:1–13.

29. Lubis AR, Nasution MKM, Sitompul OS, Zamzami EM. The feature extraction for classifying words on social media with the naïve Bayes algorithm. IAES Int J Artif Intell. 2022;11(3):1041.
30. Zaw HT, Maneerat N, Win KY. Brain tumor detection based on naïve Bayes classification. In: 2019 5th international conference on engineering, applied sciences and technology (ICEAST). IEEE; 2019, July. p. 1–4.
31. Salmi N, Rustam Z. Naïve Bayes classifier models for predicting the colon cancer. IOP Conf Ser Mater Sci Eng. 2019;546(5):052068.
32. Govindu A, Palwe S. Early detection of Parkinson's disease using ML. Proc Comput Sci. 2023;218:249–61.
33. Jackins V, Vimal S, Kaliappan M, Lee MY. AI-based smart prediction of clinical disease using random forest classifier and naive Bayes. J Supercomput. 2021;77:5198–219.
34. Noshad Z, Javaid N, Saba T, Wadud Z, Saleem MQ, Alzahrani ME, Sheta OE. Fault detection in wireless sensor networks through the random forest classifier. Sensors. 2019;19(7):1568.
35. Wang X, Zhai M, Ren Z, Ren H, Li M, Quan D, et al. Exploratory study on classification of diabetes mellitus through a combined random Forest classifier. BMC Med Inform Decis Mak. 2021;21(1):1–14.
36. Challa KNR, Pagolu VS, Panda G, Majhi B. An improved approach for prediction of Parkinson's disease using ML techniques. In: 2016 international conference on signal processing, communication, power and embedded system (SCOPES). IEEE; 2016. p. 1446–51.
37. Polat K. Freezing of gait (fog) detection using logistic regression in parkinson's disease from acceleration signals. In: 2019 scientific meeting on electrical-electronics & biomedical engineering and computer science (EBBT). IEEE; 2019. p. 1–4.
38. Prashanth R, Roy SD. Early detection of Parkinson's disease through patient questionnaire and predictive modelling. Int J Med Inform. 2018;119:75–87.
39. Celik E, Omurca SI. Improving Parkinson's disease diagnosis with ML methods. In: 2019 scientific meeting on electrical-electronics & biomedical engineering and computer science (EBBT). IEEE; 2019. p. 1–4.
40. Saberian MJ, Vasconcelos N. Boosting classifier cascades. NIPS. 2010;23:2047–55.
41. Ferreira AJ, Figueiredo MA. Boosting algorithms: a review of methods, theory, and applications. In: Ensemble ML: methods and applications; 2012. p. 35–85.
42. Enireddy V, Gunda K, Kalyani NL, Prakash KB. Nature inspired binary Grey wolf optimizer for feature selection in the detection of neurodegenerative (Parkinson) disease. Int J. 2020;9(3):3977.
43. Ali L, Zhu C, Golilarz NA, Javeed A, Zhou M, Liu Y. Reliable Parkinson's disease detection by analyzing handwritten drawings: construction of an unbiased cascaded learning system based on feature selection and adaptive boosting model. IEEE Access. 2019;7:116480–9.
44. Yadav S, Singh MK. Hybrid ML classifier and ensemble techniques to detect Parkinson's disease patients. SN Comput Sci. 2021;2:1–10.
45. Jalal S, Alkhafaji S. Empirical mode decomposition-based method for Parkinson detection from voice signal. Eur Chem Bull. 2023;
46. Sekhar C, Rao MS, Bhattacharyya D. ML algorithms for Parkinson's disease detection. Asia-Pacific J Neural Netw Appl. 2020;4(1):29–36.
47. Byeon H. Exploring the predictors of rapid eye movement sleep behavior disorder for Parkinson's disease patients using classifier ensemble. Healthcare. 2020, May;8(2):121.
48. Magesh PR, Myloth RD, Tom RJ. An explainable ML model for early detection of Parkinson's disease using LIME on DaTSCAN imagery. Comput Biol Med. 2020;126:104041.
49. Loh HW, Hong W, Ooi CP, Chakraborty S, Barua PD, Deo RC, et al. Application of deep learning models for automated identification of Parkinson's disease: a review (2011–2021). Sensors. 2021;21(21):7034.
50. Wang W, Lee J, Harrou F, Sun Y. Early detection of Parkinson's disease using deep learning and ML. IEEE Access. 2020;8:147635–46.
51. Johri A, Tripathi A. Parkinson disease detection using deep neural networks. In: 2019 twelfth international conference on contemporary computing (IC3). IEEE; 2019, August. p. 1–4.

52. Asmae O, Abdelhadi R, Bouchaib C, Sara S, Tajeddine K. Parkinson's disease identification using KNN and ANN algorithms based on voice disorder. In: 2020 1st international conference on innovative research in applied science, engineering and technology (IRASET). IEEE; 2020, April. p. 1–6.
53. Taleb C, Likforman-Sulem L, Mokbel C, Khachab M. Detection of Parkinson's disease from handwriting using deep learning: a comparative study. Evol Intel. 2020:1–12.
54. Rizvi DR, Nissar I, Masood S, Ahmed M, Ahmad F. An LSTM based deep learning model for voice-based detection of Parkinson's disease. Int J Adv Sci Technol. 2020;29(8)
55. Ahsan MM, Luna SA, Siddique Z. Machine-learning-based disease diagnosis: a comprehensive review. Healthcare. 2022;10(3):541.
56. Quan C, Ren K, Luo Z. A deep learning based method for Parkinson's disease detection using dynamic features of speech. IEEE Access. 2021;9:10239–52.
57. Varalakshmi P, Bhuvaneaswari R. Parkinson disease detection based on speech using various ML models and deep learning models. In: 2021 international conference on system, computation, automation and networking (ICSCAN). IEEE; 2021. p. 1–6.
58. Singh M, Renji NM. A comparative study of different ML algorithms in detection of Parkinsons disease. Int J Res Appl Sci Eng Technol. 2021;9(1):755–66.
59. Nissar I, Rizvi DR, Masood S, Mir AN. Voice-based detection of Parkinson's disease through ensemble ML approach: a performance study. EAI Endorsed Trans Pervasive Health Technol. 2019;5(19):e2.
60. Mathur R, Pathak V, Bandil D. Parkinson disease prediction using ML algorithm. In: Emerging trends in expert applications and security: proceedings of ICETEAS 2018. Singapore: Springer; 2019. p. 357–63.
61. Chowdhary MS, Paga JJR, Deepa D, GowriManohari V, Nithya S. Comprehensive and comparative analysis of different types ML algorithm for detecting Parkinson's disease. Math Stat Eng Appl. 2022;71(3s2):588–94.
62. Ali L, Chakraborty C, He Z, Cao W, Imrana Y, Rodrigues JJ. A novel sample and feature dependent ensemble approach for Parkinson's disease detection. Neural Comput Applic. 2022:1–14.
63. Borzì L, Sigcha L, Rodríguez-Martín D, Olmo G. Real-time detection of freezing of gait in Parkinson's disease using multi-head convolutional neural networks and a single inertial sensor. Artif Intell Med. 2023;135:102459.
64. Rahman W, Lee S, Islam MS, Antony VN, Ratnu H, Ali MR, Al Mamun A, Wagner E, Jensen-Roberts S, Waddell E, Myers T, Pawlik M, Soto J, Coffey M, Sarkar A, Schneider R, Tarolli C, Lizarraga K, Adams J, et al. Detecting parkinson disease using a web-based speech task: observational study. J Med Internet Res. 2021;23(10):1–15. https://doi.org/10.2196/26305.
65. Habets JGV, Heijmans M, Leentjens AFG, Simons CJP, Temel Y, Kuijf ML, Kubben PL, Herff C. A long-term, real-life Parkinson monitoring database combining unscripted objective and subjective recordings. Data. 2021;6(2) https://doi.org/10.3390/data6020022.
66. Alzubaidi MS, Shah U, Dhia Zubaydi H, Dolaat K, Abd-Alrazaq AA, Ahmed A, Househ M. The role of neural network for the detection of Parkinson's disease: a scoping review. Healthcare. 2021;9(6):740.
67. Biswas SK, Nath Boruah A, Saha R, Raj RS, Chakraborty M, Bordoloi M. Early detection of Parkinson disease using stacking ensemble method. Comput Methods Biomech Biomed Engin. 2023;26(5):527–39.
68. Yadav S, Singh MK. Hybrid machine learning classifier and ensemble techniques to detect Parkinson's disease patients. SN Comput Sci. 2021;2:1–10.
69. Ul Haq A, Li J, Memon MH, Ali Z, Abbas SZ, Nazir S. Recognition of the Parkinson's disease using a hybrid feature selection approach. J Intell Fuzzy Syst. 2020;39(1):1319–39.
70. Xiong Y, Lu Y. Deep feature extraction from the vocal vectors using sparse autoencoders for Parkinson's classification. IEEE Access. 2020;8:27821–30.
71. Alissa M, Lones MA, Cosgrove J, Alty JE, Jamieson S, Smith SL, Vallejo M. Parkinson's disease diagnosis using convolutional neural networks and figure-copying tasks. Neural Comput Applic. 2022:1–21.

72. Karaman O, Çakın H, Alhudhaif A, Polat K. Robust automated Parkinson disease detection based on voice signals with transfer learning. Expert Syst Appl. 2021;178:115013.
73. Papadopoulos A, Iakovakis D, Klingelhoefer L, Bostantjopoulou S, Chaudhuri KR, Kyritsis K, et al. Unobtrusive detection of Parkinson's disease from multi-modal and in-the-wild sensor data using deep learning techniques. Sci Rep. 2020;10(1):21370.
74. Goyal J, Khandnor P, Aseri TC. Classification, prediction, and monitoring of Parkinson's disease using computer assisted technologies: a comparative analysis. Eng Appl Artif Intell. 2020;96:103955.
75. Gupta R, Kumari S, Senapati A, Ambasta RK, Kumar P. New era of artificial intelligence and machine learning-based detection, diagnosis, and therapeutics in Parkinson's disease. Ageing Res Rev. 2023:102013.
76. Chandrabhatla AS, Pomeraniec IJ, Ksendzovsky A. Co-evolution of machine learning and digital technologies to improve monitoring of Parkinson's disease motor symptoms. NPJ Dig Med. 2022;5(1):32.
77. Lepakshi VA. Machine learning and deep learning based AI tools for development of diagnostic tools. In: Computational approaches for novel therapeutic and diagnostic designing to mitigate SARS-CoV2 infection. Academic Press; 2022. p. 399–420.

Chapter 9
Kidney Disease Prediction by Machine Learning Techniques

Tanupriya Choudhury, Ayan Sar, Avita Katal, Sumit Aich, and Purvika Joshi

9.1 Introduction

Temporal kidney disease (CKD) is a worrisome health issue on a global scale, affecting over 600 million people, and the number has risen over the years. Early detection and intervention are the key elements in the fight against the progressive course of the disease, which increases the possibility of the desired end. Conventional diagnostics, although adequate on many occasions, still do not possess the level of sensitivity as well as the precision required for early detection and making an accurate prediction of a kidney disease. In this situation, a set of machine learning techniques has played a major role in the field of paradigm-shifting predicting kidney diseases. This introduction provides a general description of the use of machine learning to forecast kidney disease, which includes its benefits, difficulties, and innovations alongside its future outlook.

9.1.1 Challenges in Traditional Diagnosis

The usual methods of diagnosing kidney disease mostly involve laboratory tests, imaging studies, and physical examinations as the main diagnostic tests. Despite the fact that these methods are effective for getting diagnosed when the harm to the

All authors are first authors for this chapter.

T. Choudhury (✉) · A. Sar · A. Katal · S. Aich · P. Joshi
School of Computer Science, University of Petroleum and Energy Studies (UPES), Dehradun, Uttarakhand, India
e-mail: tanupriya@ddn.upes.ac.in; akatal@ddn.upes.ac.in

T. Choudhury, A. Katal (eds.), *Machine Learning for Disease Detection, Prediction, and Diagnosis*, https://doi.org/10.1007/978-981-96-4241-0_9

kidneys is so established that it could not even be discovered earlier, the early signs of the development and missing clues about the pace of the progression of the disease with the required accuracy are impossible to get. In addition, the interpretation of diagnostic results can be subjective, hence sensitive to variability, which may delay the diagnosis as well as the strategic treatment planning. The ever-increasing requirement for precision and accuracy of predictive models in kidney disease has stoked the desire to capitalize on machine learning technology, which, armed with the plethora of data, would be of great utility.

9.1.2 Machine Learning in Kidney Disease Prediction: Machine Learning in Kidney Disease Prediction

Machine learning fits kidney disease prediction into multifaceted data analysis by means of the computer processing complex data at a high rate and extracting main features from it. Through the assistance of computational algorithms and statistical models, machine learning can identify attributes that signify a meaningful pattern from many diverse datasets, including the patient's demographic profile, previous medical history, laboratory results, and imaging outcomes. These algorithms use the results to develop predictive models that could accurately diagnose and assess the due seriousness of kidney disease. Machine learning combines multiple data sources and gathers from mass databases. The algorithms, hence, look for hidden correlations and biomarkers that doctors fail to recognize using traditional diagnostics.

9.1.3 Recent Advances and Applications

Recent advancements in machine learning have propelled the development of innovative approaches for kidney disease prediction.

9.1.4 Risk Stratification

Machine learning model algorithms can be helpful in classifying individuals into different categories of risk depending on the probability of developing kidney renal disease or experiencing the progression of renal disease. As a result of these categorization schemes, the risk factors will be classified according to which intervention or resource utilization can be prioritized by healthcare providers with a view to optimizing patient healthcare and resource allocation.

9.1.5 Early Detection

Machine learning models are superior in that they can highlight those "weak signals" in the form of biomarkers that signify minor renal abnormalities, which may not be visible with usual test procedures. Through the early detection of kidney diseases, these models facilitate timely interventions and preventive measures; therefore, they can help to avert the progression of the disease and may play a major role in reducing the volume of complications.

9.1.6 Personalized Medicine

Machine learning algorithms create unique risk profiles for patients based on their clinical features, taking into account their genetic predispositions, environmental hazards, and social determinants. These personalized risk assessments thus give clinical teams a chance to find ways of amending the therapeutic goals for each unique patient and improve the treatment efficacy and patient-taking care.

9.1.7 Future Directions

The future of kidney disease prediction should thus depend on further progress in areas for machine learning techniques, including deep learning, reinforcement learning, and ensemble methods. In addition, the combination of polymodal information streams, electronic health records, genomics, proteomics, and wearable sensing data is a huge promise for the improvement of predictive power and clinical significance. Integrating researchers, professionals, and data scientists is vital in order to test out models and iron out mistakes for the prediction of kidney diseases in different patient groups and settings. Through the use of machine learning, we will be able to accomplish a quantum leap in kidney disease prediction and hence the introduction of the era of precision medicine and patient care precisely directed toward each one of the patients.

9.2 Literature Review

Chronic kidney disease (CKD) has become a global health problem because of the huge number of incidences and the mortality and morbidity rates observed. Early discovery and prediction of CKD are the first things that manage to seem the intervention. ML techniques have been adopted as prospective approaches in regard to CKD prognosis and risk development. Different research that includes ML models shows the prognosis of CKD, using a variety of data sources containing but not limited to EHR, clinical records, biomarkers, and genetic information. In fact, Tomasev et al.'s [1] scientific paper revealed the potential of deep learning approaches to forecasting CKD progression using electronic health records. However, other investigations have been designed to optimize the predictive capability of certain CKD features, such as demographic and biomedical data, by omitting those that do not add further discriminatory power. In a study by Shieh et al. [2], a genetic algorithm-based hybrid feature selection method (GAHFS) was employed, which involves combinations of genetic algorithm and support vector machine classifiers to predict potential risk factors of CKD. First, among ensemble learning techniques, we take the convention of using multiple base learners to improve prediction accuracy. Khan et al. [3] experimented with the ensemble of classifiers to predict the stages of CKD, and the results were shown to be superior and better than those made by individual classifiers. Rani et al. [4] proposed diabetes risk prediction using a Gradient Boosting Model fine-tuned; likewise, Sachdeva et al. [5] presented heart failure diagnosis the heart diagnosis using an efficient method. In their approach, Rani et al. [6] developed machine learning algorithms optimized by Bayesian method for predicting heart disease. Jeyalakshmi and colleagues [7] include machine learning algorithms to have renal function declining in chronic kidney disease patients, and Revathy and colleges are [8] essential for using a deep neural network to predict heart diseases. Chen et al. [9] proposed a deep learning and Bayesian Personalized Ranking (BPR) method for the acute complication prediction. Al-Jamimi [10] carried out the synergistic features engineering as well as the ensemble learning for a prompt chronic disease prediction. Chowdary et al. [11] set forth the machine learning-based prognostic model of the disease. Thacker and Vaghela [12] conducted a comparative analysis on how smart AI and intelligent machine learning are used for supplying means for chronic kidney disease prediction models. Tripathy et al. [13] argue whether supervised and unsupervised learning algorithms are more effective than deep learning algorithm in predicting diabetes index. Other studies include Jiang et al. [14] on urine routine tests-based machine learning for predicting acute kidney injury, Yashwante et al. [15] on meta-heuristic feature selection and feature extraction approaches for chronic kidney disease prediction, Sneha et al. [16] on the early detection of chronic kidney disease using different machine learning algorithms, Wani et al. [17] on enhancing disease prediction through ensemble learning techniques, Jayaprabha and Priya [18] on the early prediction of chronic kidney disease through a comprehensive survey, and Shukla et al. [19] and Sneha and Justus [20] on chronic kidney disease prediction using ML techniques and supervised algorithms, respectively.

Other researchers concentrate on the contribution of multimodal data sources such as clinical, genetic, and imaging data, which are then used to design predictive models of CKD that are comprehensive. Machine learning techniques have also been widely utilized for risk assessment of each given individual patient and disease

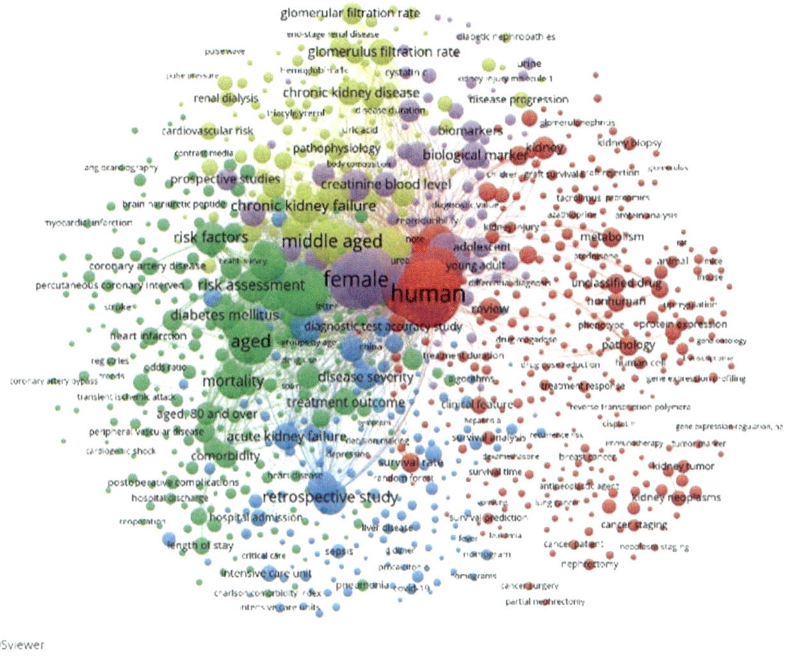

Fig. 9.1 Bibliometric study

outcome, specifically looking at their common patient characteristics and individual disease track. However, some issues, including heterogeneity of data, imbalanced datasets, and also the interpretability of ML models, are still major problems that limit the development of models for CKD prediction. The current age of medical research is devoted to adequately responding to setbacks that limit the adoption of ML-based prognostic mechanisms for CKD that could thus be used for developing therapeutic and preventive measures aimed at decreasing CKD-related healthcare expenditures across the globe. The bibliometric study is provided in Fig. 9.1.

9.3 Methodology

The predictive approach to the enterprise of machine learning in order to detect a condition of CKD involves multiple stages. Initially, researchers obtain and process heterogeneous datasets that are composed of medical records of the same people and those who have not been affected by CKD. Techniques such as feature engineering could be used to identify essential features or remedy the issue of missing values. Having done that, scientists choose the most accurate machine learning algorithms, considering factors like the nature and amount of data, specific prediction task, and amount of access to computing resources. The most frequently applied algorithms cover classifications, such as logistic regression, support vector machines, decision trees, random forests, gradient boosting, as well as deep learning models. The complexity of the data itself could be a major reason for drop in performance. One approach to improve prediction is to use ensemble learning techniques where multiple base learners are combined. Some of the methodologies for feature selection, like genetic algorithms and recursive feature elimination, can be used to zero the redundant predictors. The dataset is usually broken down into training, validation, and testing sets in order to train and evaluate the model. Performance parameters such as accuracy, sensitivity, specificity, and area under the receiver's operating curve (AUCROC) are used to evaluate model performance. Lastly, a model that is nearly ready to be utilized is validated on nonindependent datasets to assess its performance in terms of generalizability and clinical utility. In this context, the researchers' purpose is to design models possessing the ability to predict CKD risk scores correctly and progress in an accurate manner, which shall culminate in quality care and effective patient management. In addition, researchers may involve multimodal (clinic, genetics, multi-imaging, and multi-omics data) data to express the overall predictive models for CKD. The integration undertakes a delicate preprocessing and feature engineering that are designed to make sure that the data being integrated and the layouts are compatible and could be represented in a meaningful way. Along with such ones, individualized risk assessment can be taken into account because it might include features that individual patients have, such as their disease trajectories, thus leading to the prediction that could be suitable for the individual application. Data variety in the heterogeneous nature, class imbalance in the datasets, and model understanding are the intrinsic issues that must be addressed while coming up with a methodology. Techniques in this advanced area, like transfer learning, domain adaptation, and explainable AI, may become important tools to help address the challenges and improve the transparency and credibility of these models. Collaborations between clinicians, data scientists, and domain experts are core to the success of the clinical relevance and acceptable proliferation of machine learning, or AI, in healthcare. The application of this methodology is intended to be progressive and corrected by way of validation. The first stage is the emergence of machine learning-based CKD prediction tools into clinical practice that, in turn, will improve patients' outcomes and cut down the CKD-associated healthcare costs. A detailed diagram is shown in Fig. 9.2.

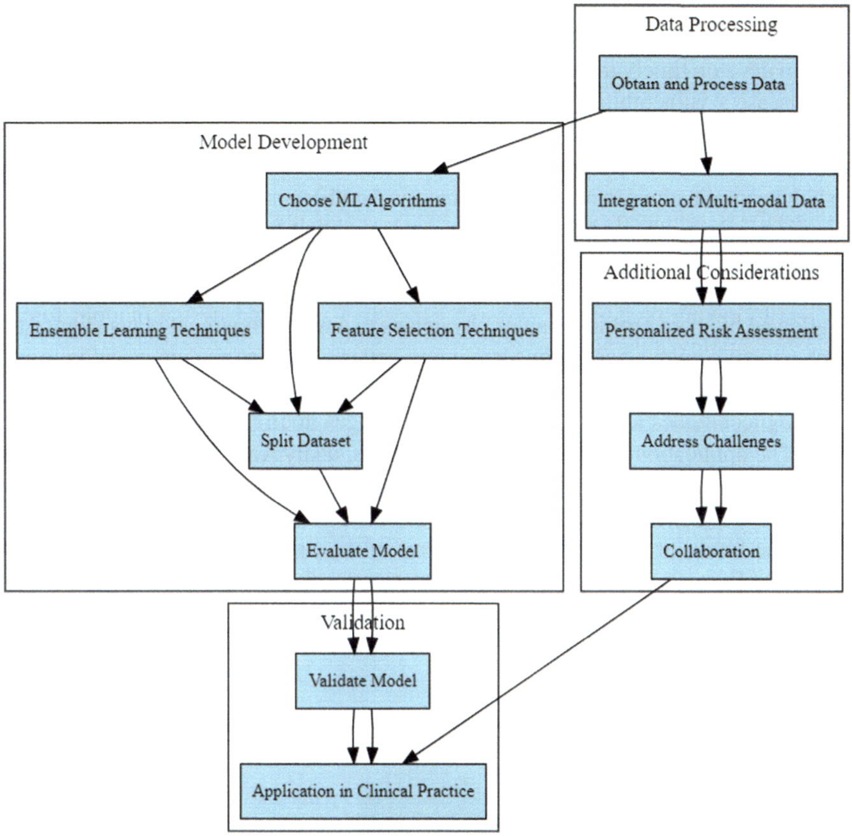

Fig. 9.2 Methodological approach to the system

9.4 Machine Learning in Healthcare

Machine learning methods are becoming more used in the healthcare sector. In particular, these methods support the early detection and prognosis of a wide spectrum of medical conditions like kidney disease. In the framework of machine learning for kidney failure prediction, different types of algorithms use many sources of data from patient records, including demographics, medical history, lab findings, and medical images. They can pick out patterns and relationships that are difficult to see by the physicians, not because they are unaware of it but because it just takes them much time. Such discoveries by methods of machine learning allow for forecasting the risk of developing kidney disease or diagnostic decisions on the nature of its development and the severity of the disease with a high accuracy. One of the noteworthy strengths of machine learning applications in kidney disease prediction is the fact that they are highly efficient in aiding the early detection of disease as well. Conventional diagnostics might not be able to identify some early signs of dysfunction, being very professionally assessed. Nevertheless, ML algorithms demonstrate unparalleled effectiveness in early anomaly detection. Hence, health workers can develop timely treatments and preventive maneuvers straightaway to reduce the disease's vulnerability and impact. Early detection is an extremely powerful anticancer weapon because it enables doctors to start the treatment before the disease has been able to progress too far, so it becomes difficult to destroy it, and the risk the complications increases significantly. Another machine learning application is customizing medical interventions in renal condition forecasting. Machine learning through individual patient features like genotypes and comorbidities is a better option that can generate profiles of risk expectancy. The individuated treatment that is allowed by these profiles enables physicians to develop treatment plans aligned to each patient's unique needs, thus leading to better treatment outcomes and improving the general quality of care. In short, machine learning facilities for machine learning techniques lead to comprehensive changes in disease prediction in healthcare. The application of advanced data analytics, analytical processes, and predictive modeling through machine learning spots the early signs, identifies individuals at high risk, and tailors treatment plans to individual patients. These technologies are at the forefront of development in terms of their potential to enhance patient's quality of life as well as their clinical outcomes for those experiencing kidney disease. The pillars of machine learning in the healthcare sector can be visualized in Fig. 9.3.

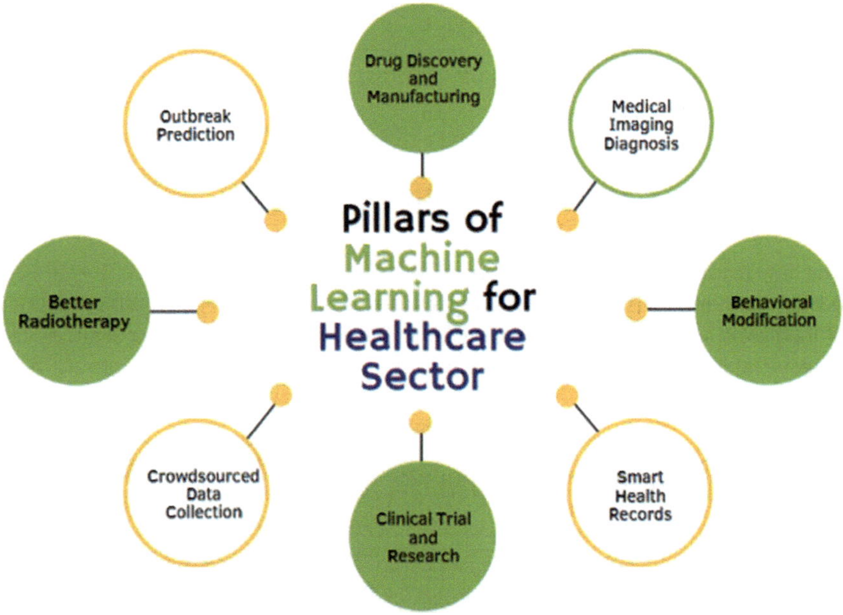

Fig. 9.3 Pillars of machine learning in the healthcare sector

9.5 Data Collection and Preprocessing

Data collecting and preprocessing seek the core stepping in ML techniques to kidney disease prediction, which evaluates the quality and accuracy of predictive models.

9.5.1 Data Collection

The masks of choosing the diverse real datasets that can include data like electronic health records (EHRs), clinical databases, public repositories, genetic data, and wearable sensor data are very important. EHRs (electronic health records) are rich sources of data about patients like patient demographics, summaries of medical history, laboratory results, and imaging findings. Clinical databases and public repositories provide access to services allowing for datasets sharing among researchers and clinicians focusing on renal disease. Genetic conjugates may uncover genetic prehabilitation and biomarkers connected with the risk of kidney disease. Simultaneously, continuous tracking of relevant renal function parameters via wearable data sensors is made possible, which provides real-time updates.

9.5.2 Data Preprocessing

Once collected, the data undergo preprocessing to ensure its quality and suitability for machine learning analysis, as shown in Fig. 9.4.

Handling missing values: in place of missing data, we have techniques like imputation or deletion to grab our attention and completeness, ensuring that we do not lose anything.

Feature selection: hence, the relevant characteristics that appear to contribute to generating a prediction of kidney disease are turned into features so as to reduce dimensionality and computational complexity.

Data normalization/scaling: the numerical parameters scale is scaled and adjusted to correspond with the training of the models for the sake of getting rid of bias in training.

Handling categorical variables: categorical variables are encoded into vector covers so that the model can learn from them.

Addressing class imbalance: Approaches including upsampling, downsampling, or employing synthetic data are some of the techniques used to equalize the classes' distributions.

Data splitting: The set of data is clustered into three sets, namely for training, validation, and testing, which work in harmony for effective training and performance evaluation of the model, thus avoiding overfitting.

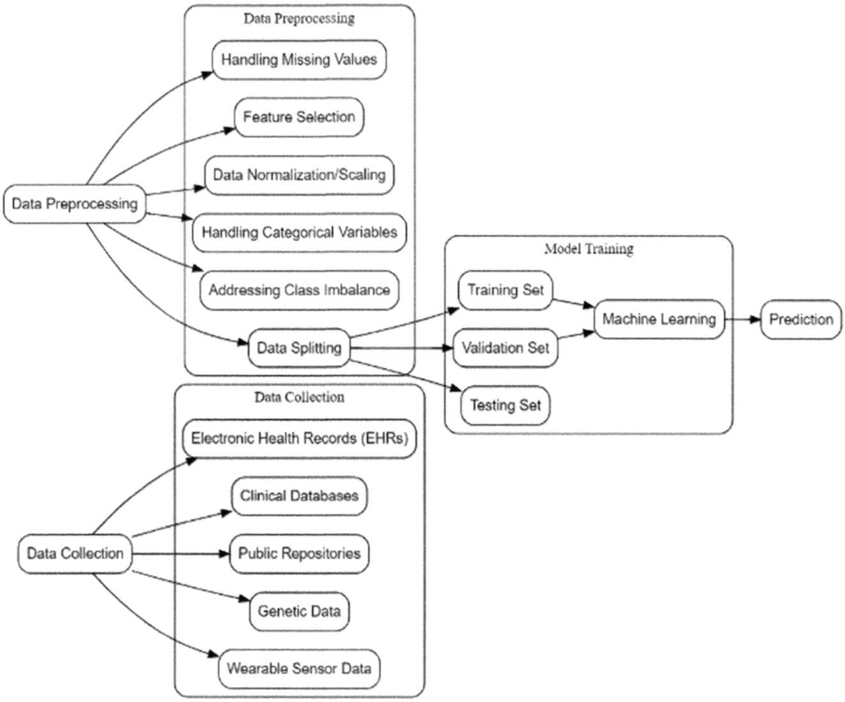

Fig. 9.4 Data preprocessing and collection diagrammatic representation

By painstakingly taking data collection and preprocessing, researchers and physicians come to possess robust machine learning algorithms that then are able to accurately predict kidney disease, allowing for early detection and yielding individualized patient care. Such programs have been the light of hope in fighting off the effects of kidney disease, with the ultimate goal of saving lives globally.

9.6 Feature Selection and Engineering

Feature selection and engineering are the key components in the construction of machine learning models for kidney disease screening. They determine the accuracy of the model, its ease of interpretation and its generalizability for all patients in general. Here is how these processes are approached.

9.6.1 Feature Selection

The feature selection operation is oriented toward finding the most relevant variables among the dataset that have a significant impact on predicting kidney disease. The objective here is to get rid of unimportant features without losing the meaningful information from the dataset. Several techniques are commonly used for feature selection.

Univariate feature selection: statistical tests like chi-square or ANOVA (for each feature) or mutual information can be used to measure their significance individually (APPROVED).

Recursive feature elimination: this interactive way of training allows for the model to lack in less important parameters and adds as many as necessary based on the impact of such on the model performance.

Feature importance ranking: algorithms such as decision trees or ensemble methods may evaluate a measure of feature importance and help to select the most suited ones.

Dimensionality reduction techniques: the AI can employ a technique known as principal component analysis (PCA) or t-distributed stochastic neighbor embedding (t-SNE) to make the transition from the original to the lower dimensional space while preserving a maximum amount of variance.

9.6.2 Feature Engineering

Feature engineering encompasses the creation of new features from the established ones or the transformation of any useless features to boost the model's performance. Through that, health workers would have to rely on their experience and a full clear picture of the complexity of kidney diseases. Some common techniques for feature engineering in kidney disease prediction include

Creating interaction terms: integrating two or more omics methods in parallel or in sequence to observe how they may work together or interact in order to form features predictive of kidney disease.

Time-series analysis: extracting temporal patterns or trends from dates to picture disease proliferation on the line of time.

Domain-specific transformations: using the transformation domain with the built-in transformations or the custom normalization can result in a feature dataset within proper scale and representation.

Incorporating external data: integrating any other datasets, such as genetic or environmental ones, into the feature space and also using them to take advantage of further predictive probabilities.

Selective, purposeful, as well as smart feature engineering can help the developers make machine learning models that are powerful enough to predict kidney disease without delay, targeted interventions, and the best of the results. Such approaches are therefore not only indispensable to the improvement of personalized medicine and the augmentation of the efficacy of machine learning in health, but they also lay the ground for the better use of these technologies in medicine.

9.7 Machine Learning Models for Kidney Diseases Prediction

A wide range of machine learning models were used by developers in the field of kidney disease prediction, and as a result, they have their benefits and challenges. Here is an overview of some commonly used models in this domain.

9.7.1 Logistic Regression

Logistic regression is a line-based model that is employed for binary classification tasks so that it allows us to tell whether or not kidney disease is present. It offers the possibility of sparse coefficients that are easy to understand, and they show the influence of one feature on the likelihood of a disease. Nevertheless, it supposes that there is a linear link between covariates and features, which is why it ignored potential multifactorial interactions that may exist in the dataset.

9.7.2 Random Forest

Random restoration is an ensemble procedure that sums up multiple trees and lifts the predictive ability of the model. It can bear the nonlinear relationships due to the fact it is a black box and handles interactions between the features that would make a very robust prediction model for chronic kidney disease. Random forest models generally perform better in tasks that involve nonlinear dependencies and large datasets, which can be divided into multiple subsets correlating to independent variables. Besides, they also give the significance of characteristics, thus, in the feature selection and interpretation.

9.7.3 Support Vector Machines (SVM)

Such persistently downward-biased economic indicators are a potential harbinger of turmoil, and the continuation of the current situation is not something that Ukraine's economic managers can afford. Support vector machine (SVM) uses a supervised learning algorithm to make both classification and regression predictions. It operates by optimizing partitioning of all classed hyperplanes in a feature space. SVMs have shown their efficacy in predicting the onset of renal diseases, especially when the data is not linearly separable. On the one hand, these machines have difficulties in making decisions because of the dependence on kernel function and the need for tuning hyperparameters of the model.

9.7.4 Gradient Boosting Machines (GBM)

By leveraging industry networking platforms such as LinkedIn, job boards, and professional networking events, I have been able to proactively seek opportunities and connect with individuals with expertise in marketing, business operations, and entrepreneurship. Another ensemble technique is GBM, which is comprised of consecutive and sequential decision trees in an order where the front one tries to take care of the errors of the previous. GBM models perform well in various input structures and can force different enterprises to represent complex relations in the data. Thus, GBM is a good choice as a tool for kidney disease prediction. They also provide robustness against the outliers and handle the missing value issue well. Nevertheless, both GBM models and the choice of hyperparameters to be used can potentially consume much time on computing and can be quite tricky to tune.

9.7.5 Deep Learning Models

The representation of models, which are convolutional neural networks and recurrent neural networks (CNNs and RNNs), provides hope in predicting. Recently, convolutional neural networks (CNNs) may process spatial features from images of pathology, such as renal ultrasound images or MRI scans. Recurrent neural networks (RNNs) maybe go through longitudinal data, seeing serial information. AI's deep learning models require vast amounts of labeled data and substantial computing resources to train. However, they can reach such a level of performance in the task of kidney disease prediction that they can even be considered better than specialists.

To sum up, the application of various machine learning models to kidney disease prediction is a viable practice, yet it has pros and cons specific to each. Whether one chooses a model that is data-driven or interpretable or if there are enough computational resources available are some of the factors that affect how a model is selected. Through the adoption of these models, researchers and clinical staff can diagnose the disease early, stratify risk, and customize the management program, which will ultimately lead to a better outcome for the patients.

9.8 Performance Evaluation Metrics

Evaluation metrics for the performance of models in machine learning are vital to evaluate the efficacy and accuracy of models used in kidney disease diagnosis. Such metrics give us information about the model's predictability, durability, and ability to address a broad set of issues. Here are some commonly used performances evaluation metrics.

9.8.1 Accuracy

Accuracy measures the probability of correct predictions over the number of cases correctly classified. Truthfulness may be required overall and is not even suitable for the imbalanced datasets that, on the other hand, have an unequal class partition.

9.8.2 Precision and Recall

Each of those measures senses the coverage of the prediction in terms of achieved predictions and recognized samples in a dataset. Accuracy of positive predictions, denoted by precision, and the number of positive instances captured, determined by recall, emphasize recall. The contexts where one type of class is highly prevalent compared to the other, precision and recall are advantageous metrics for the assessment of the model.

9.8.3 F1 Score

In the case of balanced performance, the F1 score takes the harmonic mean between precision and recall. It takes both the false positives and the false negatives into account, specifically for datasets imbalanced with precision and recall, where they might as well trade off each other.

9.8.4 Receiver Operating Characteristic (ROC) Curve and Area Under the Curve (AUC)

Whether it is through conservation practices, community development, or educational efforts, environmental organizations, from local grassroots movements to large-scale global campaigns, contribute to the overall health and sustainability of our planet.

It is drawn by the rate of true positive (TPR) on the x-axis and false positive rate (FPR) with the compensating threshold settings on the y-axis. AUC is a measure that is achieved by drawing the line seen on the graph. This line is known as the ROC curve. So, AUC is a metric for the model's ability to separate between positive and negative instances. A higher AUC means the model discriminates data as accurately as possible.

9.8.5 Confusion Matrix

The confusion matrix converts into a tabular format of actual and mistake classifications, which clearly denotes TP (true positive), FP (false positive), TN(true negative), and FN (false negative). It is the grounding for deriving other quantitative indicators such as accuracy, precision, recall, and F1 score for each classification type.

Therefore, the evaluation of performance metrics is an essential element in assessing the expectations and dependability of machine learning models during kidney disease prediction. Researchers and healthcare professionals would be able to achieve a holistic perception of the performance of the model by considering multi-metrics, so it would be possible for them to make an ideal decision on the deployment of the model in clinical settings.

9.9 Challenges and Limitations

While machine learning techniques hold promise for kidney disease prediction, several challenges and limitations need to be addressed to ensure their effective application in clinical settings.

9.9.1 Data Quality and Availability

One of the major issues in this regard is the lack of very good coverage as well as the quantity of datasets for machine learning model training. The problem of not having a diverse corpus of well-labeled data (especially the longitudinal one for different stages of kidney disease) could hinder using the hypothesis of the predictive modeling strictly.

9.9.2 Data Imbalance

The problem of imbalanced datasets, where the examples for one class (e.g., patients with kidney disease) are significantly more common than the other class, creates new training and evaluation difficulties for model makers. Unbalancing data can create unfavorable model performance metrics, and it can be difficult to properly predict the values of a minority class.

9.9.3 Interpretability

Fundamentally, numerous machine learning models, particularly sophisticated ones like deep learning algorithms, are not interpretable; thus, it is very hard for doctors to grasp the logic and reasoning behind machine predictions. Human-understandable models are essential for the healthcare practitioners' trust and also their availability and acceptance.

9.9.4 Generalization and External Validation

Therefore, it is of utmost importance to model predictive mechanisms capable of generalizability among multiform patient populations and clinical situations. Models trained with data covering the population or healthcare facilities of a particular area may fail to play well for other communities regardless of the demography or the settings. This observation shows the significance of external validation.

9.9.5 Integration into Clinical Workflow

Along with the technical aspects, bringing the AI models into the clinical processes must be considered from the practical and logistics point of view. Doctors may be unclear about interpreting and even interpret output production or be unwilling to trust predictions that are not explained by algorithms.

9.9.6 Ethical and Regulatory Considerations

Ethical considerations, such as patient privacy, data security, and algorithmic bias, must be carefully addressed when deploying machine learning models in healthcare settings. Regulatory frameworks and guidelines for the development and deployment of medical AI technologies need to be established and adhered to.

Addressing these challenges requires collaboration among researchers, clinicians, policymakers, and industry stakeholders. Strategies such as data-sharing initiatives, transparent model development processes, explainable AI techniques, and interdisciplinary training programs can help overcome these limitations and harness the full potential of machine learning in kidney disease prediction and management.

9.10 Real-Time Case Studies

We use diverse clinical cases in real time to illustrate the usefulness of machine learning techniques for the prediction of chronic kidney disease and kidney failure, which is a vivid showcase of the application of the method in the clinical setting.

Here are two examples.

9.10.1 Early Detection and Risk Stratification

At the same time, the very nature of mobile devices presents challenges as portability can encourage multitasking, loss of focus, and information overload.

That had a real-world case study at the nephrology clinic where machine learning algorithms had decided to predict the risk of developing chronic kidney disease (CKD) for patients with diabetes. Through the use of electronic health records (EHRs), which we achieved by integrating demographic data, laboratory results, medication history, and comorbidities into one database, the research was conducted. Through this analysis using logistic regression and random forest algorithms, the algorithms accurately identified those with a higher probability of the disease develop in a specific time frame. Hospitals, doctors, and other service providers (clinicians) took the prediction into account when they made their decisions for treatment, such as making lifestyle adjustments, modifying the drugs, and closely monitoring signs of kidney dysfunction. Machine learning-based risk stratification realization was very effective in the healthcare system, and it led to a positive impact on patient results, better disease control, efficient resource allocation, and other benefits.

9.10.2 Personalized Treatment Planning

An example of a real-time case study is that patients with advanced kidney disease cannot be neglected. However, they are undergoing dialysis, and they need to have personalized treatment planning. The machine learning models were trained on a number of longitudinal patient data, such as information from the dialysis, laboratories, and clinical assessments. The predictive models, which were based on gradient boosting machines (GBM) and recurrent neural networks (RNN), would be used to forecast the personal medication trajectories, including the changes in kidney function, responses to medication, and complication risk. Clinicians implemented these predictions, thus treating the patients differently, tailoring the treatment regimens to each of them, adjusting dialysis prescriptions, and advising on the next follow-up appointments through their prognoses. The integrated access to machine learning-based knowledge in clinical practice facilitated the physicians to get better treatment outcomes, reduce the hospitalization rate, and improve the quality of life for patients on the maintenance therapies of advanced kidney disease.

These practical examples show how these machine learning algorithms become useful for early detection, better stratification risks, and individualized treatment plans in day-to-day clinical practice and set a benchmark for prediction accuracy. Through the use of techniques such as advanced analytics and predictive modeling, physicians can ensure maximum utilization of the available medical resources, which in turn can lessen the impact of kidney disease on the people as well as the whole healthcare system.

9.11 Conclusion

Machine learning as a technique is capable of paving a unique pathway for future kidney disease prediction, accelerating early detection, and creating patient-centered treatment plans and better outcomes. Indeed, the progress made in deploying machine learning algorithms to a number of datasets for developing predictive models for kidney disease overshadows the challenges and limitations it has faced so far.

Machine learning capitalizes on computational algorithms and statistical model skills to bring out the hidden patterns, signatures, features, and associations in the massive amount of patient data sources, which comprises electronic health records, clinical databases, genetic data, and wearable sensor data. Such predictive models will not only predict the chance of getting kidney disease but will also determine disease severity, stratify further patient risk, and inform patient-based treatment choices.

Additionally, machine learning models are reported to have impressive accuracy in different aspects of the prediction of renal disease, namely, the detection of some small signatures, mapping into time-series information in longitudinal data, and the integration of multiple data sources for a thorough risk assessment. The ensemble learning technique, the deep learning architecture, and the model interpretable are the three technology trends that are known to have improved the predictive accuracy and clinical usefulness.

Nevertheless, the quality and quantity of data, which may be imbalanced, model interpretability, generalizability, incorporating machine learning in routine workflows, and ethical issues must be generally resolved to obtain the utmost use of machine learning in predicting kidney disorders. The introduction of machine learning applications in healthcare demands close working relationships among researchers, clinicians, policymakers, and industry players to overcome the challenges this brings and enhance the ethical use of machine learning tools in health settings.

As for the future, conducting more research, open data flows, model development with transparency, and cross-border collaboration will be key to improving OMKD prediction. Through the scientific acquisition of these problems, along with the right usage of machine learning, the early diagnosis of this condition will be made easier, risk stratification will be given attention, and finally, personalized management in the patient will be achieved. There would be an improvement in patients' prognosis as well as a reduction in global kidney disease burden.

Acknowledgments We acknowledge VosViewer for providing us with the visualization of the literature survey and bibliometric study.

References

1. Tomasev N, Glorot X, Rae JW, Zielinski M, Askham H, Saraiva A, et al. A clinically applicable approach to continuous prediction of future acute kidney injury. Nature. 2019;572(7767):116–9.
2. Shieh CS, Lin SH, Chiang TY, Tseng VS. A hybrid feature selection approach for chronic kidney disease prediction. Expert Syst Appl. 2020;140:112906.
3. Khan J, Zubair M, Khan A. Chronic kidney disease stages prediction using ensemble learning methods. In: Proceedings of the international conference on advances in computing and communication engineering; 2019. p. 1–5.
4. Rani P, Lamba R, Sachdeva RK, Jain A, Choudhury T, Kotecha K. Diabetes risk prediction through fine-tuned gradient boosting. In: Garg D, Rodrigues JJPC, Gupta SK, Cheng X, Sarao P, Patel GS, editors. Advanced computing: 13th international conference, IACC 2023, vol. 2054. Cham: Springer; 2024. https://doi.org/10.1007/978-3-031-56703-2_11.
5. Sachdeva RK, Singla A, Bathla P, Jain A, Choudhury T, Kotecha K. An efficient method for heart failure diagnosis. In: Garg D, Rodrigues JJPC, Gupta SK, Cheng X, Sarao P, Patel GS, editors. Advanced computing: 13th international conference, IACC 2023, vol. 2054. Cham: Springer; 2024. https://doi.org/10.1007/978-3-031-56703-2_23.
6. Rani P, Lamba R, Sachdeva RK, Jain A, Choudhury T, Kotecha K. Heart disease prediction using Bayesian optimized classification algorithms. In: 2023 7th international symposium on multidisciplinary studies and innovative technologies (ISMSIT). Ankara; 2023. p. 1–5. https://doi.org/10.1109/ISMSIT58785.2023.10304966.
7. Jeyalakshmi G, Lloyd FV, Subbulakshmi K, Vinudevi G. Integration of machine learning algorithms in predicting renal function decline in chronic kidney disease patients. In: 2024 international conference on distributed computing and optimization techniques (ICDCOT). Bengaluru; 2024. p. 1–5. https://doi.org/10.1109/ICDCOT61034.2024.10515631.
8. Revathy G, Kalaiselvi DMT, Priya PM. GUI based heart disease prediction using deep neural networks. In: 2024 5th international conference on intelligent communication technologies and virtual mobile networks (ICICV). Tirunelveli; 2024. p. 22–4. https://doi.org/10.1109/ICICV62344.2024.00011.
9. Chen X, Cheng Q. Acute complication prediction and diagnosis model CLSTM-BPR: a fusion method of time series deep learning and Bayesian personalized ranking. Tsinghua Sci Technol. 2024;29(5):1509–23. https://doi.org/10.26599/TST.2023.9010103.
10. Al-Jamimi HA. Synergistic feature engineering and ensemble learning for early chronic disease prediction. IEEE Access. 2024;12:62215–33. https://doi.org/10.1109/ACCESS.2024.3395512.

11. Chowdary MK, Prasad RJC, Kumar KA, Sapthami I, Krishna BM, Kiran A. Machine learning based disease prognostic model. In: 2024 international conference on advancements in smart, secure and intelligent computing (ASSIC), vol. 2024. Bhubaneswar. p. 1–5. https://doi.org/10.1109/ASSIC60049.2024.10508032.
12. Thacker S, Vaghela R. Assessing chronic kidney disease prediction models: a comparative analysis of smart AI and intelligent machine learning approaches. In: 2024 international conference on advancements in smart, secure and intelligent computing (ASSIC), vol. 2024. Bhubaneswar. p. 1–6. https://doi.org/10.1109/ASSIC60049.2024.10507811.
13. Tripathy N, Moharana B, Balabantaray SK, Nayak SK, Pati A, Panigrahi A. A comparative analysis of diabetes prediction using machine learning and deep learning algorithms in healthcare. In: 2024 international conference on advancements in smart, secure and intelligent computing (ASSIC). Bhubaneswar; 2024. p. 1–6. https://doi.org/10.1109/ASSIC60049.2024.10508008.
14. Jiang T, Sang Y, Tan H. Urine routine tests-based machine learning and urinalysis to predict acute kidney injury: URT-based machine learning and urinalysis to predict AKI. In: 2024 4th international conference on consumer electronics and computer engineering (ICCECE), vol. 2024. Guangzhou. p. 362–8. https://doi.org/10.1109/ICCECE61317.2024.10504188.
15. Yashwante P, Patil Y, Nadar K, Khade A. Comparative analysis of meta-heuristic feature selection and feature extraction approaches for enhanced chronic kidney disease prediction. In: 2024 IEEE international conference on interdisciplinary approaches in technology and management for social innovation (IATMSI). Gwalior; 2024. p. 1–6. https://doi.org/10.1109/IATMSI60426.2024.10502980.
16. Sneha SZR, Chakraborty P, Haque P. Early detection of chronic kidney disease using different machine learning algorithms. In: 2024 international conference on advances in computing, communication, electrical, and smart systems (iCACCESS), vol. 2024. Dhaka. p. 1–6. https://doi.org/10.1109/iCACCESS61735.2024.10499497.
17. Wani SR, Attri SH, Setia S. Enhancing disease prediction through ensemble learning techniques. In: 2024 11th international conference on computing for sustainable global development (INDIACom). New Delhi; 2024. p. 974–80. https://doi.org/10.23919/INDIACom61295.2024.10498596.
18. Jayaprabha MS, Priya VV. Early prediction of chronic kidney disease: a comprehensive survey. In: 2024 5th international conference on Mobile computing and sustainable informatics (ICMCSI). Lalitpur; 2024. p. 45–51. https://doi.org/10.1109/ICMCSI61536.2024.00013.
19. Shukla A, Robin F, Narvaria H, Joshi A. Prediction of chronic kidney disease using ML techniques. In: 2024 IEEE international conference on computing, power and communication technologies (IC2PCT). Greater Noida; 2024. p. 469–75. https://doi.org/10.1109/IC2PCT60090.2024.10486621.
20. Sneha A, Justus JJ. Development of supervised algorithm for predicting chronic kidney disease. In: 2024 IEEE international conference on computing, power and communication technologies (IC2PCT). Greater Noida; 2024. p. 342–8. https://doi.org/10.1109/IC2PCT60090.2024.10486505.

Chapter 10
Prediction of Diabetes Using Different Machine Learning Algorithms

Akshatha Bhat, Abhishek Kumar, Saurabh Agarwal, and Ishan Srivastava

Nomenclature

ANN	Artificial neural network
AUC	Area under the receiver operating characteristic curve
DT	Decision tree
ML	Machine learning
RF	Random forest
SVM	Support vector machine

A. Bhat
Department of Electronics and Communication Engineering, SOET, CMR University, Bengaluru, Karnataka, India

A. Kumar (✉)
Department of Electronics and Communication Engineering, Presidency University, Bengaluru, India
e-mail: abhishek.kumar@presidencyuniversity.in

S. Agarwal
Department of Information and Communication Engineering, Yeungnam University, Gyeongsan, South Korea

I. Srivastava
Electrical Engineering Department, UIET Babasaheb Bhimrao Ambedkar University, Lucknow, Uttar Pradesh, India

T. Choudhury, A. Katal (eds.), *Machine Learning for Disease Detection, Prediction, and Diagnosis*, https://doi.org/10.1007/978-981-96-4241-0_10

10.1 Introduction

Diabetes is a chronic disease with potential complications affecting multiple organs and tissues. Diabetes can result in imbalances in insulin levels in the blood, leading to episodes of hyperglycemia and hypoglycemia. Specific clinical signs and treatment modalities are associated with different classifications of diabetes, such as type 1 and type 2. When the immune system unintentionally kills the pancreatic cells responsible for producing insulin, type 1 diabetes results. Instead of being absorbed by the cells for energy, glucose accumulates in the bloodstream when there is insufficient insulin. Type 1 diabetes frequently causes increased thirst, inexplicable weight loss, exhaustion, frequent urination, and blurred eyesight, among other symptoms. People with type 2 diabetes either produce less insulin to uphold normal blood sugar levels or become immune to insulin [1]. Diabetes is frequently linked to lifestyle factors such as food, physical inactivity, and obesity. Some common effects of type 2 diabetes include frequent urination, vision problems, recurring infections, and slow recovery of wounds. The global impact of diabetes is significant. The number of adults with diabetes is projected to increase to 700 million by 2045 [2]. Approximately 4.2 million fatalities globally in 2019 were related to diabetes [3], making it one of the major causes of death.

Machine learning (ML) methods can assist in the preliminary diagnosis of diabetes by analyzing patients' physical examination data and providing valuable reference information for doctors. In ML-based diabetes prediction, careful consideration of feature selection and choosing the appropriate classifier are crucial. Several algorithms have been utilized, including support vector machine (SVM), decision tree (DT), logistic regression (LR), principal component analysis, neuro-fuzzy inference, and ensemble methods [4]. Medical professionals can use data mining algorithms and decision support systems to predict and diagnose the disease accurately. Early prediction of diabetes is vital for effective management and improved patient outcomes. With the use of a dataset including the characteristics of diabetics, this work attempts to predict diabetes using a variety of ML methods, including LR, artificial neural network (ANN), DT, SVM, and others [5]. These methods leverage computational algorithms to analyze data and make accurate predictions. ML methods for diabetes prediction show potential for applications such as wearable health tracking devices and medication delivery with automation.

For diabetics, ML algorithms can be very helpful in automating and optimizing drug administration. By analyzing patient data, such as their blood sugar levels, insulin dosage, and other relevant factors, algorithms can make accurate predictions and provide personalized recommendations for medication dosage adjustments. Diabetes sufferers may benefit from this as it can enhance treatment results and lower their risk of complications [6]. Another area of advancement is the use of wearable devices to track health. Wearable devices, such as glucose monitors, can continuously monitor glucose levels in individuals with diabetes [7]. By integrating ML algorithms with these devices, real-time data can be analyzed to detect patterns and identify potential risks or abnormalities. Moreover, researchers use advanced Markov chains to predict diabetes [8–10]. This can provide individuals with timely feedback and alerts, enabling them to take proactive measures to manage their condition effectively.

However, there are challenges associated with predicting diabetes using ML algorithms. One major challenge is data collection and its authenticity. Inaccurate or incomplete data can lead to unreliable predictions and incorrect recommendations. Thus, ensuring the gathering of trustworthy and high-quality data is essential for both algorithm testing and training. Another challenge is maintaining privacy. Diabetes prediction requires access to sensitive health data, which raises concerns about privacy and data security. Implementing robust privacy measures and adhering to regulatory guidelines is essential to protect individuals' health information and maintain their trust in the system [11].

Predicting diabetes accurately is a goal in the healthcare sector [12], and various data mining algorithms and decision support systems can assist medical professionals. Sifting through massive amounts of healthcare data to find hidden patterns and knowledge requires the use of big data analytics. Predictive analysis, incorporating ML algorithms and statistical methods, can aid in making informed decisions, optimizing resources, and improving clinical outcomes. Some advantages of utilizing ML algorithms for diabetes prediction are accuracy, personalized forecasts, efficiency, scalability, and automation, making them valuable tools for diabetes prediction and management.

10.2 Data Gathering and Preliminary Processing

Data gathering and preliminary processing are essential actions in predicting diabetes using ML algorithms. They enable researchers to gather pertinent information and prepare the data for analysis, ultimately leading to the development of accurate prediction models for this chronic condition. Detailed explanations of the importance and steps of data pre-processing can be found in works by authors [13, 14]. Additionally, the authors [15, 16] have emphasized the significance of balanced data to avoid any form of biasing.

Data collection on diabetes involves gathering information on various aspects, including demographics, medical history, lifestyle factors, diagnostic tests, treatment, and outcomes. Multiple data sources can be collected for predicting diabetes using ML algorithms. Clinical trials, surveys, wearable technology, electronic health records, public health databases, and research studies are some examples of these sources [17]. Analyzing this data can help identify trends, patterns, and outcomes related to diabetes care and management.

To make sure that the data is appropriate for prediction, it goes through a preprocessing step when it is gathered. Cleaning, addressing outliers, resolving missing values, and formatting the data into an analysis-ready format are all part of data pre-processing. Different approaches to data pre-processing have been used by researchers in the context of diabetes prediction [18]. These techniques include regression imputation, mean imputation, and mode imputation, among others, to handle missing variables. Statistical techniques such as the z-score and interquartile range can be used to detect and handle outliers. The most pertinent features for diabetes prediction can be found using feature selection methods including lasso regression, forward selection, and backward elimination. Preparing the data ensures that it is in a format and quality suitable for analysis, which improves the precision and.

10.2.1 Data-Informed Method for Choosing Features

Data-informed method for choosing features is a process in which the most relevant and informative methods are automatically selected [19] from a given dataset for a specific task, such as diabetes prediction. The objective is to decrease the number of dimensions in the data by eliminating superfluous or unnecessary characteristics and enhancing the interpretability, computational efficiency, and performance of the model. In the context of diabetes research, a data-informed method for choosing features procedures can be used to identify the most essential features that contribute to diabetes prediction. Figure 10.1 shows a general description of the data-informed method for choosing features procedure [20].

Data-informed methods for choosing feature selection procedures help identify the most informative methods for diabetes prediction, dimensionality reduction in the data, and enhanced interpretability and accuracy of the models. They enable researchers and healthcare professionals to focus on the most relevant factors contributing to diabetes and guide further research and interventions.

10.2.2 Prediction Model

A prediction model related to diabetes is a statistical or ML model that uses various input features, such as patient demographics, medical history, and clinical measurements, to estimate the probability that a person has diabetes. The input variables and corresponding diabetes outcomes are derived from historical data, which is used to create these models.

Figure 10.2 illustrates the steps involved in creating a diabetes prediction model [21]. It is significant to highlight that careful consideration of data quality, feature selection, model selection, and evaluation metrics is necessary while designing a diabetes prediction model.

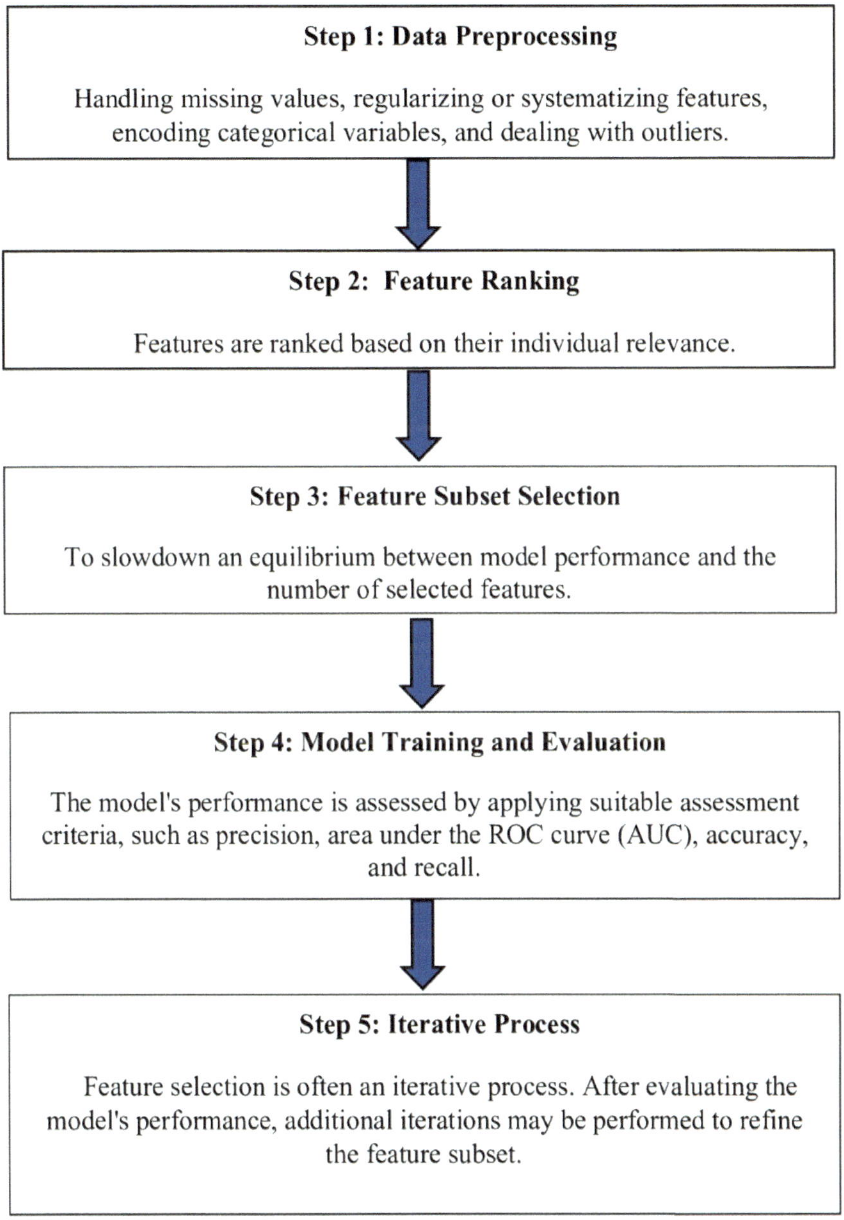

Fig. 10.1 Data-informed method for choosing features

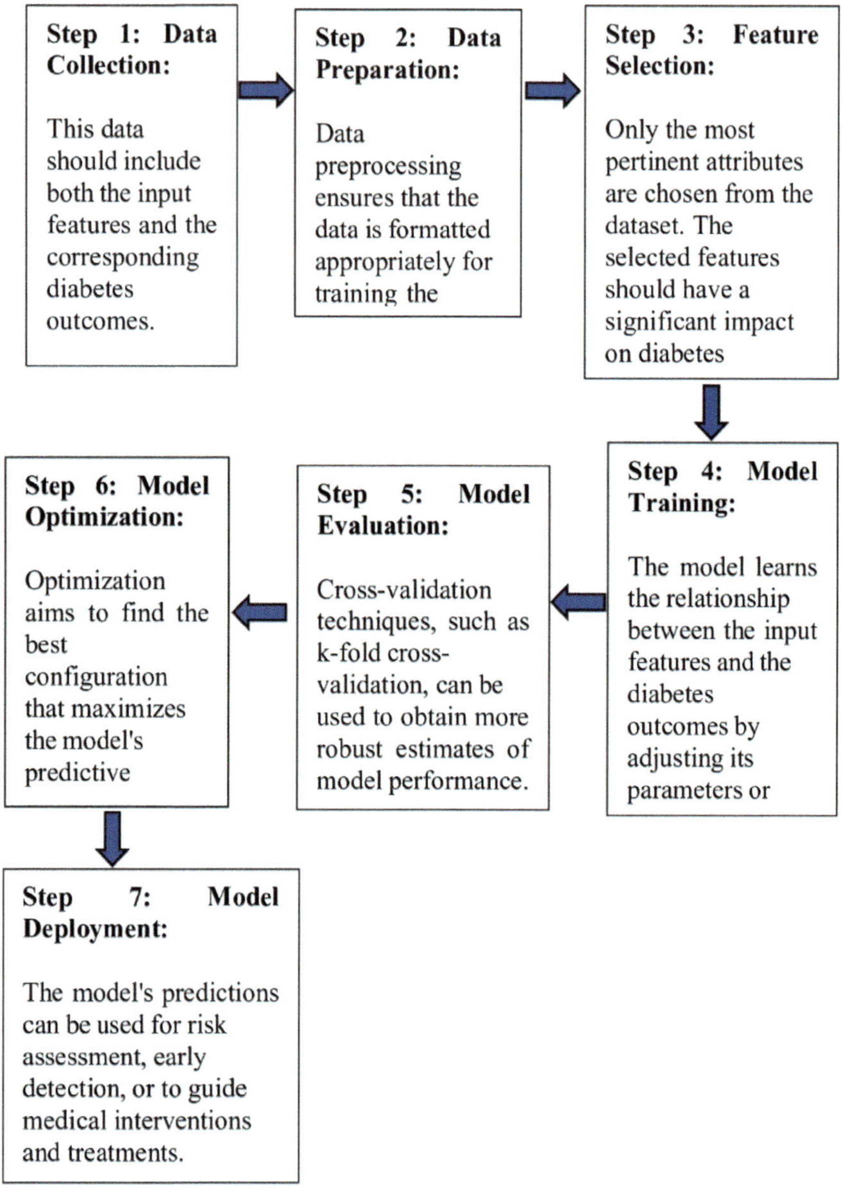

Fig. 10.2 Process of building a prediction diabetic model

10.3 Prediction Using Machine Learning

ML is crucial in developing computer systems that learn from experiences with minimal programming. It can be applied to predict and diagnose diseases like diabetes with high accuracy, enhancing patient care and clinical outcomes. By utilizing **ML** algorithms and data mining techniques, predictive models can be built to aid in diabetes prediction. This approach is faster and more efficient than traditional lab tests.

10.3.1 Support Vector Machine

SVM is frequently utilized in the context of diabetes prediction and diagnosis. SVM can used regarding challenges involving regression and classification. Within the case of diabetes, SVM can be utilized to predict the possibilities of diabetes in a person from a set of input features [22]. These features may include medical history, demographic information, lifestyle factors, and clinical measurements. To use SVM for diabetes prediction, a dataset with labeled instances (diabetic or non-diabetic) is required for training the model. The SVM algorithm learns from these labeled instances to create a decision boundary to classify new, unseen samples accurately. Researchers have applied SVM to diabetes prediction using various datasets and have achieved promising results [23, 24].

10.3.2 Artificial Neural Network

ANN is a widely used algorithm for ML that simulates how neural networks in the human brain work. It is a computational model made of artificial neurons that are networked together and collaborate to process and learn from input data. ANNs are widely utilized for many different types of tasks, including pattern recognition, prediction, regression, and classification. In diabetes, ANN has been extensively applied to predict and diagnose the disease based on input features such as patient demographics, medical history, and clinical measurements. The architecture of an ANN can be found in [25]. Backpropagation is a technique used in the training phase of an ANN to modify the weights of its connections iteratively. Researchers have explored various architectures and configurations of ANNs for diabetes prediction, experimenting with different activation functions, network topologies, and training algorithms [4, 26, 27]. The objective is to enhance the functionality of ANNs and improve their accuracy in predicting diabetes.

10.3.3 Decision Tree

DT is a supervised algorithm for learning. It can be very effective in the context of diabetes prediction and diagnosis. A DT represents a structure resembling a flowchart. DTs are particularly useful for diabetes prediction as they can handle numerical and categorical features. In addition to managing nonlinear relationships between targets and features, they can handle outliers and missing values. Choosing the optimal characteristic to divide the data at each node of a DT requires selection. This selection is according to the information gain, Gini impurity, or entropy. The simplicity and efficiency of DTs make them suitable for real-time and online prediction scenarios. Researchers have applied DTs to diabetes prediction tasks, achieving promising results [28–30]. DTs can handle various types of features, handle missing values and outliers, and capture nonlinear relationships. They are easy to interpret and visualize, making them valuable tools in diabetes research and healthcare.

10.3.4 Logistic Regression

LR aims to approximate the likelihood, given the input qualities, that the result variable will belong to a particular class. A logistic function is used in LR to translate the input features to a probability value between 0 and 1, whereas linear regression predicts continuous values. In diabetes prediction, LR can calculate the probability of an individual's diabetes, given their demographic information, medical history, and clinical measurements. LR assumes a linear relationship and might not be able to grasp sophisticated nonlinear relationships and interactions between features. To overcome this limitation, researchers often use feature engineering, To improve the prediction performance of LR models, regularization and feature selection strategies are used. LR is simple, interpretable, and efficient, making it a valuable tool in diabetes research and healthcare [31–33].

10.3.5 Random Forest

Ensemble methods, such as random forest (RF), are powerful ML algorithms commonly used in diabetes research for prediction and diagnosis tasks. To increase forecast accuracy and robustness, ensemble approaches mix several separate models. To estimate the risk that a person will have diabetes, a variety of input features, including patient demographics, medical history, and clinical measures, can be investigated using RF. The RF model considers the average probabilities or votes of each DT when combining all of the forecasts to produce the final prediction. RF has several advantages for diabetes prediction [34, 35]. It can handle both numerical and categorical features without necessitating extensive data pre-processing. It can effectively handle missing values and outliers. RF is also robust against noise and can handle nonlinear relationships between features and the target variable. RF is also computationally efficient and can be parallelized, allowing faster training and prediction.

Table 10.1 briefly gives an insight into the latest implementation of various **ML** algorithms in diabetes prediction.

Table 10.1 Brief overview of ML algorithm diabetes prediction

Techniques	Authors	Key point
SVM	Tasin et al. [22]	The paper presents the development of an automatic diabetes forecasting system using a dataset of female patients to reduce the risk of consequences including diabetes retinopathy, renal disease, nerve damage, and heart problems associated with long-term diabetes
	Assegie and Nair [23]	Diabetes is caused by a deficiency of insulin, which regulates blood sugar amounts. When the pancreas is unable to produce the required amount of insulin, complications occur, leading to an inability to control sugar levels effectively
	Ahmed et al. [24]	The study employed various feature selection approaches to identify and prioritize several risk factors associated with diabetes. The developed model, utilizing two distinct datasets, was analyzed
ANNs	Bukhari et al. [26]	The emergence of ANNs has revolutionized the medical industry, particularly in the healthcare domain, as they offer data-driven solutions for various applications
	Bano et al. [4]	The analysis aims to create a more accurate diabetes prediction system by identifying intricate connections between input variables and the output, uncovering unique patterns in the process
	Massari et al. [27]	Early detection and timely diagnosis of diabetes pose significant challenges. However, the emergence of various **ML** techniques, including ontology-based ML techniques, has revolutionized medical science by enabling the development of automated systems capable of accurately detecting patients with diabetes
Dt	Soni and Varma [28]	Implemented a methodology that focuses on early diabetes prediction in individuals or patients, aiming for higher accuracy using the DT algorithm
	Khanwalkar and Soni [29]	If left untreated and undetected, polymer disorders can lead to various complications. Seeking medical attention and diagnosis from a healthcare professional is crucial in managing the condition effectively. A critical aspect of addressing polymer disorders is early detection, which serves as a primary focus of this study. The research involves conducting a comprehensive survey and analyzing multiple parameters to diagnose poly-infected disorders. The methodology employed in this study includes utilizing classification algorithms and data collection techniques to gather and analyze
	Gandhi and Pandi [30]	The DT algorithm utilizes questions about an object's qualities found at nodal nodes to classify them. Based on the answers, the algorithm selects one of the branches and poses a fresh query at the intersection that follows. Until the algorithm arrives at the final response, which is symbolized by a tree leaf, this procedure is repeated

(continued)

Table 10.1 (continued)

Techniques	Authors	Key point
LR	Rajendra and Latifi [31]	LR, a widely used classification model in ML, finds extensive application in clinical analysis. LR utilizes probabilistic estimations to connect the independent and dependent variables
	Joshi and Dhakal [32]	Utilizing an LR model, we aim to among Pima Indian women, with type 2 diabetes can be predicted, allowing for prompt diagnosis and early detection of the condition. Consequently, this improves our knowledge of the related risk factors and gives patients the power to choose the best preventive and treatment plans
	Alghamdi [33]	The LR algorithm is utilized to predict a binary outcome's likelihood by using a sigmoid function to translate the output into a probability after fitting a regression model to the data
RF	Jiby [34]	Multiple DTs of different sizes make up the tree-based classifier RF. By taking into account the combined predictions of these trees, it establishes the class of dependent variables. During the construction of each tree, random sampling is performed on the training data. Ultimately, the decision made by the RF algorithm is based on the most commonly predicted outcome across the DTs
	Deberneh and Kim [35]	The authors created a ML model to estimate cases of type 2 diabetes in the next year if we have information for the current year. Hence, proactive steps to stop or slow the disease's course can be taken by those who are at risk
	Nikhila [36]	Predicting diabetes using ML techniques is crucial, as it relies on a comprehensive understanding of the disease's symptoms. ML algorithms use a variety of techniques to accurately predict diabetes, such as naive Bayes, RFs, LR, and DTs

These case studies illustrate the efficacy of different ML algorithms in diabetes prediction. The selection of an appropriate algorithm depends on factors such as the dataset's characteristics, interpretability requirements, and computational complexity. Evaluating and comparing different algorithms' performance is vital to choosing the most suitable one for a specific prediction task.

10.4 Model Performance

Model performance plays a vital role in designing a prediction model. Using this model performance only, we can identify the efficacy of the developed model. There are some standard matrices and parameters that all artificial intelligence (AI)/ML researchers widely use to evaluate prediction models' efficacy and performance related to diabetes. Assessment metrics provide quantitative measures of how well the model predicts the likelihood of an individual having diabetes. Table 10.2 shows commonly used evaluation metrics [14]:

Table 10.2 Brief of model performance parameters

Evaluation matrix	Key point
Accuracy	When dealing with unbalanced datasets, with a notably lower proportion of instances with diabetes than without it relying solely on accuracy can be misleading Should have a high accuracy and correctly classify both diabetic and non-diabetic cases
Precision	Precision is functional when the focus is on minimizing false positives, as in the case of reducing the risk of misdiagnosing a non-diabetic individual as diabetic The objective is to correctly identify diabetic cases with a high recall rate while minimizing false positives with a high precision rate
Recall (sensitivity or true positive rate)	Remember is necessary when the emphasis is on minimizing false negatives, as in the case of ensuring that all diabetic individuals are correctly identified
F1-score	Given its ability to account for false positives and false negatives, the F1 score is a useful indicator when there is an imbalance in the dataset between cases with and without diabetes
Area under the receiver operating characteristic (ROC) curve (AUC)	A higher AUC worth signifies a stronger discrimination ability of the model in effectively distinguishing between diabetic and non-diabetic cases
Specificity	Specificity is useful when the goal is to minimize false positives, as in the case of reducing unnecessary medical interventions or treatments for non-diabetic individuals A good model should have high specificity, minimizing false positive predictions and reducing unnecessary medical interventions or treatments for non-diabetic individuals

The evaluation metrics mentioned offer diverse viewpoints on the performance of prediction models concerning diabetes. Selecting the appropriate evaluation metrics relies on the application's specific goals, priorities, and requirements. It is commonly advised to consider multiple evaluation metrics concurrently to understand the model's performance comprehensively.

A model's performance can be affected by many elements, such as the features used as input, the selected ML algorithm, the caliber and representativeness of the training data, and optimized model parameters. It is crucial to evaluate model performance on the training data, independent test data, or cross-validation techniques to ensure its generalization ability. Depending on the particular demographic or situation it is used to, the model may change. Therefore, ongoing monitoring and evaluation of model performance using real-world data are essential for assessing effectiveness and making necessary adjustments.

10.5 Future Scope and Challenges

Diabetes prediction with ML methods holds promising prospects in several areas. Two notable areas of advancement are automated medication and the use of wearable devices to track health.

1. Automated Medication: ML algorithms can have a significant part in automated medication management for individuals with diabetes. These algorithms can analyze various data sources, such as blood glucose levels, physical activity, and dietary information, to provide personalized recommendations on medication dosage, timing, and adjustments. This can help individuals improve diabetes management which can lower the chance of complications. One such algorithm is reinforcement learning [37, 38] and its variant Markov game [39], where an agent learns to select an appropriate action (e.g., amount of insulin) by itself via trial and error. Researchers now focus on applying reinforcement learning [40–42] in automated medication.
2. Wearable Devices to Track Health: Wearable technology, including fitness trackers and smartwatches, has seen a sharp increase in popularity in recent years. People are using these devices more and more because they make it possible to track and monitor a variety of health metrics, such as heart rate, sleep patterns, and physical activity. By integrating ML algorithms, wearable devices can provide real-time feedback and personalized insights on diabetes management. For example, they can alert users about abnormal glucose levels, suggest physical activities, or provide reminders for medication adherence. When combined with wearable devices, ML algorithms have proved to be very useful in healthcare [43], apart from diabetes detection and management.

However, along with these prospects, specific challenges need to be addressed:

1. Data Collection and Authenticity: Collecting accurate and reliable data is crucial for developing effective prediction models. However, data collection can be challenging due to missing data, data quality issues, and inconsistencies. Maintaining the accuracy of prediction models relies on ensuring the authenticity and reliability of the collected data.
2. Maintaining Privacy: With the increasing use of personal health data for diabetes prediction, ensuring privacy and data security is a significant concern. Protecting sensitive health information and adhering to privacy regulations is vital to gaining the trust of individuals and maintaining ethical standards. Robust data anonymization techniques, secure data storage, and strict access controls are necessary to address privacy challenges.
3. Interpretability and Explainability: ML algorithms, especially complicated models like deep learning, have a black-box character that makes it difficult to grasp and explain their predictions, which makes them challenging to interpret in the context of diabetes prediction. However, because transparency is important, attempts are being made to create explainable AI methods that can reveal important information about how these algorithms make decisions.

Addressing these challenges will be crucial for successfully implementing and adopting ML-based diabetes prediction systems. To improve diabetes management and patient outcomes, researchers, healthcare providers, and policymakers must collaborate and prioritize transparency in developing robust, privacy-preserving solutions. This is particularly important as ML algorithms, including complex models like deep learning, tend to be opaque, hindering the interpretability of their predictions.

10.6 Conclusion

Various studies have demonstrated promising outcomes in predicting diabetes using different ML algorithms. SVM, ANN, RF, and LR are among the algorithms utilized to forecast diabetes based on various patient characteristics and risk factors. These algorithms offer several advantages: accuracy, personalized predictions, efficiency, scalability, and automation. They can analyze extensive datasets, detect patterns, and provide timely forecasts, enabling early detection and intervention in diabetes management.

Accurate prediction models heavily rely on effective data collection and pre-processing. Multiple sources, such as clinical trials, surveys, and electronic health records, can be utilized to gather diabetes-related data. Pre-processing techniques, including addressing missing values, managing outliers, and employing feature selection methods, aid in preparing the data for analysis. Comparative analyses have been conducted to assess the performance of various algorithms. SVM, ANN, RF, and LR have shown promising outcomes in accuracy, sensitivity, specificity, and AUC-ROC values when predicting diabetes.

Eventually, the use of ML algorithms in diabetes prediction holds potential for automated medication management and the integration of wearable devices to monitor health. However, certain challenges must be addressed for successful implementation, including ensuring accurate and authentic data collection, maintaining privacy, and improving the interpretability of algorithmic predictions. ML algorithms offer valuable tools for medical personnel to make well-informed choices and improve patient outcomes when managing diabetes. To further enhance the precision, dependability, and usefulness of diabetes prediction models, more research and cooperation are essential.

References

1. Sonar P, JayaMalini K. Diabetes prediction using different machine learning approaches. In: 2019 3rd international conference on computing methodologies and communication (ICCMC). IEEE; 2019. p. 367–71.
2. Muthukumaran N, Aiswarya R, Sankari SA, Bharathi KD. Deep learning neural network based human emotion classification with ANFIS algorithm. Irish Interdiscip J Sci Res. 2020;4(3):105–11.

3. Kopitar L, Kocbek P, Cilar L, Sheikh A, Stiglic G. Early detection of type 2 diabetes mellitus using machine learning-based prediction models. Sci Rep. 2020;10(1):11981.
4. Farhana B, Munidhanalakshmi K, Mohana RM. Predict diabetes mellitus using machine learning algorithms. J Phys Conf Ser. 2021;2089(1):012002.
5. Faruque MF, Sarker IH. Performance analysis of machine learning techniques to predict diabetes mellitus. In: 2019 international conference on electrical, computer and communication engineering (ECCE). IEEE; 2019. p. 1–4.
6. Ye J, Yao L, Shen J, Janarthanam R, Luo Y. Predicting mortality in critically ill patients with diabetes using machine learning and clinical notes. BMC Med Inform Decis Mak. 2020;20(11):1–7.
7. Zhu T, Li K, Herrero P, Georgiou P. Deep learning for diabetes: a systematic review. IEEE J Biomed Health Inform. 2020;25(7):2744–57.
8. Srikanth P. Using Markov chains to predict the natural progression of diabetic retinopathy. Int J Ophthalmol. 2015;8(1):132.
9. Lubis AR, Prayudani S, Lubis M. Analysis of the Markov chain approach to detect blood sugar level. J Phys Conf Ser. 2019;1361(1):012052.
10. Gill NS, Mittal P. A novel hybrid model for diabetic prediction using hidden markov model, fuzzy based rule approach and neural network. Indian J Sci Technol. 2016;9(35):192–9.
11. Birjais R, Mourya AK, Chauhan R, Kaur H. Prediction and diagnosis of future diabetes risk: a machine learning approach. SN Applied Sciences. 2019;1:1–8.
12. Herman WH, Ye W, Griffin SJ, Simmons RK, Davies MJ, Khunti K, Rutten GE, Sandbaek A, Lauritzen T, Borch-Johnsen K, Brown MB. Early detection and treatment of type 2 diabetes reduce cardiovascular morbidity and mortality: a simulation of the results of the Anglo-Danish-Dutch study of intensive treatment in people with screen-detected diabetes in primary care (ADDITION-Europe). Diabetes Care. 2015;38(8):1449–55.
13. García S, Luengo J, Herrera F. Data preprocessing in data mining, vol. 72. Cham: Springer International Publishing; 2015. p. 59–139.
14. Saar-Tsechansky M, Provost F. Handling missing values when applying classification models. J Mach Learn Res. 2007;8:1625–57.
15. Rahman MM, Davis DN. Addressing the class imbalance problem in medical datasets. Int J Mach Learn Comput. 2013;3(2):224.
16. Noor NSEM, Ibrahim H, Lah MHC, Abdullah JM. Improving outcome prediction for traumatic brain injury from imbalanced datasets using RUSBoosted trees on electroencephalography spectral power. IEEE Access. 2021;9:121608–31.
17. Sisodia D, Sisodia DS. Prediction of diabetes using classification algorithms. Proc Comput Sci. 2018;132:1578–85.
18. Larabi-Marie-Sainte S, Aburahmah L, Almohaini R, Saba T. Current techniques for diabetes prediction: review and case study. Appl Sci. 2019;9(21):4604.
19. Theis J, Galanter WL, Boyd AD, Darabi H. Improving the in-hospital mortality prediction of diabetes ICU patients using a process mining/deep learning architecture. IEEE J Biomed Health Inform. 2021;26(1):388–99.
20. Dargan S, Kumar M, Ayyagari MR, Kumar G. A survey of deep learning and its applications: a new paradigm to machine learning. Arch Computat Methods Eng. 2020;27:1071–92.
21. Hou J, Sang Y, Liu Y, Lu L. Feature selection and prediction model for type 2 diabetes in the Chinese population with machine learning. In: Proceedings of the 4th international conference on computer science and application engineering; 2020. p. 1–7.
22. Tasin I, Nabil TU, Islam S, Khan R. Diabetes prediction using machine learning and explainable AI techniques. Healthcare Technol Lett. 2023;10(1–2):1–10.
23. Assegie TA, Nair PS. The performance of different machine learning models on diabetes prediction. Int J Sci Technol Res. 2020;9(01)
24. Ahmed N, Ahammed R, Islam MM, Uddin MA, Akhter A, Talukder MA, Paul BK. Machine learning based diabetes prediction and development of smart web application. Int J Cogn Comput Eng. 2021;2:229–41.

25. Saha S, Das S, Acharya A, Kumar A, Mukherjee S, Pan I, Gupta A. Identification of nonlinear systems from the knowledge around different operating conditions: a feed-forward multi-layer ANN based approach. In: 2012 2nd IEEE international conference on parallel, distributed and grid computing. IEEE; 2012, December. p. 413–8.
26. Bukhari MM, Alkhamees BF, Hussain S, Gumaei A, Assiri A, Ullah SS. An improved artificial neural network model for effective diabetes prediction. Complexity. 2021;2021:1–10.
27. El Massari H, Sabouri Z, Mhammedi S, Gherabi N. Diabetes prediction using machine learning algorithms and ontology. J ICT Stand. 2022;10(2):319–37.
28. Soni M, Varma S. Diabetes prediction using machine learning techniques. Int J Eng Res Technol. 2020;9
29. Khanwalkar A, Soni R. A survey on prediction of diabetes using classification algorithms. J Achiev Mater Manuf Eng. 2021;104(2):77–84.
30. Gandhi P, Pandi GS. Diabetes prediction using machine learning techniques. Int J Res Publ Rev. 2022;3(2):77–82.
31. Rajendra P, Latifi S. Prediction of diabetes using logistic regression and ensemble techniques. Comput Methods Prog Biomed Update. 2021;1:100032.
32. Joshi RD, Dhakal CK. Predicting type 2 diabetes using logistic regression and machine learning approaches. Int J Environ Res Public Health. 2021;18(14):7346.
33. Alghamdi T. Prediction of diabetes complications using computational intelligence techniques. Appl Sci. 2023;13(5):3030.
34. Jiby TC. A study on various machine learning classification algorithms for diabetes prediction. Int J Eng Res Technol. 2021;10(8):425–7.
35. Deberneh HM, Kim I. Prediction of type 2 diabetes based on machine learning algorithm. Int J Environ Res Public Health. 2021;18(6):3317.
36. Nikhila G. Diabetes prediction using RF based classifier using machine learning: a review. Int J Res Publ and Rev. 2022;3(11):591–8.
37. Kumar A, Sharma R. Neural/fuzzy self learning Lyapunov control for non linear systems. Int J Inf Technol. 2018:1–14.
38. Kumar A, Sharma R. Linguistic Lyapunov reinforcement learning control for robotic manipulators. Neurocomputing. 2018;272:84–95.
39. Kumar A, Sharma R, Varshney P. Lyapunov fuzzy Markov game controller for two link robotic manipulator. J Intell Fuzzy Syst. 2018;34(3):1479–90.
40. Javad MOM, Agboola SO, Jethwani K, Zeid A, Kamarthi S. A reinforcement learning–based method for management of type 1 diabetes: exploratory study. JMIR Diab. 2019;4(3):e12905.
41. Zohora MF, Tania MH, Kaiser MS, Mahmud M. Forecasting the risk of type ii diabetes using reinforcement learning. In: 2020 joint 9th international conference on informatics, electronics & vision (ICIEV) and 2020 4th international conference on imaging, vision & pattern recognition (icIVPR). IEEE; 2020. p. 1–6.
42. Tejedor M, Woldaregay AZ, Godtliebsen F. Reinforcement learning application in diabetes blood glucose control: a systematic review. Artif Intell Med. 2020;104:101836.
43. Kumar A, Anand JJ, Hemanth Kumar BN. Intrusive video oculographic device: an eye-gaze-based device for communication. Innov Emerg Technol. 2022;9:2250002.

Chapter 11
Investigation of Machine Learning Algorithms in Detecting Chronic Kidney Disorder

Manoj Ishwar Hegde, N. Dhanvina, N. Dharshan, and Snigdha Sen

11.1 Introduction

The human body is an intricately designed system comprising several vital organs, including the heart, liver, and kidneys. These organs work in harmony to maintain normal bodily functions. The kidneys, which are about the size of an electronic mouse, are the most vital organ for maintaining general health. These bean-shaped organs tirelessly filter our entire blood supply approximately every 30 minutes, ensuring the efficient removal of waste products, toxins, and maintaining a delicate balance of blood pressure, production of red blood cells, bone health, and essential blood chemicals. However, it is alarming to note that over 13% of adults in India are considered to have chronic kidney disease (CKD). CKD is a complex condition characterized by progressive kidney damage. This results in a diminished capacity to filter blood effectively. This impairment results in the buildup of excess fluid and waste in the body, subsequently increasing the risk of various health complications, including heart disease and stroke. Notably, CKD can often remain asymptomatic, with individuals not experiencing noticeable symptoms, making early detection critical. To confirm a diagnosis of CKD, specific blood and urine tests are required. These tests involve assessing creatinine levels in the blood and measuring protein levels in the urine.

CKD presents different stages of severity, typically worsening over time. However, early detection and proactive actions can considerably impede its progression. If left untreated, CKD may progress to kidney failure, markedly increasing the

M. I. Hegde · N. Dhanvina · N. Dharshan
Department of Computer Science, Global Academy of Technology,
Bangalore, Karnataka, India

S. Sen (✉)
Department of Information Technology, Manipal Institute of Technology Bengaluru,
Manipal Academy of Higher Education, Manipal, India

T. Choudhury, A. Katal (eds.), *Machine Learning for Disease Detection, Prediction, and Diagnosis*, https://doi.org/10.1007/978-981-96-4241-0_11

risk of premature cardiovascular issues. When kidney function reaches a complete standstill, survival hinges on dialysis or a kidney transplant, marking the stage known as end-stage renal disease. Facts show that CKD is a global health concern. The World Health Organization reports that CKD is prevalent worldwide, affecting people of all ages and backgrounds. Among the main causes of CKD are elevated blood pressure and diabetes, both of which have become more prevalent recently. Additionally, CKD can result from infections, kidney stones, and genetic factors. It is essential to understand that CKD does not discriminate based on age, gender, or ethnicity, affecting millions of people across the globe. To mitigate the risk of CKD and the dire consequences of kidney failure, it is essential to manage the associated risk factors. Regular annual checkups can aid in early detection, allowing for timely interventions. Furthermore, adopting a healthier lifestyle by incorporating a balanced diet and regular exercise can contribute to kidney health. Adhering to prescribed medications when necessary and maintaining consistent visits with healthcare providers are vital steps in preventing or managing CKD. By heeding these precautions and actively managing kidney health, individuals can reduce their susceptibility to CKD and its potentially devastating consequences.

However, the real challenge arises when CKD reaches stage 5, known as end-stage kidney disease or renal failure. At this point, patients require renal replacement therapy (RRT), a costly and resource-intensive treatment that not only places a significant financial burden on individuals but also highlights disparities in healthcare access, particularly in developing countries like Ethiopia. Unfortunately, RRT is not readily available in many regions, making the management of renal failure and its associated complications a complex and daunting task. The shortage of facilities and medical personnel, along with the high expenses associated with RRT, underscores the crucial importance of timely detection. Early diagnosis is vital for enhancing patient outcomes as well as minimizing the economic strain associated with advanced CKD treatment. Timely intervention can significantly impact the trajectory of the disease and enhance the effectiveness of therapies, potentially alleviating some of the challenges faced in resource-limited settings.

11.2 Machine Learning in CKD Diagnosis

The use of machine learning (ML) techniques and data-driven methodology [1, 2] has emerged in recent years as a promising approach to enhance the detection of chronic renal disease. Support vector machines (SVMs), naïve Bayes, and decision tree approaches are a few examples of ML techniques that have the ability to deliver quick and accurate evaluations of CKD risk, enabling more targeted and timely interventions. This work focuses on using these ML techniques to more accurately diagnose CKD. The study has worked with large datasets to test and train these models, and in the end, it has assessed how accurate these models are at identifying CKD. The goal of the study is to determine which method of collecting data about

CKD is most useful for diagnosing the condition. This could have noteworthy implications for clinical practices.

The remaining structure of this manuscript is structured in the following manner:

- Section 11.3 offers a thorough examination of the current literature, shedding light on the current state of CKD diagnosis, the challenges faced, and how ML might be able to help with these problems.
- In Sect. 11.4, the suggested methodology is presented, providing details on the development of ML models that are trained and tested using datasets related to CKD.
- Section 11.5 delves into the practical implementation of these models, illustrating their real-world applicability in a clinical context.
- Section 11.6 provides a thorough analysis of the outcomes arising from different ML methodologies, helping us understand which approach is the most effective for CKD diagnosis.
- Finally, the article concludes on the broader implications of ML in the field of healthcare.

11.3 Related Work and Background Study

Pal et al. [3] carried out a study to use ML techniques to create a prediction model for the prompt identification of chronic kidney illness. A dataset with 25 unique features was taken from the University of California (UCI) ML storage facility, and classifiers such as SVM, decision tree, and logistic regression (LR) were used by the researchers. The outcomes demonstrated a noteworthy 97.23% accuracy rate, demonstrating the efficacy of the suggested methodology. The primary focus of the research was to emphasize the crucial significance of early CKD diagnosis. To achieve this goal, the study explored a range of methods, including both traditional and emerging approaches, to enhance CKD detection. The successful application of ML techniques demonstrated the potential for enhancing the precision of early identification in chronic renal disease.

Zhang et al. [4, 5] introduced an innovative approach to predict CKD at an early stage, leveraging advanced ML techniques. They conducted their research utilizing an extensive dataset obtained from the National Health and Nutrition Examination Survey, which encompassed 30 clinical and demographic variables. Utilizing a varied range of ML models, which encompass random forest, K-nearest neighbors (KNNs), and gradient boosting machine (GBM), the researchers crafted a predictive model that demonstrated an impressive accuracy rate of 94.5% through a meticulous evaluation process. The study underscored the paramount importance of timely and accurate CKD prediction, along with effective intervention strategies. In addition to random forest, support vector, and decision tree techniques, the researchers explored the data using various unsupervised algorithms. Significant accuracy ratings of 82.56% for five-class classification and 99.8% for binary classification were obtained from this investigation. The findings suggested that experimenting with diverse unsupervised algorithms could further enhance outcomes and improve the accuracy of CKD detection, contributing valuable insights to the field.

Ilyas et al. [6] employed a pair of algorithms, J48 and random forest algorithms to forecast various stages of CKD. Their study achieved notable accuracy levels, specifically 85.5% using the J48 algorithm within a rapid 0.03 seconds and 78.25% with the random forest approach in a slightly longer 0.28 seconds. Despite the time constraints, the researchers concluded that J48 outperforms random forest in real-time predictions. This research contributes insights into the trade-off between accuracy and computational speed in predicting CKD stages. By favoring the J48 algorithm for its superior real-time performance, the study sheds light on the practical considerations essential for implementing predictive models in a time-sensitive clinical setting.

The researchers [7] explored the random forest technique for identifying CKD. Their findings indicated that the random forest algorithm demonstrated the highest accuracy in identifying CKD. However, they also noted certain drawbacks associated with its application, including issues related to time complexity and the potential for overfitting with noisy datasets. This insight highlights a common challenge in utilizing powerful algorithms like random forest in real-world scenarios. Despite its high accuracy, the researchers acknowledge the importance of addressing practical limitations such as time complexity and potential overfitting, underscoring the need for a balanced consideration of algorithmic performance and feasibility in clinical applications.

Kunwar et al. [8] conducted an analysis of current techniques in the study of CKD. They highlighted several approaches, including decision trees, artificial neural networks (ANNs), SVM, fuzzy expert systems, and Naïve Bayes. Notably, they found that Naïve Bayes outperforms ANN in their evaluation.

Chetty et al. [9] emphasized the good accuracy of Naïve Bayes, ranking it second to sequential minimal optimization and Instance-Based K-nearest neighbor (IBK) in CKD studies. This observation underscores Naïve Bayes as a competitive choice for CKD applications. On a different note, Vijayarani et al. [10] noted the superior performance of SVM over Naïve Bayes in their study. The collective insights from these studies indicate varying perspectives on the effectiveness of different algorithms in CKD prediction. While Naïve Bayes has shown promising results in certain analyses, there is a call for further exploration and improvement of this relatively simple algorithm, as highlighted by Vijayarani et al. This underscores the dynamic nature of research in the field, with ongoing efforts to optimize and refine methodologies for CKD detection.

In [11], the authors conducted a thorough investigation into diverse ML algorithms with the aim of detecting early-stage CKD diagnosis. The research employed CKD patient data for constructing, training, and validating models based on input parameters discussed earlier in the study. To enhance model efficiency, the dataset underwent preprocessing, study explored relationships among various factors to streamline features and eliminate redundant data. Hemoglobin, albumin, and specific gravity were determined to be the most significant factors in predicting CKD when a filter technique was used for feature selection. This finding highlighted key biomarkers crucial for accurate CKD detection. Principal component analysis was also used in the study to identify important aspects of the dataset and improve the CKD identification procedure. Furthermore, correlations among various factors were scrutinized to streamline features and remove redundant data. Employing a filter approach for feature selection, it was found that hemoglobin, albumin, and specific gravity were pivotal in predicting CKD.

Accuracy was the primary criterion taken into account when assessing the ML algorithms' efficacy, stressing the significance of precise forecasts for the diagnosis of CKD in its early stages. Through a systematic process involving preprocessing, feature selection, and model evaluation, the research aimed to enhance comprehension and utilization of ML methods in CKD detection [12].

The research tackles the difficulty of predicting and preventing CKD, which is often challenging to detect because it typically lacks noticeable symptoms. Utilizing ML techniques for prediction and analysis, the chapter presents eight ML methods: KNNs, SVMs, LR, Naïve Bayes, extra tree classifiers, AdaBoost, Xgboost, and LightGBM.

The study seeks to determine the most efficient classifier for forecasting CKD, utilizing a dataset comprising 14 attributes and 400 records obtained from Kaggle.com. The proposed ML models are systematically compared to assess their performance in predicting chronic kidney illness. Notably, the LightGBM model [13] stands out, achieving an impressive accuracy level of 99.00%, providing a promising solution for accurate CKD prediction.

The goal of the research is to build a predictive model for CKD by utilizing the SVM, decision tree, and KNN classification techniques. Notably, the decision tree method predicted CKD with an astounding 97.85% accuracy rate. The body of research on the subject highlights the value of ML in healthcare, especially in the early diagnosis and treatment of patients with CKD. Because SVM, KNN, and decision tree algorithms are effective at managing medical datasets, they are frequently utilized. This study [14] adds to the growing corpus of research that highlights how ML is improving healthcare diagnostics, particularly in the area of chronic disease prediction such as CKD.

The research utilizes a dataset on CKD sourced from Kaggle, utilizing random forest and XGBoost classifiers to construct a predictive model. Assessment is conducted through the confusion matrix, emphasizing both categorical and non-categorical attributes. The results show that random forest achieves 95.83% accuracy, while XGBoost outperforms with 98.33%. Literature supports ML in CKD prediction, emphasizing classifier metrics and the significance of early detection. The comparative analysis aligns with existing research on classifier effectiveness. Future work includes refining the model through feature selection methods. Keywords such as chronic kidney disease, random forest, XGBoost, and prediction model [15, 16] ensure relevance in ML applications in CKD research.

As ML algorithms have huge potential to deliver tremendous performance already in many domains [17–19], we are motivated to apply in this present study of CKD prediction.

11.4 Algorithms Under Study

In this section, we describe briefly multiple algorithms used for this work.

11.4.1 Decision Tree

The decision tree method is one of the commonly employed supervised ML techniques for categorization. It addresses the challenge of ML by organizing feature values into a tree-like structure. Every leaf node in a decision tree represents the class to which the examples belong, and every node of a decision tree symbolizes characteristics (features) in an instance to be classed. Figure 11.1 describes the working of the decision tree algorithm.

This study employs multi-class J48 and random forest classification algorithms to categorize CKD into several stages. Subsequent sections provide explanations of these algorithms and their functioning.

11.4.2 J48Algorithm

C4.5 is a renowned classification algorithm, widely recognized as the successor to Ross Quinlan's ID3 approach [11]. It is commonly referred to as J48 in Weka, and its implementation is in Java, making it accessible and versatile. This algorithm excels in generating informative-theoretic decision trees, which serve as valuable statistical classifiers, aiding in the grouping and categorization of data.

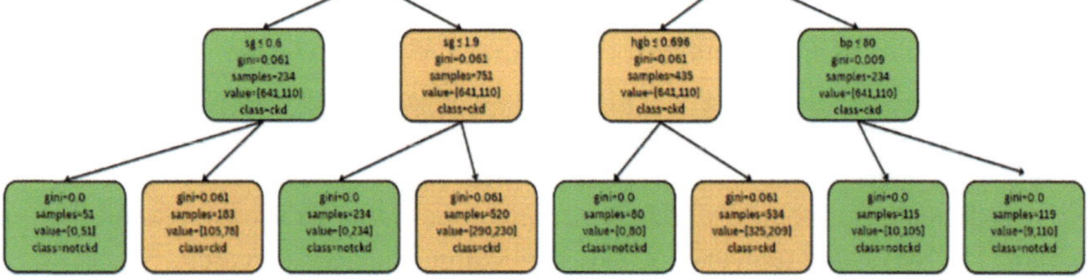

Fig. 11.1 Working of decision tree algorithm

In the C4.5 algorithm, the training data, represented as D, comprises categorized samples, each sample being depicted as an m-dimensional vector (n_1, i, n_2, i, ..., n_p,i). These vectors encompass not only the attribute values but also crucial class information, which is vital for the classification process. To improve the accuracy of categorization of C4.5, the algorithm employs a strategic approach. It selects the attribute with the highest information gain, as determined by measures like entropy, to act as the pivotal dividing factor in the decision tree construction.

The capacity of C4.5 to independently choose the most informative attributes for categorization makes it an invaluable instrument in the ML space. It empowers the algorithm to craft decision trees that effectively differentiate and classify data, ultimately contributing to the automation of the feature selection process in addition to improving the classification procedure's accuracy. This adaptability and inherent intelligence in attribute selection are key factors in C4.5's widespread popularity and effectiveness in various classification tasks.The output generated by this study's J48 algorithm is dependent on the following stages.

1. Select the input dataset. Classification attributes are separated according to the ID3 algorithm using the J48 approach.
2. Evaluate each feature's normalized informative content. To build a decision tree, choose the attribute that has the most informative content as the root node.
3. Determine each attribute's informative information, then add it to the tree as a child node. Iteratively repeat the preceding two stages until the stopping condition is satisfied.

11.4.3 Random Forest Algorithm

Random forest, a supervised classification technique, is an effective tool for making predictions based on randomly selected data samples. Its strength lies in the generation of a multitude of decision trees, from which forecasts are obtained, and then a collective "vote" is cast to pick the best option. One significant aspect to consider in the concept of random forest is the count of overall trees used in this classifier. This parameter directly influences the model's accuracy. Unlike some ML algorithms that require meticulous fine-tuning of hyperparameters, random forest exhibits a robust ability to adapt and produce reliable results even without extensive hyperparameter adjustments. Figure 11.2 describes the working of the random forest algorithm.

What sets random forest apart is its easiness of use and effectiveness, especially when dealing with vast datasets. This makes it a priority for data scientists and analysts working with big data. The adaptability and accuracy of random forest are particularly advantageous in scenarios where huge amounts of data need to be processed efficiently. Furthermore, random forest has the added benefit of effectively identifying outliers and abnormalities within the data. This capability is crucial in

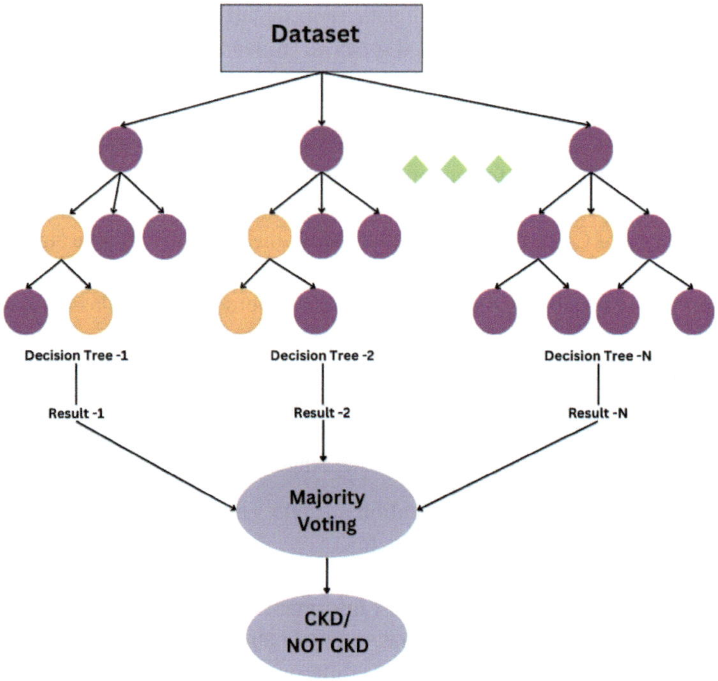

Fig. 11.2 Working of the random forest algorithm

data analysis, as it helps in pinpointing points in data that significantly take a diversion from the norm, potentially signaing errors or anomalies that require further investigation.

The following processes determine the output produced by the random forest classifier employed in this study:

1. Select randomized samples from the initial dataset. These kinds of randomly selected samples are sometimes referred to as "boot-strapped dataset set."
2. Take into account any particular set of factors while creating a decision tree for the bootstrapped data collection.
3. To the greatest extent possible, repeat the above steps 1 and 2.
4. Run the recently collected data through each decision tree that has already been constructed in order to predict the result for a new data point. As per votes received, the anticipated class is assessed.
5. As a way to ascertain the final class, assess the model using the data set's out-of-bag (OOB) instances. OOB examples are those instances that are excluded from the bootstrapped data.

11.4.4 Support Vector Machines

SVMs [20] are frequently employed for problems involving both regression and classification. SVM is a model of supervised learning used in categorization that may identify patterns in data by analyzing it and categorizing fresh data. Selecting the most effective hyperplane to split the data points is the aim of SVM, with the maximum margin between classes. SVM is very helpful for handling difficult decisions boundaries and high-dimensional data. Figure 11.3 illustrates the proposed workflow with SVM.

1. Collect labeled data.
2. Preprocess Data: Prepare the data for SVM. In this stage, missing values are handled, features are scaled or normalized, and the testing and training sets of the data are separated.
3. Select Kernel: Choose an appropriate kernel function for the SVM. Common choices include the following:

   - Linear Kernel: Suitable for data that can be separated linearly.
   - Polynomial Kernel: Ideal for data with non-linear boundaries.
   - Radial Basis Function (RBF) Kernel: For complicated, non-linear data, the RBF kernel is an effective method.

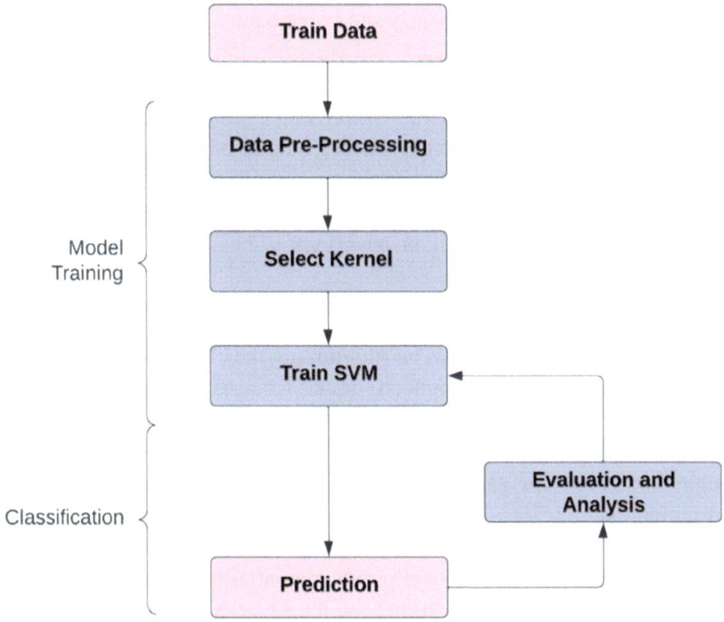

Fig. 11.3 The proposed workflow with SVM

4. Train SVM: Train the SVM using the training data. During training, the SVM learns to determine which decision boundary best divides the data's several classifications.
5. Predict: Using SVM to forecast fresh, untainted data. This is done by feeding the SVM with the features of the new data points.
6. Evaluate Model: Analyze the SVM's performance utilizing a variety of metrics, such as recall, accuracy, precision, and F1-score. This stage aids in evaluating the model's performance and determining whether any modifications are required.
7. Make Predictions: Following the model's evaluation, predictions for fresh, unseen data can be made. This is how your trained SVM model is used in real life.
8. Analyze Results: Analyze the results to gain insights into your model's performance. Understand its strengths, weaknesses, and areas for improvement. You may need to fine-tune parameters, consider different kernels, or gather more data to enhance accuracy.

Key Components of SVM

- Kernel Functions:
  - Linear Kernel: This kernel is appropriate for data that is linearly separable points. It represents the simplest form of SVM and works well for relatively straightforward classification tasks.
  - Polynomial Kernel: Non-linearly separable data is handled using the polynomial kernel. The SVM may carry out classification in the modified space once it maps the original data into a higher-dimensional space.
  - RBF Kernel: The RBF kernel is effective in capturing complex relationships within the data. It is capable of mapping data into infinite-dimensional space, making it a popular choice for nonlinear classification.
  - Sigmoid Kernel: While the sigmoid kernel is less frequently employed in SVM than other kernels, it is nevertheless beneficial for neural network applications.
- Regularization Parameter (C): The regularization parameter in SVM known as "C" regulates the trade-off between minimizing the classification error and optimizing the margin. While a higher C seeks for a narrower gap and can cause overfitting, a lower C permits a larger margin and might result in more misclassifications.
- Gamma Parameter: The gamma parameter determines how much of an influence one training example has. A low gamma value implies a broader reach, meaning that points farther away are considered in the decision boundary, while a high gamma value focuses on the closer points, potentially leading to overfitting.
- Shrinking Heuristic: The shrinking heuristic is an optimization technique employed by SVM to speed up the training process. It helps in discarding non-support vectors from the working set, reducing the computational load, and improving the training speed without significantly impacting the model's performance.

Significance of Chosen Parameters

- C Parameter: The C parameter influences the soft margin, impacting the balance between achieving a wider margin and allowing more margin violations. Choosing an appropriate C value is crucial for controlling the bias-variance trade-off and preventing overfitting. It is shown in Fig. 11.4.
- Kernel Parameter: Selecting the appropriate kernel is crucial for capturing the underlying data patterns accurately. The kernel's choice depends on the data's linearity and complexity. It enables the SVM to handle both linearly separable and nonlinear data effectively.
- Gamma Parameter: The gamma parameter plays a vital role in defining the impact of a single individual training sample. A low gamma value implies a more generalized decision boundary, whereas a high gamma value leads to a more precise but potentially overfitted decision boundary.

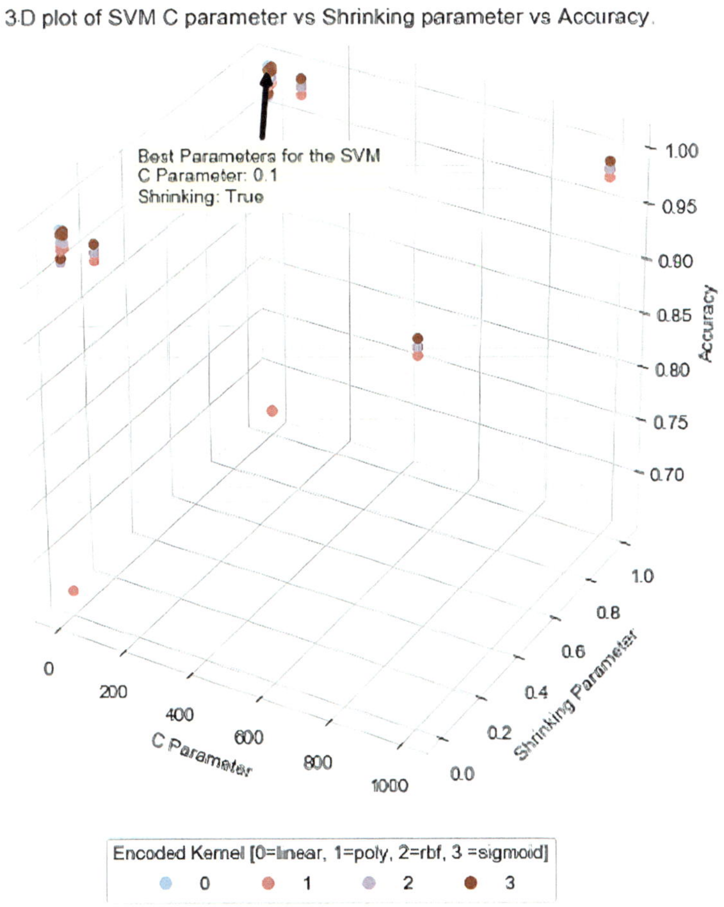

Fig. 11.4 3D plot of SVM C parameter

- Shrinking Parameter: The shrinking heuristic is a beneficial optimization technique that accelerates the training process. By discarding non-support vectors, the SVM algorithm focuses on the most critical data points, resulting in quicker convergence and reduced computational complexity.

Three three-dimensional (3D) graphs of SVM accuracy for various kernel functions as a function of the shrinking and C parameters are displayed in the figure. Additionally displayed for each kernel function are the optimal SVM settings.

The figures demonstrate how important the C parameter and the shrinking parameter are to SVM accuracy. The optimal SVM accuracy for each of the three kernel functions is obtained with a high C value and a false shrinking parameter.

However, it is crucial to remember that the best SVM parameters can vary depending on the specific dataset. Therefore, it is important to perform a grid search over the C parameter and the shrinking parameter to search the best parameters for your dataset.

11.4.5 Naïve Bayes

Naive Bayes shown in Fig. 11.5 is a fundamental algorithm for classification which is based on Bayes' theorem, a fundamental principle in probability theory. It is used in various domains, which include natural language processing, email filtering, sentiment analysis, and medical diagnosis. Naive Bayes estimates the probabilities required for classification using a labeled dataset, which includes a features set and a class label. It calculates the prior probabilities of each class (P(A)) and the likelihoods of observing every characteristic for every class (P(B|A)):

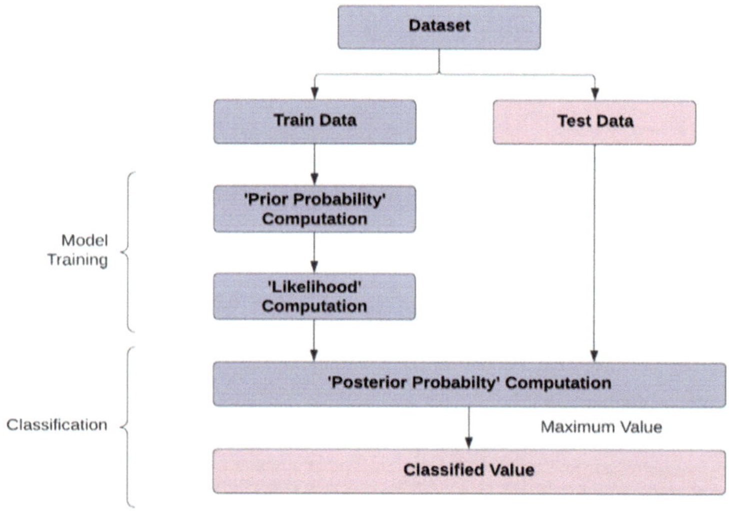

Fig. 11.5 Working with Naïve Bayes algorithm

$$P(A \mid B) = \frac{P(B \mid A) \cdot P(A)}{P(B)}$$

- P(A|B)—P(A|B) is how likely it is that event A will happen in regard to event B's occurrence.
- P(B|A)—P(B|A) is the likelihood that event B will occur in relation to the happening of event A.
- P(A)—P(A) and P(B)P(B) are the individual chances of occurrences A and B.

The "naive" part of Naive Bayes is its assumption of feature independence, which simplifies calculations, but they need not hold true in the dynamic or real world. For example, in text analysis, the words "cat" and "kitten" are often correlated, but Naive Bayes treats them as independent.

The classification phase involves using the model to classify new, unseen data. For each class, Naive Bayes calculates the posterior probability (P(A|B), which expresses the likelihood, based on the observable features. The class with the highest posterior probability is given the data point.

11.4.6 Insights From Implementation

Dataset Source The UCI Machine Learning Repository is the source of the dataset used in this investigation [12], specifically designed to diagnose chronic renal disease. It comprises 24 diverse features aimed at facilitating CKD prognosis, ranging from numerical attributes like age and blood pressure to categorical attributes like specific gravity and the presence of red blood cells. The dataset description is shown in Fig. 11.6.

Attribute Information This dataset includes 24 varied features kept track of to help with CKD prognosis:

1. Age (numerical) age described in years
2. Blood Pressure (numerical) BP in mm/Hg

   …

25. Class (nominal) class—(ckd, notckd)

Preprocessing
Categorical attributes such as Albumin (Al), Specific Gravity (Sg), Sugar (Su), Hypertension (Htn), and Red Blood Cell (Rbc) were processed to ensure compatibility with the ML algorithms utilized.

Numeric attributes including Blood Pressure (Bp), Blood Urea (Bu), Serum Creatinine (Sc), Sodium (Sod), Potassium (Pot), White Blood Cell Count (Wbcc), Hemoglobin (Hemo), and Red Blood Cell Count (Rbcc) were retained for analysis.

Correlation analysis was conducted to comprehend the connections between various characteristics and how they affect the likelihood of developing CKD.

The numeric attributes recorded are shown in Fig. 11.7.

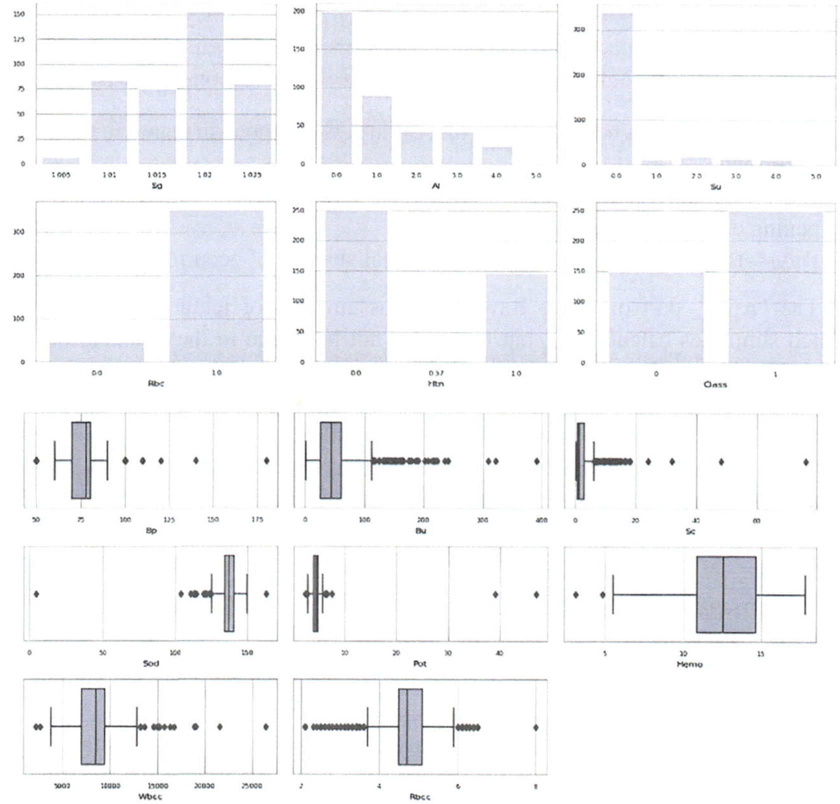

Fig. 11.6 Insights of the dataset

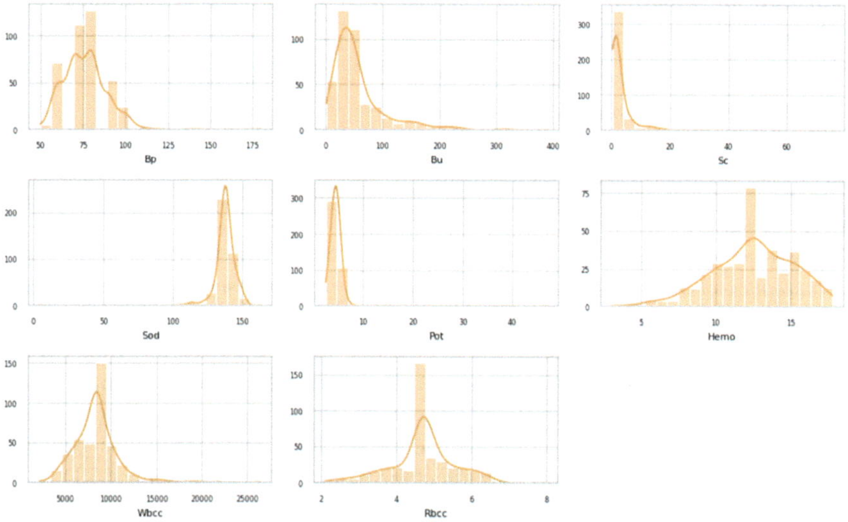

Fig. 11.7 Distribution of numeric attributes

Bp—Blood Pressure
Sc—Serum Creatinine
Wbcc—White Blood
Cell Count
Sod—Sodium
Pot—Potassium
Bu—Blood Urea
Hemo—Hemoglobin
Rbcc—Red Blood Cell Count

In Fig. 11.8, we present the correlation matrix among features of the dataset.

Correlation

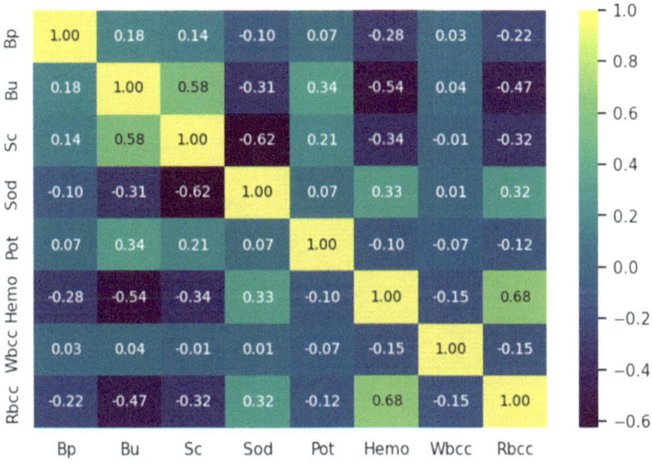

Fig. 11.8 Correlation among features of the dataset

11.5 Experimental Results

The study was conducted using Google Colab, providing a cloud-based framework for running ML experiments. Four models were evaluated: SVM with RBF (SVM RBF), random forest tree, J48 algorithm, and Naive Bayes. Precision, accuracy, recall, and F1-score were among the performance evaluation criteria used. Figure 11.9 and Table 11.1 display the results.

Analysis of Results

SVM (RBF): It achieved exceptional performance across all metrics, demonstrating perfect precision and high scores in accuracy, recall, and F1-score. The SVM RBF model effectively distinguished data points with a high degree of accuracy.

Random Forest Tree: While demonstrating strong performance with high precision and accuracy, the model displayed slightly lower scores in recall and F1-score compared to SVM RBF. This indicates a relatively higher tendency to misclassify positive instances.

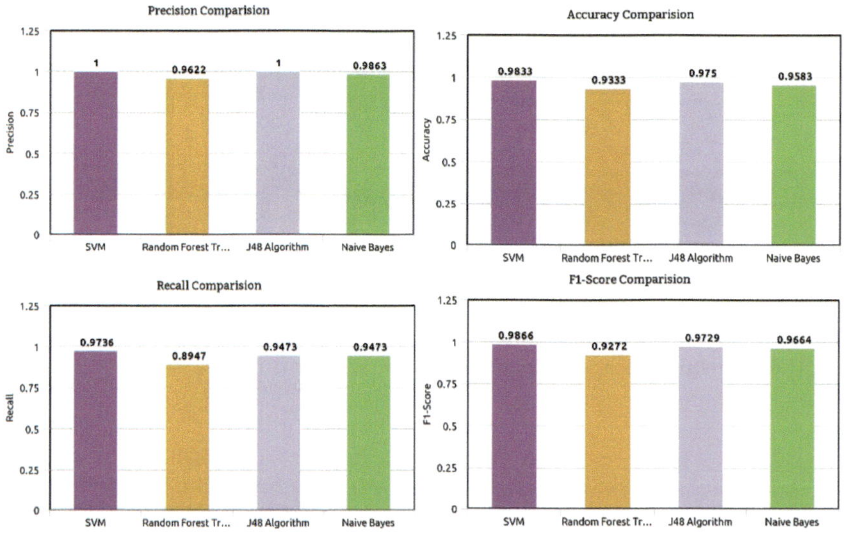

Fig. 11.9 Comparative performance graphs of experimented algorithms

Table 11.1 Comparative analysis of four algorithms

Models	Precision	Accuracy	Recall	F1-score
SVM (RBF)	1	0.9833	0.9736	0.9866
Random forest tree	0.9622	0.9333	0.8947	0.9272
J48 algorithm	1	0.975	0.9473	0.9729
Naive Bayes	0.9863	0.9583	0.9473	0.9664

J48 Algorithm: It exhibited perfect precision and commendable results in accuracy, recall, and F1-score. The model effectively classified instances with high precision and recall, resulting in a robust F1-score.

Naive Bayes: It showcased impressive performance with high precision and accuracy. Competitive scores in recall and F1-score indicate the model's ability to accurately classify instances with balanced precision and recall.

11.6 Conclusion

The importance of ML algorithms in enhancing the diagnosis and prognosis of CKD is highlighted by this work. We assessed the effectiveness of four different models using a varied dataset that we obtained from the UCI Machine Learning Repository: SVM RBF, random forest tree, J48 algorithm, and Naive Bayes.

Key Findings

Our analysis revealed that the SVM RBF model exhibited exceptional performance, achieving perfect precision and high scores in accuracy, recall, and F1-score. This underscores the robust classification capabilities of SVM RBF in accurately distinguishing CKD instances.

While the random forest tree model demonstrated strong overall performance, it displayed a slightly higher tendency to misclassify positive instances compared to SVM RBF.

The J48 algorithm showcased commendable results, with perfect precision and effective classification of instances, leading to a robust F1-score.

Naive Bayes also showed impressive performance, accurately classifying instances with high precision and competitive scores in recall and F1-score.

References

1. Amrita I, Sen S, Ashwini K. (2022). Analysis of Trend in COVID-19, World Vaccination and Its Side Effects Using Machine Learning. In: Rout RR, Ghosh SK, Jana PK, Tripathy AK, Sahoo JP, Li KC. (eds) Advances in Distributed Computing and Machine Learning. Lecture Notes in Networks and Systems, vol 427. Springer, Singapore. https://doi.org/10.1007/978-981-19-1018-0_27.
2. Khasnis NS, Sen S, & Khasnis SS. (2021, August). A machine learning approach for sentiment analysis to nurture mental health amidst COVID-19. In Proceedings of the international conference on data science, machine learning and artificial intelligence (pp. 284–289).
3. Pal S. Chronic kidney disease prediction using machine learning techniques. Biomed Mater Devices. 2022;1(3) https://doi.org/10.1007/s44174-022-00027-y.
4. Almansour N. Neural network and support vector machine for the prediction of chronic kidney disease: a comparative study. Comput Biol Med. 2019;109:101–11.
5. Chen Z, Zhang X, Zhang Z. Clinical risk assessment of patients with chronic kidney disease by using clinical data and multivariate models. Int Urol Nephrol. 2023;48(12):2069.

6. Ilyas H, Ali S, Ponum M, Hasan O, Mahmood MT, Iftikhar M, Malik MH. Chronic kidney disease diagnosis using decision tree algorithms. BMC Nephrol. 2021;22:273.
7. Lakshmi SV, Meena MK, Kiruthika NS. Diagnosis of chronic kidney disease using random forest algorithms. Int J Res Eng Sci Manage. 2019;2(3):559. ISSN (Online): 2581-5792
8. Kunwar V, Chandel K, Sabitha AS, Bansal A. Chronic kidney disease analysis using data mining classification techniques. In: 6th international conference - cloud system and big data engineering (Confluence); 2016.
9. Chetty N, Vaisla KS, Sudarsan SD. Role of attributes selection in classification of chronic kidney disease patients. In: International conference on computing, communication and security (ICCCS); 2015.
10. Vijayarani S, Dhayanand S. Data mining classification algorithms for kidney disease prediction. Int J Cybern Inf Sci. 2015;4(4):13–25.
11. Salzberg SL. C4. 5: Programs for machine learning by J. Ross Quinlan. Morgan Kaufmann Publishers, Inc., 1993. Mach Learn. 1994;16:235–40.
12. Rubini L, Soundarapandian P, Eswaran P. Chronic_Kidney_Disease. UCI Machine Learning Repository; 2015.
13. Md. Ariful Islam. Ziaul Hasan Majumder Md. Alomgeer Hussein, Chronic kidney disease prediction based on machine learning algorithms. In PMC, Published online 2023 Jan 12. https://doi.org/10.1016/j.jpi.2023.100189.
14. Farjana A, et al. Predicting chronic kidney disease using machine learning algorithms. In: 2023 IEEE 13th Annual Computing and Communication Workshop and Conference (CCWC), Las Vegas, NV, USA, vol. 2023. p. 1267–71. https://doi.org/10.1109/CCWC57344.2023.10099221.
15. Muatu S. Predicting chronic kidney diseases using machine learning algorithms. Int J Emerg Technol Innov Res (www.jetir.org). 2024;11(1):b534–48. Available: http://www.jetir.org/papers/JETIR2401161.pdf. ISSN:2349-5162.
16. Dhanush M, Raj H, Bothra P, Zanwar R, Nagraj S. Chronic kidney disease prediction system using machine learning. International Journal for Research in Applied Science & Engineering Technology. 12(V):1994. (https://www.ijraset.com/), ISSN: 2321-9653.
17. Mayank K, Sen S, Chakraborty P. Implementation of cascade learning using apache spark. In: 2022 IEEE international conference on electronics, computing and communication technologies (CONECCT). IEEE; 2022.
18. Kumar A, et al. Plant leaf disease detection using ResNet. In: Soft computing: theories and applications: proceedings of SoCTA 2022. Singapore: Springer; 2023. p. 867–77.
19. Sahana D, et al. A CNN-based approach for facial emotion detection. In: Soft computing: theories and applications: proceedings of SoCTA 2022. Singapore: Springer; 2023. p. 1–10.
20. Ray S. Understanding support vector machine with an example code. In Analytics Vidhya. Link: https://www.analyticsvidhya.com/blog/2017/09/understaing-support-vector-machine-example-code/. September 2017.

Chapter 12
Skin Disease Prediction Using Machine Learning Techniques

Mrunal Patil, Nikita Shinde, Rakesh Kumar Sharma, Sachin Kadam, and Shivaji Kashte

12.1 Introduction

Skin diseases, collectively known as dermatoses, are serious worldwide health issues impacting millions of individuals of all ages and backgrounds. Skin, which is the most significant organ in the body, acts as an essential defense against external threats and environmental strains. Nonetheless, it can be prone to numerous disorders resulting from various causes, including genetics, environmental factors, autoimmune responses, and infectious agents [1].

There is a substantial effect of dermatological conditions on public health in terms of prevalence and the associated economic burden. Common conditions such as scabies, onchodermatitis, superficial mycoses, pyoderma, psoriasis, leprosy, eczema, acne, filarial lymphedema, pediculosis, endemic treponematoses, and Buruli ulcers have far-reaching impact on millions of people worldwide, significantly affecting their quality of life. Skin diseases profoundly and complexly impact individuals' lives, influencing their mental well-being, daily functioning, and social interactions [2, 3].

Moreover, skin cancers, notably melanoma, pose a significant concern due to their propensity for metastasis and high mortality rates in the absence of timely

M. Patil · N. Shinde · S. Kashte (✉)
Department of Stem Cell and Regenerative Medicine, Centre for Interdisciplinary Research, D. Y. Patil Education Society (Deemed to be University), Kolhapur, MS, India
e-mail: ShivajiKashte.cir@dypgroup.edu.in

R. K. Sharma
D.Y. Patil Medical College, D. Y. Patil Education Society (Deemed to be University), Kolhapur, India

S. Kadam
Sophisticated Analytical and Technical Help Institute, Indian Institute of Technology, Delhi, India

T. Choudhury, A. Katal (eds.), *Machine Learning for Disease Detection, Prediction, and Diagnosis*, https://doi.org/10.1007/978-981-96-4241-0_12

diagnosis and intervention. Epidemiological analysis of global cancer data underscores a worrisome surge in melanoma cases. Around 325,000 new cases with an estimated 57,000 casualties were reported in 2020. If current patterns persist, projections indicate a concerning future, with an anticipated 96,000 casualties and 510,000 instances of skin cancer before the year 2040. This underscores the vital need to address the growing health challenge [4, 5].

Dermatologists have traditionally played a crucial role in diagnosing and treating skin diseases. Their expertise is built on years of clinical experience, visual examination, and occasionally invasive procedures like skin biopsies. However, the current dermatology practice has its challenges. Dermatological diagnoses can be subjective and vary among practitioners, potentially leading to misdiagnoses and delayed treatments. Additionally, the rising occurrence of skin diseases and restricted availability of dermatologists in specific regions stress the healthcare system's capacity to deliver prompt care [6, 7].

In this context, integrating machine learning (ML) techniques into dermatology offers a transformative approach to improving the precision and effectiveness of detecting skin diseases. A subset of artificial intelligence (AI), ML, employs computational algorithms to identify patterns in records and generate predictions despite the need for written code. Numerous fields have experienced notable achievements, such as natural language processing, computer vision, and healthcare, through the application of this technology [8].

Within dermatology, ML has acquired substantial momentum in recent times. The accessibility of ample dermatological datasets, combined with advances in computational power, has facilitated the development of models capable of analyzing images [9]. These models have the potential to provide consistent, objective, and rapid diagnostic assessments, thereby revolutionizing the field of dermatology [10].

Once trained, ML models can detect trends and attributes in wounds that may be imperceptible to the naked eye. They can help dermatologists make more accurate and timelier diagnoses, especially when dealing with uncommon or intricate conditions. Additionally, these models can be employed to conduct preliminary screenings for skin diseases in areas where access to dermatological expertise is limited, thus extending healthcare services to underserved populations [11].

This chapter aims to provide a comprehensive review of the application of ML techniques in dermatology, with a specific focus on skin disease prediction and diagnosis. We will cover the basics of ML, the preparation and refinement of dermatological data, the exploration of various ML algorithms, and the presentation of case studies demonstrating the practical application of these techniques (Table 12.1). Furthermore, we will discuss the ethical considerations and challenges associated with using ML in dermatology and outline prospective avenues for upcoming investigations and therapeutic uses. The convergence of dermatology and ML shows significant potential to enhance patient results, lower medical expenses, and alleviate the impact of skin conditions on people and communities worldwide. By harnessing the capabilities of AI, we can improve access and precision of dermatological testing, ultimately bringing about a new era of data-driven healthcare in the discipline of dermatology.

Table 12.1 Case studies for skin disease prediction using ML

Sr. No.	Type of skin disease	Type of algorithm/ model	Findings/accuracy	References
1.	Digital hair removal technique	DT, SVM, KNN, GLCM	95%	[25]
2.	Melanoma and basal cell carcinoma detection	DT	79.2%	[29]
3.	Skin cancer	SVM	96.25%	[32]
4.	Major skin disease detection	DL	92.62%	[36]
5.	Pre-classification of SL	DT	90%	[37]
6.	Multi-disease detection system	CNN	98.23%	[42]
7.	Eczema disease recognition	RF	88%	[43]
8.	SL classification—Benign and malignant	SVM	85.19%	[48]
9.	Acne object detection	RNN	85%	[49]
10.	NTDs	SVM	70%	[50]
11.	Tinea pedis and tinea corporis	CNN	93.3%	[51]
12.	Diagnosing of oral lesions	CNN	88%	[52]
13.	Differentiation of nevus and melanoma SLs	RNN	95%	[56]
14.	Melanoma detection	DL	85%	[66]
15.	Differentiation between malignant and non-malignant	CNN	0.874 (AUC)	[70]
16.	Skin cancer detector	CNN	91%	[71]
17.	Diagnosis of skin cancer	LinkNet-B7	97.80%	[79]
18.	Skin cancer detection	SVM	92.79%	[80]
19.	Classification of skin diseases	CNN	85.34%	[83]
20.	Skin cancer prediction	SVM	80–100%	[33]
21.	Skin diseases prediction	SVM	98%	[34]
22.	COVID-19	DT	–	[38]
23.	Dermatological diagnosis	DT	93.98%	[36]
24.	Skin disorder	RF	96.1%	[44]
25.	Melanoma lesions	CNN	70%	[53]
26.	Cutaneous Leishmaniasis	DT	–	[106]
27.	Nevi and melanoma detection	CNN	–	[53]
28.	Skin cancer detection	RNN	95%	[57]
29.	Skin disease detection	RNN	–	[58]
30.	SL detection	RNN, CNN	RNN—97.3% CNN—94%	[59]
31.	Skin cancer classification	Squeeze-MNet	99%	[67]
32.	Skin disease classification	CNN	98%	[68]
33.	Benign, melanoma, and keratosis classification	Naïve Bayes classifier	Benign—94.3 Melanoma—91.2% Keratosis—92.9%	[69]

(continued)

Table 12.1 (continued)

Sr. No.	Type of skin disease	Type of algorithm/ model	Findings/accuracy	References
34.	Skin cancer diagnosis	CNN	–	[72]
35.	Melanoma detection	CNN	–	[77]
36.	Melanoma detection	Proposed algorithm	–	[81]
37.	MPXV infection	CNN	–	[84]
38.	Skin disease classification	SVM, DT	95%	[25]
39.	Skin disease recognition	ANN	80%	[87]
40.	Psoriasis	Developed algorithm	–	[90]

12.1.1 Review of Literature

Traditional skin disease diagnosis relies on clinicians visually examining skin lesions (SLs) and conducting tests like dermoscopy and histopathological examination. However, these methods have limitations. There are not enough dermatologists with specialized knowledge, leading to delays in diagnosis. Also, interpretations can vary between clinicians and can be influenced by factors like lighting, affecting accuracy. Additionally, the complexity of skin disease images makes it challenging to distinguish between different conditions, resulting in misdiagnosis or missed diagnosis. This traditional approach lacks efficiency and accuracy, impacting timely and effective patient care [12].

AI holds immense promise in transforming patient care and healthcare systems, offering a spectrum of techniques from ML to deep learning (DL). Its applications span individual risk assessment, medication discovery, and disease identification. For precise illness detection using AI, integrating many medical data sources, including computed tomography scans, genomics, magnetic resonance imaging, and ultrasound, is crucial. Moreover, AI innovations substantially enhance patient experiences within healthcare facilities and streamline the transition to home-based rehabilitation [13].

Heinlein et al. compared the performance of an open-source ensemble algorithm, "All Data are Ext," with dermatologists for melanoma detection using a heterogeneous dataset. The AI showed higher sensitivity (0.921 vs. 0.734) and balanced accuracy (0.798 vs. 0.781) but lower specificity (0.673 vs. 0.828), suggesting its potential as a supportive tool for diagnosing challenging melanoma cases in clinical practice [14].

Abbasi et al. utilized high-throughput ML ensemble methods to predict skin cutaneous melanoma cancers and analyze overall survival probability. Employing eight baseline classifiers, including random forest (RF) and gradient boosting, the proposed ensemble techniques outperformed traditional methods, with the RF classifier achieving an excellent accuracy level of 99%. The research introduces four techniques, namely, stacking, bagging, boosting, and voting, with the voting method achieving a noteworthy accuracy level of 99% [15].

Yadav et al. propose a novel Sailfish-based Gradient Boosting Framework (SbGBF) for precise skin lesion (SL) prediction and segmentation, achieving a high accuracy of 98% with a minimal error rate of 0.02%. The framework effectively addresses challenges related to noise and poor segmentation accuracy while demonstrating superior computational efficiency (3 s) compared to existing models in medical imaging [16].

Lyakhov and colleagues present a novel multimodal neural network system incorporating customized loss functions for skin cancer identification, achieving an impressive 85.20% accuracy across ten diagnostic categories. Their approach, including weighting factors to reduce false negatives, notably improves accuracy by 1.99–4.28 percentage points compared to traditional multimodal systems [17].

Therefore, AI has the potential to revolutionize skin disease prediction with significant improvement over traditional methods.

12.2 Machine Learning in Dermatology

The term AI is a machine, specifically a system of computers, that emulates human cognitive functions. In today's age, the sheer complexity and abundance of data essential for comprehending and predicting clinical outcomes have exceeded the scope of human comprehension. This is where AI steps in, offering valuable assistance within the medical industry. AI includes an extensive set of computational operations that conventionally require human input, such as speech recognition, translation of languages, visual perception, and making decisions.

ML, a component of AI, is dedicated to empowering machines to carry out smart actions by gaining knowledge from their mistakes and enhancing their capabilities through practice. ML operates iteratively, autonomously identifying patterns in data and deriving meaningful insights using statistical models.

DL is a subfield of ML that uses different architectures of artificial neural networks (ANNs) to gather and process intricate attributes from data. This evolution gradually reduces to the particular domain of DL from the more general arena of AI [18, 19].

12.2.1 Fundamentals of Machine Learning

As stated by Arthur Samuel, ML is a "Field of study that gives computers the ability to learn without being explicitly programmed." In simpler terms, it concentrates on developing computer programs that can autonomously improve their performance as they accumulate more experience. In the context of dermatology, ML has emerged as a potent tool to aid dermatologists in achieving precise and efficient diagnoses of skin diseases [20, 21].

12.2.2 Types of Machine Learning

12.2.2.1 Supervised or Predictive Learning in Dermatology

Supervised learning is an ML paradigm wherein a computer system undergoes training using a dataset with well-defined and labeled examples. In dermatology, for instance, when determining the malignancy of an SL, the computer system sees many pictures of SLs, all of which have been classified as normal or cancerous. Following this training, the system's proficiency is assessed by presenting it with fresh, unlabeled images, requiring it to decide whether to label them as cancerous or not based on the information it learned during the learning stage [22].

The most notable application of supervised learning in dermatology is SL classification. Dermatologists have historically relied on visual inspection; however, this process can be subjective and prone to errors [23].

In clinical practice, skin cancer diagnoses are determined using the ABCDE guideline:

- A—Asymmetry.
- B—Border irregularity.
- C—Color of lesion.
- D—Diameter of lesion.
- E—Enlarging lesion.

The ABCDE model and traditional clinical procedures for skin cancer diagnosis have limitations and a relatively high error rate. Dermoscopy has emerged as a critical tool that captures high-resolution images, offering improved accuracy in SL evaluation and helping to mitigate diagnostic errors [24].

Ahammad et al. utilized a supervised learning methodology, applying ML classifiers such as decision tree (DT) and support vector machines (SVMs) to categorize dermatological photographs into distinct groups, such as melanomas, carcinoma of the squamous cells, carcinoma of the basal cells, melanocytic nevus, dermatofibroma, vascular lesion, and a condition called actinic keratosis or benign keratosis. Thanks to this supervised learning technique, they were able to instruct the models with labeled data, leading to effective skin disease prediction and contributing to a more efficient and automated diagnosis in dermatology. Additionally, they introduced a digital hair removal technique, followed by an image enhancement and segmentation method (as depicted in Fig. 12.1), which shows 95% accuracy. This comprehensive approach holds significant promise for expediting treatment planning and enhancing patient care within the field of dermatology [25].

Supervised learning can also predict disease outcomes based on patient data. For instance, researchers have used ML to forecast an individual's likelihood of adverse events with chronic skin diseases like psoriasis. Through examination of clinical variables and patient records, these algorithms can determine who is more likely to experience severe symptoms or comorbidities [26].

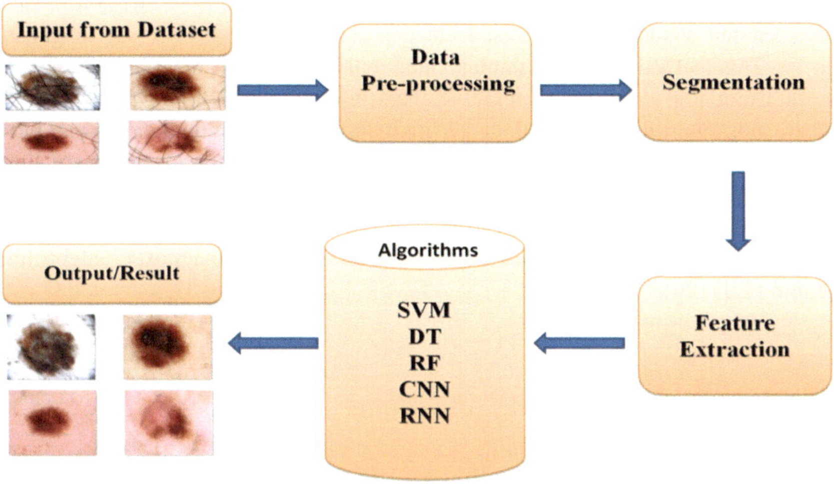

Fig. 12.1 Flow chart of Machine Learning process for skin Lesion processing. (Adapted from Ahammad et al. [26] Creative Common license 4.0)

12.2.2.2 Unsupervised or Descriptive Learning in Dermatology

Unsupervised learning refers to the ML paradigm when the result data cannot access a labeled dataset used for training. Unlike supervised learning, which categorizes data based on labeled examples, unsupervised learning classifies input data solely based on their intrinsic characteristics. This approach enables a more exploratory analysis of the underlying data distribution, uncovering patterns that might otherwise remain undiscovered [22, 27].

12.2.2.3 Reinforcement Learning (RL) in Dermatology

RL falls within the realm of ML and centers on the process by which agents acquire the ability to make choices and execute actions in response to intermittent rewards or penalties received from their surroundings. This approach is particularly applicable in scenarios where traditional data-driven methods are limited. In RL, agents learn to maximize predefined rewards by interacting with their environment. It is a valuable concept in fields like self-driving cars and cognitive computing, which aim to simulate human thought processes and, in the broader context, artificial general intelligence (AGI). Success in RL hinges on accurately modeling the environment in which the agent operates, allowing it to adapt and make informed decisions to achieve its goals [27, 28].

Barata et al. utilized reinforcement learning techniques to explore how AI-driven diagnostic decision support systems for skin cancer diagnosis could be enhanced. They introduced a unique approach by employing rewards and penalties that were not evenly distributed, drawing from expert-generated charts to strike a balance between the advantages and disadvantages associated with different diagnostic errors. This methodology was implemented within a reinforcement learning framework. Compared to supervised learning, their reinforcement learning model brought about substantial improvements in sensitivity when detecting melanoma and basal cell carcinoma. Simultaneously, it helped mitigate AI overconfidence, maintained accuracy, and improved the precision of dermatologists' diagnoses, resulting in improved decisions regarding the optimal course of action. This showcases the possibility of incorporating individual inclinations into diagnostic algorithms (Table 12.1) [29].

12.2.3 Machine Learning Algorithms

Different ML algorithms (depicted in Fig. 12.2) can be used for dermatological issues, depending on the data and the particular problem being addressed. Dermatology primarily involves categorizing data into different groups, particularly for predicting skin disease types using features extracted from images. Various classification approaches are adapted to the specific data and purpose, and ML algorithms are significantly contributing to the progress of dermatology by automating the categorization of SLs into different disease groups. Such progress substantially improves the accuracy and speed of diagnosis, representing a notable shift in the field [25].

12.2.3.1 Support Vector Machines

SVMs serve as a helpful tool for dermatologists as they help doctors determine whether a skin issue is a significant concern. They do this by drawing a line, but it's not an ordinary line; it's a special line that helps them separate different skin issues. SVMs are good at telling the difference between harmless and dangerous skin problems, especially when looking at skin images. This helps doctors catch dangerous skin issues like melanoma early, essential for saving lives. In a nutshell, SVMs are

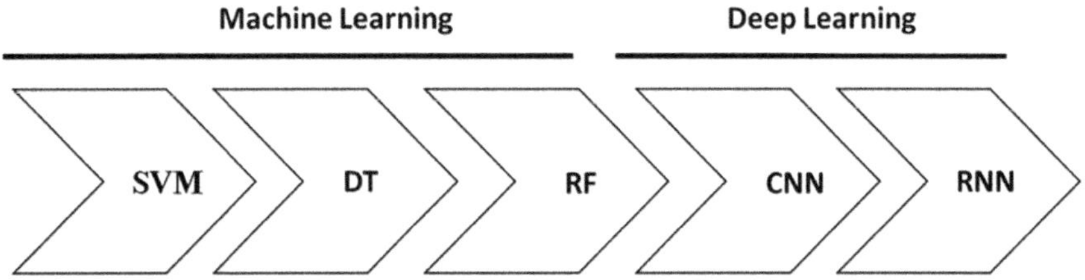

Fig. 12.2 ML Algorithms used in Dermatology

like helpful guides for dermatologists, making it easier to diagnose and classify skin problems [30, 31].

Skin cancer, driven by DNA damage, presents a significant threat, underscoring the critical importance of early detection. Recognizing the gravity of this issue, scientists have devised a method employing multi-support vector machine (MSVM) classification. This method harnesses two powerful techniques, the ABCD and the MSVM models, for feature extraction. Notably, this approach achieved an impressive accuracy rate of around 96.25% [32].

Bedair et al. introduced an innovative approach by combining a convolutional neural network (CNN) with SVM to build a Mobile Android Application for skin disease diagnosis. Their system achieves impressive accuracy rates ranging from 80% to 100% across various skin diseases, surpassing other methods such as Naive Bayes and CNN. By leveraging SVM classification, the application aims to offer early detection and treatment recommendations for skin diseases, utilizing the widespread accessibility of mobile devices to improve healthcare delivery. This research highlights the capabilities of algorithms for ML, particularly SVM, in advancing early diagnosis and treatment of skin diseases, particularly in regions with limited access to medical facilities [33].

Abinash et al. conducted a comparative study to predict skin diseases using SVM and RF algorithms, employing a sample size of $N = 10$. The SVM algorithm achieved an accuracy of 98%, while RF attained an accuracy of 97.5%. Statistical analysis, including independent sample T-tests with a confidence interval of 95%, confirmed the superior accuracy of SVM over RF for skin disease prediction, providing valuable insights for improving recognition systems in dermatology based on image datasets. Despite the slower processing rate of SVM due to large databases, its higher accuracy suggests the potential for enhanced predictive models, addressing the challenge of accurately performing algorithms in skin disease prediction applications [34].

12.2.3.2 Decision Trees

DTs are interpretable models that can be used for dermatological diagnosis. They build a structure resembling a network of branches, in which each branch stands for a choice made in response to input features. DT has been employed to diagnose various skin diseases, including eczema and psoriasis [35].

Chang et al. employed a DT method as part of their study to construct predictive models for dermatology. However, when combined with sensitivity analysis, the study found that this DT method resulted in lower accuracy (80.33%) compared to the Neural Network Model (92.62%). Despite this, the research highlights the potential of AI classification technology to improve dermatological diagnoses, aiding physicians in reducing medical waste and enhancing healthcare quality [36]. *Ozkan* et al. ML methods pre-classified SLs into normal, abnormal, and melanoma categories. The results showed a promising classification accuracy rate of 90.00% with DT. This research demonstrates the potential for a valuable dermatological decision support system, enhancing the accuracy of SL diagnoses and assisting healthcare professionals in their clinical assessments [37].

Chrimes et al. conducted a preliminary investigation highlighting the absence of a publicly available chatbot for assessing COVID-19 severity and related comorbidities. They created a DT utilizing binary classification, stratified based on age, body system, viral infection, comorbidities, and manifestations. This tree comprised 212 nodes and encompassed 63,360 different scenarios. Notwithstanding its limitations, this prototype offers significant insights into the requisite data for proficient decision-making support in COVID-19 instances, underscoring the intricate nature of severe COVID-19 manifestations and the criticality of stratified frameworks [38].

Chang et al. conducted a study comparing the effectiveness of two data mining categorization methods, DT and multi-layer neural networks, in predicting skin diseases. Utilizing the WEKA data mining tool, they performed experimental analyses using N-fold cross-validation (N = 10) for both classifiers. The results showed that multi-layer perceptron (MLP) achieved a higher accuracy level of 96.99%, while the J48 decision tree attained a 93.98% accuracy level. This highlights the superior predictive performance of MLP over J48 in diagnosing skin diseases, emphasizing its potential to enhance healthcare systems' predictive modeling capabilities [36].

Ramezankhani et al. conducted a study to investigate the influence of environmental elements on Cutaneous Leishmaniasis incidence from 2007 to 2015. Using a decision tree model, they determined that certain conditions such as wind speeds below 14 m/s, altitudes ranging from 1234 to 1810 m above sea level, Normalized Difference Vegetation Index (NDVI) below 0.12, rainfall less than 1.6 mm, and air temperatures exceeding 30 °C were associated with a seasonal CL incidence of 163.28 per 100,000 individuals. In contrast, when wind speed was below 14 m/s, altitude was below 1810 m, and NDVI was above 0.12, the average seasonal incidence decreased significantly to 2.27 per 100,000 individuals. These quantitative results highlight the importance of considering environmental variables in CL control and prevention programs [39].

12.2.3.3 Random Forests

RFs are aggregate learning methods that achieve improved accuracy and mitigate overfitting by combining multiple DTs. In dermatology, RF was previously used to improve the accuracy of dermatological condition classifiers [40].

When classifying a new object, it passes this object through all trees within the forest. Every tree generates a categorical result, and the forest makes its final decision by selecting the class that receives the most votes among the individual tree outputs [41].

Shinde et al. developed a robust multi-disease detection system for skin conditions, including psoriasis, ringworm, atopic dermatitis, and nail growth, using AI and CNN methods. Their approach, which combines feature extraction via CNN and classification through the RF algorithm, attained an exceptional precision of 98.23% following 100 epochs. This research highlights the capacity of AI-driven techniques for the enhancement of precision in skin disease diagnosis, with plans to expand the system to cover even more skin diseases [42].

Eczema is a prevalent skin disease marked by itching and inflammation of the skin. Accurate diagnosis is crucial for effective management. A study showcased a machine vision-based system for skin disease recognition (Eczema), utilizing characteristic disease features. The RF classifier achieved an 88% accuracy, with potential for improvement through enhanced image quality and expanded dataset size [43].

Kalaivani et al. conducted a study addressing the challenge of predicting skin disorders by introducing a novel ensemble technique combining two data mining approaches, significantly improving accuracy to 96.1%. Leveraging a Dermatology dataset, the ensemble DL method effectively categorized skin diseases into seven categories, outperforming single-classifier approaches and conventional methods. Furthermore, the study highlights the importance of integrating predictive heterogeneous ensemble models for multiclass dermatitis, demonstrating superior performance compared to existing literature, with the highest accuracy achieved using the RF Classifier at 96.1% [44].

The Syahab et al. study employed logistic regression (LR) and RF algorithms to forecast extreme UV conditions based on UV Index data. Using historical UV index data from Palu City, Central Sulawesi, the RF model achieved an impressive accuracy of 0.997 and an F1-score of 0.947, while the LR model achieved an accuracy of 0.958 and an F1-score of 0.996. These outcomes show the superior predictive ability of the RF model in extreme UV index prediction [45].

12.2.4 Deep Learning Architectures

DL models, including CNNs and recurrent neural networks (RNNs), are components of the larger ML domain and have significantly contributed to the recent advancements in dermatology. These advanced models possess the ability to autonomously acquire and depict intricate connections within dermoscopic images and patient information [46]. In particular, CNNs have been applied in skin disease classification, disease detection, and feature extraction. Their success in these applications has demonstrated the potential of DL to transform dermatological practice by providing highly accurate and efficient diagnostic tools [47].

12.2.4.1 Convolutional Neural Networks

The CNN algorithm ushered in a transformative era of dermatology, particularly in image-based diagnosis. These DL models excel at processing visual data and have become prominent tools for SL classification and melanoma detection tasks.

Seeja et al. employed a CNN-based algorithm for image segmentation, followed by feature extraction. This feature was then used as input for ML classifiers to diagnose plus differentiate skin images as melanoma or benign lesions. The results demonstrated the proficiency of this approach, attaining a Dice coefficient of 77.5% in the context of image segmentation, while the SVM classifier achieved an accuracy of 85.19%. U-Net segmentation was the most efficient approach in a DL environment, significantly enhancing classification performance. Accurate segmentation can improve diagnostics and treatment planning [48].

Researchers built an AI-generated system named "AcneDet," involving two models created to automatically identify acne-related objects and rate the severity of the condition based on pictures of oneself taken with digital devices. The R-CNN model attained mean average precision of 0.54 mAP for detecting acne-related objects, while the LightGBM model demonstrated a mean accuracy of 0.85 for severity grading. This study aims to advance AI systems for acne diagnosis, enhancing patient understanding and supporting medical professionals [49].

Yotsu et al. utilized a dataset comprising 1709 images from 506 patients in Ghana, collected via digital health tools and teledermatology efforts. They employed two CNNs, ResNet-50, and VGG-16, to assess the efficacy of different DL algorithms in diagnosing specific skin neglected tropical diseases (NTDs). Both models exhibited a diagnostic accuracy of over 70%, with improved performance observed as the training dataset size increased. Notably, the ResNet-50 model outperformed the VGG-16. Furthermore, training the model with polymerase chain reaction confirms that Buruli ulcer resulted in a 1–3% hike in accuracy compared to training with unconfirmed cases, highlighting the potential for enhancing diagnostic accuracy in NTDs [50].

A CNN was created by Nigat et al. to categorize common fungal skin diseases, addressing the challenge of limited access to costly diagnostic technologies in resource-limited regions like sub-Saharan Africa. Their model achieved an impressive 93.3% accuracy, outperforming other CNN architectures. This research offers a promising foundation for automated image-based screening systems in dermatology, potentially enabling timely treatment and expanding to include more skin diseases in the future, as well as knowledge-based and clinical decision support systems [51].

When identifying the distinction between oral lichen planus (OLP) and non-OLP lesions from images, Achararit and colleagues found that CNN algorithms performed well. The Xception model was the most successful of the models they tested, outperforming the others in terms of score and overall accuracy. According to their research, CNN models can attain diagnostic accuracy levels between 82% and 88%; the Xception model is the best performer in accuracy and F1-score. This implies that CNNs—specifically, the Xception model—have the potential to improve the use of image data in the diagnosis of OLP and non-OLP lesions [52].

Collenne et al. designed an ANN to classify nevi and melanoma lesions based on handcrafted expert features, achieving an AUC of 0.873, a sensitivity of 90%, and a specificity of 67%. In comparison, an array of seven cutting-edge CNNs outperformed the ANN with an AUC of 0.938, a sensitivity of 91%, and a specificity of 82%. This highlights the superior performance of CNNs in melanoma detection, emphasizing their potential in medical image analysis and collaboration with dermatological expertise [53].

CNN demonstrated significant potential in assisting dermatologists with interpreting clinical images for cutaneous tumor classification. CNN's overall precision and kappa score were impressive when classifying different tumor types. When aided by CNNs, dermatologists showed higher accuracy and kappa scores compared to unassisted dermatologists, particularly benefiting those with less experience. Additionally, CNN assistance significantly improved sensitivity and specificity at the binary classification level. These findings highlight the potential of CNNs to enhance dermatologists' accuracy and acceptance of this valuable diagnostic tool in clinical practice [54].

12.2.4.2 Recurrent Neural Networks

The RNN is part of DL and is designed to analyze sequential and time series data. These DL models find applications in tasks such as image captioning, language translation, speech recognition, natural language processing, and powering services such as Google Translate, Siri, and voice search. In contrast to CNN, RNNs possess a "memory" feature, enabling them to utilize previous inputs to shape present inputs and outputs. While CNNs assume independence between inputs and outputs, RNNs rely on prior sequence elements to determine their output, but they cannot consider future events in their predictions [55].

The DermoDeep system is presented as a novel method for differentiating between nevus and melanomas using a perceptual-oriented color space in a recent study. This system achieves impressive results in categorizing SLs without needing pre-and final processing stages by utilizing visual traits and a multi-layered DL neural network model that incorporates RNNs. With a sensitization rate of 93%, a specificity value of 95%, and an astounding area under the curve (AUC) value of 0.96, this innovative method not only outperforms current approaches in precision and effectiveness but also shows promise for further advancements in dermatology. Moreover, it emphasizes how DL algorithms can be applied in a wider range of image domains outside of dermatology [56].

In the study by Zhang et al., a robotic computer-assisted tactic utilizing a gated recurrent unit (GRU) network and an enhanced Orca Predation Algorithm (OPA) was proposed for early skin cancer detection. Experimenting with the HAM10000 dataset, the GRU/IOPA system demonstrated superior performance with sensitivity (0.95), specificity (0.97), Positive Predictive Value (0.95), Negative Predictive Value (0.96), and accuracy. The study highlights the potency of the GRU/IOPA process compared to traditional approaches, showcasing its potential for improving skin cancer diagnosis [57].

Using the RNN method, Bu et al. introduced a fresh classification system for skin diseases utilizing cytology and pathology, evaluating its predictive accuracy against the "International Classification of Diseases, Tenth Revision (ICD)-10." The novel taxonomy exhibited superior F1 and Kappa scores in predictive accuracy, notably excelling in Project 4, with statistically significant enhancements ($P = 0.025$ and 0.005, respectively). This research underscores the promising prospects of the newly proposed taxonomy in dermatological classification, demonstrating its enhanced predictive capability compared to the traditional ICD-10 framework [58].

Kumar et al. introduced a hybrid CNN and RNN model for SL detection using the HAM10000 dataset in their study. The model showcases efficient classification performance by achieving an accuracy of 94% for CNN and 97.3% for RNN. This innovative approach combines CNN's spatial data processing with RNN's temporal modeling, demonstrating promising potential for advancing SL diagnosis in dermatology. The proposed effort aims to create a precise and effective system for detecting SLs, providing dermatologists with valuable support in identifying and treating diverse skin conditions [59].

The algorithm selection is determined by data accessibility, problem intricacy, and the desired degree of interpretability. Researchers and practitioners in dermatology often experiment with different algorithms to identify the most suitable approach for a specific diagnostic task [19].

In summary, the fundamental principles of ML provide a strong framework for addressing complex challenges in the field of dermatology. The detection and forecasting of skin diseases could be made much more accurate and efficient by combining supervised learning methods with a range of ML algorithms.

12.3 Data Preparation and Feature Engineering

Data preparation and feature engineering are critical steps in harnessing the power of ML in dermatology. The standard and relevance of the data employed for training and testing significantly impact the effectiveness of ML models. Furthermore, feature engineering entails the choice and extraction of valuable characteristics from the dataset, which can significantly impact the model's capacity to provide precise predictions [60].

12.3.1 Quality Data Sources in Dermatology

High-quality data sources are imperative in training robust ML models for dermatology. Multiple dermatological datasets have become available, aiding research and model advancement. The caliber and pertinence of data employed in ML model training and testing profoundly affect their effectiveness. Access to top-notch dermatological datasets is crucial for developing and assessing ML models in dermatology, forming the foundation for training, testing, and validating algorithms aimed at predicting and diagnosing skin conditions. Recent times have witnessed the emergence of significant dermatological datasets, playing a vital role in advancing ML applications in this domain [61, 62].

Some common datasets are as follows.

12.3.1.1 Dermatology Society Dataset

The Dermatology Society dataset stands as another valuable asset in the dermatology field. It encompasses clinical data and images of diverse skin diseases, spanning a wide spectrum of conditions encountered in dermatological practice. Dermatologists have harnessed this dataset to create ML models for diagnosing diseases and recognizing SLs [47].

Clinical data alongside images in this dataset enables researchers to explore the fusion of patient history and visual information, potentially improving diagnostic accuracy. ML models trained on this dataset hold the potential to aid dermatologists in more effective and efficient identification of skin diseases. This dataset comprises clinical data and images of various skin diseases, making it valuable for both research and clinical purposes, particularly in tasks such as disease diagnosis and lesion recognition [25, 63].

12.3.1.2 The International Skin Imaging Collaboration Dataset

The International Skin Imaging Collaboration (ISIC) dataset encompasses a broad compilation of high-resolution dermoscopic images featuring various SLs, encompassing both melanoma and benign ones. It has found extensive utility in creating ML models to differentiate SLs. The ISIC dataset stands out as a widely acknowledged and substantial resource in dermatology research. It consists of diverse high-resolution dermoscopic images spanning melanoma, nevi, and other non-malignant lesions. With its numerous annotated images, this dataset emerged as essential to the development of ML models for classification tasks involving dermatological lesions, melanoma detection, and extracting significant features [64, 65].

Acute SLs, such as burns or insect bites, require prompt diagnosis and treatment. ML can aid in recognizing the category and seriousness of these lesions. Two DL frameworks, the Lesion Indexing Network (LIN) and Lesion Feature Network (LFN), were introduced by Li et al. to address important issues in the image processing of lesions, such as the case of segmentation obtaining of dermoscopic attributes and grouping. LIN combines segmentation and classification using deep convolutional networks, achieving superior results compared to existing methods. LFN, dedicated to dermoscopic feature extraction, sets a potential benchmark with its impressive average precision and sensitivity, demonstrating its efficacy in addressing this critical task. Their evaluation of the ISIC 2017 dataset underscores the great potential of these frameworks in dermatology [66].

The ISIC database was used to create Squeeze-MNet, a simple deep-learning method for categorizing skin cancer that is accurate, lightweight, and versatile. It integrates a pre-weighted MobileNet DL model with a Squeeze algorithm for preprocessing-related digital hair removal. The model achieved an average precision of 99.76% for benign and 98.02% for malignant diagnoses, with a dataset size reduction of 66%. Incorporating a hair removal algorithm increased skin cancer detection accuracy to 99.36%, with an area under the receiver operating curve of

98.9% demonstrating its potential for accurate and efficient skin cancer diagnosis [67].

Bajwa et al. emphasized the importance of disease taxonomy in improving classification accuracy, underscoring the need for comprehensive metrics in assessing AI-based classifiers in dermatology. This study explores the potential of DL in classifying numerous skin diseases, leveraging DermNet and ISIC Archive datasets. On DermNet, 80% accuracy and 98% AUC for 23 diseases and 67% accuracy and 98% AUC for more than 600 sub-classes were achieved. Additionally, on the ISIC Archive, all seven infections were classified with 93% average accuracy and 99% AUC. These findings demonstrate DL's capability to classify skin diseases with superior repeatability and accuracy close to human levels [68].

In a case study by Balaji et al., the ISIC 2017 dataset was utilized to evaluate a novel approach to skin disease detection and classification. Their method involved a dynamic graph cut algorithm for SL segmentation, followed by classification using a Naïve Bayes classifier. Focusing on improving segmentation accuracy and classification performance, their method achieved an accuracy of 94.3% for benign cases, 91.2% for melanoma, and 92.9% for keratosis, outperforming other contemporary procedures by large margins. This study underscores the importance of accurate SL segmentation and classification in aiding medical practitioners with timely diagnosis and treatment, leveraging advanced ML techniques on publicly available datasets like ISIC for impactful healthcare applications [69].

Researchers have leveraged the ISIC dataset to train CNN, which is capable of differentiating between malignant and benign lesions. For instance, Codella et al. used the ISIC dataset to develop a DL model that achieved remarkable performance in melanoma recognition. The dataset's size and diversity make it a valuable resource for tackling complex dermatological challenges [70].

12.3.1.3 HAM10000 Dataset

The HAM10000 dataset is a comprehensive collection of approximately 10,000 dermoscopic images of pigmented SLs. The dataset encompasses various dermatological conditions, including basal cell carcinoma, melanomas, nevi, and more. It has played a vital role in propelling the progress of ML models in the detection and categorization of melanoma [65].

Alam et al.'s study highlights the pressing challenge of efficient skin cancer detection in healthcare, emphasizing the critical importance of early diagnosis for better treatment outcomes. They propose a novel DL-based skin cancer detector, addressing data imbalance with data augmentation using the Skin Cancer MNIST: HAM10000 dataset. The results show the superior performance of the designed framework with an accuracy of 91%, an F1-score of 88.1%, and a receiver operating characteristic curve value of 0.95. This promising approach has the potential to significantly impact disease identification, potentially saving lives, reducing unnecessary procedures, and cutting costs in healthcare [71].

Another study tackled the global skin disease problem through a comprehensive approach involving digital hair removal, Grabcut segmentation, and feature extraction. Using ISIC 2019 and HAM10000 datasets, a model achieved high accuracies of 95%, 94%, 93%, and 97%, 95%, and 95%, respectively. Remarkably, it excelled with HAM10000 data and outperformed existing methods in balanced data scenarios [25].

The study of Alwakid et al. explores the efficacy of DL algorithms in diagnosing skin cancer using the HAM10000 dataset. The proposed CNN achieved remarkable results by employing Enhanced Super-Resolution Generative Adversarial Networks (ESRGAN) for the segmentation and enhanced images. With an accuracy of 0.86, a precision of 0.84, a recall of 0.86, and an F-score of 0.86, the CNN-based model exceeded previous studies significantly. Using ESRGAN for image enhancement and adopting various architectures, including Resnet-50, contributed to the model's success. Comparisons with existing works demonstrated the robustness and effectiveness of the proposed method, showcasing its potential for accurate and timely skin cancer diagnosis. Further experiments are suggested to validate the model's performance on a larger and more diverse dataset [72].

While these datasets provide a solid foundation for ML research in dermatology, challenges exist in collecting, curating, and annotating data. Data imbalance, where certain skin diseases are underrepresented, is common and can lead to biased models. Efforts are ongoing to address these challenges and create more diverse and representative datasets [73].

12.3.2 Data Pre-processing

Data pre-processing is crucial to guarantee that data is in a suitable format for ML algorithms. Common data pre-processing techniques in dermatology include the following.

12.3.2.1 Data Augmentation

Data augmentation encompasses various techniques to enlarge and improve training datasets, which in turn aids in developing more proficient DL models. These techniques cover a wide range of approaches, such as adversarial training, neural style transfer, generative adversarial networks (GANs), geometric transformations, color space adjustments, kernel filter application, image mixing, random erasure, and meta-learning [74].

Aggarwal et al. focused on diagnosing dermatological conditions such as rosacea, acne, impetigo, psoriasis, and atopic dermatitis using AI and ML, along with data augmentation techniques. The outcomes demonstrated noteworthy enhancements in various statistical measures, including F1-score, specificity score, and sensitivity, with the implementation of data augmentation. Additionally, the AUC elevated for all five dermatological manifestations, with the highest increase observed for atopic dermatitis (AUC increased by 0.18). This DL-based approach demonstrated the potential to differentiate dermatological images effectively and improve accuracy in disease diagnosis, with AUC values ranging from 0.87 for psoriasis to 0.97 for acne [75].

An algorithm based on fusion networks has been presented by Wan et al. to identify colored lesions of the skin, addressing the issue of low classification accuracy due to imbalanced data distribution. They conducted image pre-processing and augmentation to mitigate this problem and employed various fusion strategies for validation. Their approach attained an impressive accuracy rate of 92.1% and an AUC of 95.3%. This underscores its practical applicability in aiding clinicians in diagnosing pigmented skin diseases and affirms its versatility and accuracy [76].

Abdelhalim et al. introduced SPGGAN-TTUR, a novel GAN-based data augmentation method, to generate realistic and diverse 256 × 256 SL images for melanoma detection using CNNs. Compared to traditional GANs, SPGGAN-TTUR addresses the challenges of unstable training and varied lesion characteristics. The evaluation of the HAM10000 dataset shows statistically significant sensitivity enhancements of 5.6% and 2.5% over non-augmented and augmented counterparts, respectively, with general improvements for all classes, particularly 13.8% and 8.6% for melanoma. This approach holds promise for enhancing automated SL detection and supporting dermatologists in melanoma diagnosis [77].

12.3.2.2 Noise Reduction

Removing noise from images and data can improve the quality of input features. This is particularly important in dermatology, where image quality may vary. In dermoscopy, magnified subcuticular visualization enhances the detection of features not visible to the naked eye. While factors like hair and markings can disrupt image evaluation, removing such "noise" improves diagnostic accuracy without affecting the diagnosis itself [78].

Akyel et al. introduced an innovative approach for skin cancer diagnosis, emphasizing the importance of noise reduction in skin images containing various artifacts. They created a unique noise dataset and applied the LinkNet-B7 method, achieving a 6% higher success rate than previous methods. Their method contributed to a more precise diagnosis of skin cancer with its remarkable 95.72% precision in eliminating background noise and 97.80% precision in lesion categorization [79].

Wei et al. developed an ideal and automated methodology for diagnosing skin cancer from dermoscopy images. Their approach encompasses noise reduction, thresholding, feature extraction, and optimization using an altered thermal exchange optimization algorithm. The proposed technique achieved an impressive accuracy of 92.79%, making it a reliable and efficient tool for skin cancer diagnosis [80].

Gupta et al. proposed an imaging method for accurate lesion segmentation from dermoscopic images, crucial for precise melanoma detection. The method utilized in painting, edge detection, and intensity-based thresholding, followed by mathematical morphological operations to remove noisy pixels post-segmentation. This approach was computationally efficient, allowing for potential real-time application. Evaluation on a dataset yielded an average Jaccard index of 89.2383% and a correlation coefficient of 92.5271%, demonstrating the proposed method's effectiveness in accurately segmenting lesions. This noise reduction technique offers promising implications for improving computer-aided melanoma diagnosis [81].

12.3.2.3 Data Splitting

Data splitting involves partitioning a dataset into distinct subsets, typically comprising training, validation, and test sets. This separation is fundamental in ML to evaluate model performance accurately. Utilizing cross-validation techniques during this process enhances model robustness by repeatedly partitioning the data into subsets, training on different combinations, and validating across multiple folds, thus providing a more comprehensive assessment of a model's generalization capabilities. Ultimately, this approach helps in building models that perform well on unseen data [82].

Srinivasu et al. employed a dataset comprising seven skin diseases and used a random function to split it into training and validation sets. Data augmentation was implemented to tackle the problem of imbalanced data, resulting in an accuracy of 85.34% for their MobileNet V2 and long short-term memory-based classification model, surpassing other methods when tested with real-time Kaggle images. The model's computational efficiency and stateful information maintenance enhance its performance, making it a valuable tool for skin disease diagnosis, particularly when integrated with a mobile application for the early identification of skin conditions [83].

Thieme et al. developed an image-based deep CNN (MPXV-CNN) to recognize typical SLs produced by MPXV. They developed a dataset of 139,198 SL photos divided into evaluation and training/validation cohorts. The photographs included 676 MPXV images and 138,522 non-MPXV images. MPXV-CNN showed a sensitivity of 0.83 and 0.91, a specificity of 0.965 and 0.898, and an AUC of 0.967 and 0.966 in the validation and testing cohorts, respectively. This strong accuracy in classification throughout a range of body locations and shades of skin indicates that MPXV-CNN may be useful in mitigating the MPXV outbreak [84].

Ahammed et al. conducted a comprehensive case study on skin disease classification, employing advanced digital image processing techniques and ML algorithms to automate diagnosis and expedite treatment planning. Their multi-phased approach included customized digital hair removal, data splitting, lesion segmentation using Grabcut, and feature extraction using the gray-level co-occurrence matrix (GLCM) and statistical features. Three classifiers, SVM, K-nearest neighbor (KNN), and DT, were skilled and evaluated on two benchmark datasets, ISIC 2019 challenge and HAM10000, with data imbalance addressed through random oversampling. The study reported notable quantitative results, with SVM achieving an average accuracy of 95% for ISIC 2019 and 97% for HAM10000, demonstrating superior performance on the latter dataset. Despite limitations related to automatic lesion segmentation accuracy, the proposed model outperformed existing methods, highlighting its potential to enhance early detection and promote skin health among patients [25].

12.3.2.4 Feature Engineering

Feature engineering involves selecting relevant attributes from the data to aid in disease prediction. These features in dermatology include texture, color, and shape characteristics. Researchers have explored various feature extraction methods.

12.3.2.4.1 Texture Analysis

An automated skin disease detection system categorizes three conditions using texture analysis of medical images based on the GLCM. GLCM is a robust and efficient tool for extracting traits from photographs. It also quantifies co-occurrence probabilities among gray levels in various directions. Extracted GLCM features include the following:

1. Energy: Sums squared GLCM elements, ranging from 0 to 1, with a value of 1 for constant images.
2. Correlation: Varies between −1 and 1, signifying a positive or negative correlation.
3. Contrast: Measures intensity contrast between pixels and their neighbors across the image.
4. Homogeneity: Evaluates the proximity of element distribution in GLCM to its diagonal, ranging from 0 to 1, where 1 signifies diagonal-based color histograms [85].

Wei et al. have devised an efficient method to identify specific skin conditions, such as psoriasis, herpes, and dermatitis, using texture analysis and color features. Their approach, which involves pre-processing, GLCM-based segmentation, and SVM classification, achieved an impressive recognition rate exceeding 90%, significantly enhancing diagnosis accuracy for these skin conditions. The introduction of texture and color features, along with local vertical image segmentation, contributed to this notable improvement [86].

Islam et al. studied skin disease recognition using neural networks based on texture analysis. They demonstrated the feasibility and effectiveness of feature extraction, well-defined segmentation, and classification methods for neural network-based diagnosis of skin diseases. While significant progress has been made in automating medical diagnosis, this study aims to leverage widely available devices and historical skin disease image datasets to provide practical, simpler, and quicker diagnosis in deprived regions. Prospective areas of study involve the creation of algorithms for detecting different medical conditions, improving efficiency as a whole, and reducing computational time. The proposed computer-aided diagnosis system utilizes image processing, feature extraction, and classification steps, employing ANNs as classifiers. ANNs learn disease symptom patterns, enabling faster diagnosis and recognition compared to human physicians and facilitating immediate treatment based on detected symptoms [87].

12.3.2.4.2 Color Histogram

Color histograms represent the distribution of pixel colors in an image. They have been used to characterize SLs based on their color composition. To analyze and determine the count of colors within the SL, a sequential approach was employed using Euclidean distance calculations in the red, green, and blue (RGB), luminance, chrominance (YCbCr), and hue, saturation, and value (HSV) color hues [88].

Based on the number of colors present in the SL under examination, the color specifications of the ABCD system assign a rating ranging from 1 to 6. This rating scheme takes into account six primary colors: black, blue-gray, light brown, dark brown, white, and red [89].

Dhandra et al. proposed a system for diagnosing the type of psoriasis for treatment based on color histogram features, aiming to maintain psoriasis under medication. They converted RGB to HSV color space and used the H and S planes to represent color histograms with 11 bins each. A total of 121 features were considered and normalized by the number of pixels in each bin. They employed an SVM with an RBF kernel for multiclass classification. Testing on 180 medical images showed an efficiency of 92.2% accuracy, confirming the method's effectiveness.

Arunkumar et al. conducted a study on detecting psoriasis-affected skin using RGB color space analysis, which proved effective in distinguishing affected skin from normal healthy skin. Their algorithm differentiated affected areas from non-affected ones based on color features such as redness, greenness, and blueness. Through color histogram analysis of over 50 samples, they aimed to differentiate affected skin by analyzing erythema. Unlike subjective visual analysis by dermatologists, their objective model minimized treatment deviation influenced by external factors. Based on the RGB histogram mean and standard deviation, the model provided a reliable method to differentiate healthy skin from psoriasis-affected skin [90].

12.3.2.4.3 Shape Descriptors

Shape descriptors like circularity and symmetry can quantify the shape characteristics of SLs, aiding in their differentiation.

Chakravorty et al. investigated asymmetry as a key characteristic for early melanoma diagnosis, drawing inspiration from medical algorithms such as ABCD and CASH. They utilized Kullback–Leibler divergence of color histograms and the structural similarity metric to quantify irregularities in lesion color distribution and structure, aligning with dermatological assessments. Performance evaluations with various classifiers on the PH2 dataset showcased superior asymmetry classification compared to existing literature, highlighting the effectiveness of their approach. The proposed methodology outperformed previous results across multiple metrics, suggesting its suitability for decision support systems in melanoma diagnosis. This study introduces a new benchmark for asymmetry classification, emphasizing its potential to enhance early detection and decision-making in clinical settings [91].

Bhati et al. devised an innovative approach to distinguishing melanoma from SLs and detecting it early, aiming to improve sensitivity by 12% and specificity by 1%. They observed significant improvements in image quality by employing median filtering for hair and noise removal and canny edge detection for enhanced signal-to-noise ratio. Utilizing Otsu's segmentation algorithm enhanced overall image quality, making the system suitable for diagnosing malignant melanomas. The test criteria include features such as asymmetry (A), border (B), color (C), and diameter (D), aiding in assessing the degree of asymmetry, border irregularity, color variation, and maximum diameter of the lesion, crucial for distinguishing between normal or benign lesions and melanoma. This system not only eases the doctor's role but also benefits individuals in remote and rural areas without easy access to dermatologists. Based on total dermoscopic score calculation with preprocessing and Otsu's segmentation, the proposed approach demonstrated promising results in distinguishing benign from non-benign SLs, particularly when combined with the Attention Based Detection (ABeD) algorithm [92].

Effective feature engineering is crucial because it directly influences a model's ability to discern meaningful patterns in dermatological data. The selection of features should correspond to the particular diagnostic objective and the dataset's attributes. Using the Scikit-Image library, Chakravarty et al. performed feature extraction and categorization in Python to extract geometrical and structural features from dermatological photographs. To decrease the feature vector for each classification algorithm, they also used feature selection [91, 93].

As a result, the initial stages of data preprocessing and feature crafting play a crucial role in implementing ML techniques in dermatology. High-quality data sources, appropriate preprocessing techniques, and carefully selected features are essential for building accurate and robust ML models for skin disease prediction and diagnosis. As we delve further into this chapter, we will explore how these well-prepared data are used in various ML algorithms to enhance dermatological practice.

12.3.3 Challenges in Dermatological Data Collection

The collection of high-quality dermatological data poses various hurdles that must be overcome to support the creation and assessment of ML models in the dermatology domain. These challenges encompass issues related to data diversity, privacy, and representativeness.

12.3.3.1 Data Imbalance

The unequal distribution of data is among the significant difficulties associated with the collection of dermatological data. Dermatological datasets often exhibit an unequal distribution of different skin diseases. For instance, rare conditions or diseases less frequently encountered may be underrepresented in the data. This imbalance can pose difficulties for ML models as they may become biased toward more prevalent conditions [94].

Researchers have explored techniques such as oversampling, under-sampling, and data augmentation to mitigate the effects of data imbalance. These methods ensure that ML models can effectively learn and recognize rare or less common skin diseases, contributing to more comprehensive dermatological diagnostics.

12.3.3.2 Effective Negative Set

Another challenge we face is getting the right examples that are not what we are looking for. To strengthen the network's capacity to differentiate false positive instances, the negative dataset must comprise cases that resemble nodules but are, in fact, not positive [95].

12.3.3.3 Data Privacy and Ethical Concerns

Dermatological data collection frequently involves sensitive information, such as patient images and medical records. Data privacy and ethical concerns become paramount in handling such information. It is essential to safeguard patient privacy and adhere to ethical guidelines when collecting and using dermatological data for research. Healthcare institutions and researchers must implement robust data anonymization and protection measures to prevent the inadvertent exposure of patient identities. Ethical considerations encompass obtaining informed consent from patients and ensuring that data usage aligns with the established ethical standards. These safeguards are crucial to maintain patient trust and comply with regulatory requirements in healthcare research [96].

12.3.3.4 Data Diversity and Representativeness

Another significant challenge is ensuring that dermatological datasets are diverse and representative of the population. Skin diseases can manifest differently across various demographic groups, including age, gender, and ethnicity. Failure to account for this diversity can lead to biased ML models that do not perform effectively for all patients [97].

Efforts are underway to collect more diverse and representative datasets encompassing a broader spectrum of skin types and conditions. Such datasets better reflect real-world scenarios and improve the generalizability of ML models. Researchers are also exploring the impact of demographic factors on skin disease diagnosis to develop more equitable and unbiased algorithms [98].

12.3.3.5 Data Annotation and Standardization

Data annotation and standardization are crucial for ensuring the quality and consistency of dermatological datasets. Annotating SLs accurately, including providing information on lesion types, stages, and other relevant attributes, is a labor-intensive task that requires expertise [46].

To address this challenge, researchers investigate automated annotation methods and leverage expert knowledge to create high-quality labeled datasets. Standardization efforts involve defining consistent protocols for data collection, image acquisition, and diagnostic criteria, ensuring that dermatological data remains reliable and comparable across different studies [65].

Therefore, collecting dermatological data for ML applications faces several challenges related to data imbalance, privacy, diversity, and standardization. Addressing these challenges is essential to ensuring the development of robust and equitable ML models for skin disease prediction and diagnosis.

12.4 Challenges and Opportunities

12.4.1 Data Quality and Quantity

Developing effective skin disease prediction models in ML is hindered by the scarcity of high-quality and diverse dermatological datasets, which should ideally cover various skin types, ethnicities, age groups, and disease severities for model generalizability. Size and diversity are frequently limited in publicly accessible datasets. Accurate and standardized annotations for dermatological images are crucial for reliable model training. Neglecting data quality can undermine model performance despite efforts to improve model architecture and features. Thoroughly understanding and assessing data quality through metrics and transformations are essential for enhancing ML model performance [99].

12.4.2 Interpretability and Explainability

In dermatological ML models, achieving interpretability and explainability is crucial. Complex algorithms such as DNN are often regarded as "black boxes," posing challenges in understanding their decision-making processes. Dermatologists and healthcare providers necessitate transparent and interpretable models they can trust and seamlessly incorporate into their practice. To bridge this gap, explainable AI techniques are emerging as a solution, offering valuable insights into the reasoning behind model predictions. This enhanced transparency is vital for fostering confidence in the use of ML in dermatological diagnosis and treatment [100].

12.4.3 Class Imbalance and Rare Diseases

Skin diseases exhibit uneven prevalence, with rare dermatological conditions often underrepresented in datasets. This imbalance can affect how well the model works, making it better at predicting the more common conditions and possibly missing the rarer ones. To ensure comprehensive skin disease detection, employing effective strategies to tackle class imbalance is crucial. These strategies may include oversampling techniques and anomaly detection, helping address the challenges posed by imbalanced data and improving the model's ability to detect common and rare dermatological conditions [101].

12.4.4 Real-Time Application and Mobile Health (mHealth)

Integrating ML into dermatology-focused mHealth apps presents substantial early disease detection and telemedicine opportunities. With smartphones' widespread use and advanced cameras, users and healthcare providers can capture skin images conveniently. However, ensuring data privacy and security remains crucial. Overcoming challenges in data privacy, quality assessment, AI reproducibility, and user engagement is vital for widespread AI-based mHealth app adoption. Ongoing initiatives like the European project "Watching the risk factors (WARIFA)" aim to surmount these obstacles, paving the way for AI-driven mHealth apps in disease prevention and health promotion [102].

12.4.5 Generalization and Cross-Domain Adaptation

ML models often struggle to perform effectively on diverse data sources or domains, posing a challenge in generalizing to new dermatological conditions, imaging devices, and clinical settings. Guan et al. suggested the significance of research in domain adaptation and transfer learning in addressing this persistent issue. In summary, this comprehensive overview of domain adaptation techniques in medical image analysis, with its categorization of methods and discussions on challenges and future directions, serves as an invaluable resource for researchers navigating this evolving field [103].

12.4.6 Collaborative Research and Clinical Integration

The fusion of ML capabilities with dermatology holds promise for enhancing both diagnosis and long-term treatment outcomes. Du et al. underscores the need for collaborative efforts among dermatologists, computer scientists, and data scientists to harness the full potential of ML in this field. Such interdisciplinary collaboration, aligned with ethical and regulatory standards, can lead to transformative advancements in dermatological care, benefiting both healthcare professionals and patients alike [22].

12.4.7 Ethical Considerations

The integration of ML in dermatology brings about important ethical considerations. These encompass safeguarding patient privacy, mitigating bias, ensuring accountability, and enhancing transparency. During the evolution of the field, ethical guidelines must be not only established but also rigorously adhered to, ensuring that these transformative advancements ultimately benefit both patients and healthcare practitioners [104].

12.4.8 Continuous Learning and Model Updates

In conclusion, dermatology, in conjunction with the rapid advancement of AI technology, is poised for significant progress. Continuous learning and model updates are essential to ensure the ongoing effectiveness of ML models in this dynamic medical discipline. National policies and strategic plans worldwide reflect the importance of AI in healthcare, including dermatology, with China, in particular, emphasizing its leadership in this domain. While dermatological AI faces challenges such as data scarcity, talent shortages, legal and ethical issues, and commercialization hurdles, it enjoys substantial support in funding, research, education, and market growth. The future promises enhanced diagnostic and treatment capabilities through AI, ultimately benefitting both medical professionals and patients [105].

In summary, while ML offers significant promise in dermatology for skin disease prediction, several challenges related to data, interpretability, class imbalance, real-time application, generalization, collaboration, ethics, and continuous learning must be addressed. Overcoming these challenges presents opportunities to enhance patient care, enable early disease detection, and improve dermatological practice by integrating ML into clinical workflows.

12.5 Future Directions

In dermatology, the future of ML holds great promise. Continued advancements in ML techniques are expected to lead to even greater precision and accuracy in predicting and diagnosing skin disorders. Research and development efforts will likely focus on refining algorithms, enhancing model performance, and exploring innovative applications of ML in dermatological care. Additionally, addressing ethical considerations will remain a critical aspect of future developments, with ongoing efforts needed to ensure data privacy, mitigate bias in algorithms, and maintain transparency and accountability in decision-making processes. As ML continues to evolve, its integration into clinical practice is expected to expand, with more healthcare institutions leveraging these technologies to improve patient outcomes and optimize dermatological care pathways.

12.6 Conclusion

Integrating ML into dermatology represents a significant milestone in advancing healthcare. By exploring ML principles and practical applications, this convergence has demonstrated the ability to revolutionize the prediction and diagnosis of skin disorders. Despite the challenges and ethical considerations involved, the future outlook is promising. Continued refinement of ML techniques and diligent navigation of ethical concerns will pave the way for more precise, accessible, and data-driven dermatological healthcare. As ML becomes more ingrained in clinical settings, its revolutionary effect enhances patient results, lowers healthcare expenses, and mitigates the worldwide burden of skin conditions. In summary, the integration of ML into dermatology heralds a new era of innovation and possibility, promising a brighter future for dermatological healthcare worldwide.

References

1. Swaney MH, Kalan LR. Living in your skin: microbes, molecules, and mechanisms. Infect Immun. 2021;89:1–18.
2. Hay R, Bendeck SE, Chen S, Estrada R, Haddix A, McLeod T, AM. Bookshelf_NBK11733. In: Jamison DT, Breman JG, Measham AR, Alleyne G, Claeson M, Evans DB, Jha P, Mills A, PM, editors. Dis control priorities dev countries. 2nd ed. Washington (DC), New York: The International Bank for Reconstruction and Development / The World Bank/Oxford University Press; 2005. p. 707–22.
3. Seth D, Cheldize K, Brown D, Freeman EE. Global burden of skin disease: inequities and innovations. Curr Dermatol Rep. 2017;6:204–10.
4. Martinez JC, Otley CC. The management of melanoma and nonmelanoma skin cancer: a review for the primary care physician. Mayo Clin Proc. 2001;76:1253–65.

5. Arnold M, Singh D, Laversanne M, Vignat J, Vaccarella S, Meheus F, et al. Global burden of cutaneous melanoma in 2020 and projections to 2040. JAMA Dermatol. 2022;158:495–503.
6. Heibel HD, Hooey L, Cockerell CJ. A review of noninvasive techniques for skin cancer detection in dermatology. Am J Clin Dermatol [Internet]. 2020:1–12. Available from: https://doi.org/10.1007/s40257-020-00517-z
7. Kshirsagar PR, Manoharan H, Shitharth S, Alshareef AM, Albishry N, Balachandran PK. Deep learning approaches for prognosis of automated skin disease. Life. 2022;12:1–16.
8. Adegun A, Viriri S. Deep learning techniques for skin lesion analysis and melanoma cancer detection : a survey of state - of - the - art. Artif Intell Rev. Springer, Netherlands; 2020.
9. Meng H, Liu Y, Lee BP. Model polymer system for investigating the generation of hydrogen peroxide and its biological responses during the crosslinking of mussel adhesive moiety. Acta Biomater [Internet] 2017;48:144–56. Available from: https://doi.org/10.1016/j.actbio.2016.10.016.
10. da Silva Souza Campanholi K, Combuca da Silva Junior R, da Silva C I, Said dos Santos R, Vecchi CF, Bruschi ML, et al. Stimulus-responsive phototherapeutic micellar platform of rose Bengal B: a new perspective for the treatment of wounds. J Drug Deliv Sci Technol. 2021:66.
11. Li Z, Koban KC, Schenck TL, Giunta RE, Li Q, Sun Y. Artificial intelligence in dermatology image analysis: current developments and future trends. J Clin Med. 2022:11.
12. Li L, Wang XU, Hu W, Du Y, Li B, Xiong NN. Deep learning in skin disease image recognition : a. Review. 2020:8.
13. Kumar Y, Koul A, Singla R, Ijaz MF. Artificial intelligence in disease diagnosis: a systematic literature review, synthesizing framework and future research agenda. J Ambient Intell Humaniz Comput [Internet] 2023;14:8459–86. Available from: https://doi.org/10.1007/s12652-021-03612-z.
14. Heinlein L, Maron RC, Hekler A, Haggenmüller S. Clinical melanoma diagnosis with artificial intelligence : insights from a prospective multicenter study.
15. Abbasi EY, Deng Z, Magsi AH, Ali Q, Kumar K, Zubedi A. Optimizing skin cancer survival prediction with ensemble techniques. Bioengineering. 2024;11:1–26.
16. Yadav AK, Mehta R, Kumar V, Medikondu NR. An optimized boosting framework for skin lesion segmentation and classification. Multimed Tools Appl [Internet]. 2023:1–22. [cited 2024 Feb 16]. Available from: https://link.springer.com/article/10.1007/s11042-023-17042-w
17. Lyakhov PA, Lyakhova UA, Kalita DI. Multimodal analysis of unbalanced dermatological data for skin cancer recognition. IEEE Access. 2023;11:131487–507.
18. Pettit RW, Fullem R, Cheng C, Amos CI. Artificial intelligence, machine learning, and deep learning for clinical outcome prediction. 2021;0:729–45.
19. Murphree DH, Puri P, Shamim H, Bezalel SA, Drage LA, Wang M, et al. Deep learning for dermatologists: part I.Fundamental concepts. J Am Acad Dermatol. 2022;87:1343–51.
20. Samuel AL. Some studies in machine learning using the game of checkers. 1959;3:535–54.
21. El Bouchefry K, de Souza RS. Learning in big data: introduction to machine learning. In: Knowl Discov Big data from Astron. Earth Obs Astrogeoinformatics. Elsevier Inc.; 2020.
22. Du AX, Emam S, Gniadecki R. Review of machine learning in predicting dermatological outcomes. Front Med. 2020;7:1–6.
23. Goyal M, Knackstedt T, Yan S, Hassanpour S. Artificial intelligence-based image classification methods for diagnosis of skin cancer: challenges and opportunities. Comput Biol Med. 2020;127:104065.
24. Charan DS, Nadipineni H, Sahayam S, Jayaraman U. Method to Classify Skin Lesions using Dermoscopic images; 2020. p. 1–16.
25. Ahammed M, Al MM, Uddin MS. A machine learning approach for skin disease detection and classification using image segmentation. Healthc Anal [Internet]. 2022;2:100122. Available from: https://doi.org/10.1016/j.health.2022.100122
26. Yu K, Syed MN, Bernardis E, Gelfand JM. Machine learning applications in the evaluation and Management of Psoriasis: a systematic review. J Psoriasis Psoriatic Arthritis. 2020;5:147–59.

27. Murphy KP. Machine learning: a probabilistic perspective. In: Murphy KP, editor. Google books [internet]. 1st ed. Cambridge, Massachusetts, London, England; 2012 [cited 2023 Sep 28]. Available from: https://books.google.co.in/books?hl=en&lr=&id=RC43AgAAQBAJ&oi=fnd&pg=PR7&ots=umozhEMrYa&sig=6fesvu2HvRiJrq0NP-ztgYxiGsc&redir_esc=y#v=onepage&q&f=false.
28. Eapen BR. Artificial intelligence in dermatology: a practical introduction to a paradigm shift. Indian Dermatol Online J. 2020;11:881.
29. Barata C, Rotemberg V, Codella NCF, Tschandl P, Rinner C, Akay BN, et al. A reinforcement learning model for AI-based decision support in skin cancer. Nat Med [Internet]. 2023;29:1941–6. Available from: https://doi.org/10.1038/s41591-023-02475-5
30. Tan TY, Zhang L, Jiang M. An intelligent decision support system for skin cancer detection from dermoscopic images. 2016 12th Int Conf Nat Comput Fuzzy Syst Knowl Discov ICNC-FSKD. 2016;2016:2194–9.
31. El-dahshan EA, Mohsen HM, Revett K, Salem AM. Expert systems with applications computer-aided diagnosis of human brain tumor through MRI : a survey and a new algorithm. Expert Syst Appl [Internet] 2014;41:5526–5545. Available from: https://doi.org/10.1016/j.eswa.2014.01.021.
32. Krishna Monika M, Arun Vignesh N, Usha Kumari C, Kumar MNVSS, Laxmi LE. Skin cancer detection and classification using machine learning. Mater Today Proc [Internet]. 2020;33:4266–70. Available from: https://doi.org/10.1016/j.matpr.2020.07.366
33. Bedair K, Samak AH, Nisar KS, Elrashidi A, Toumi A, Alkhazaleh S, et al. Neutrosophic fuzzy logic based SVM approach for enhanced skin cancer prediction. Int J Neutrosophic Sci. 2024;23:317–34.
34. Abinash MD, Velmurugan J. Enhancing the accuracy of skin disease prediction using image support vector machines technique over random Forest. EurchembullCom [Internet]. 2022;12:3985–92. Available from: https://www.eurchembull.com/uploads/paper/cef2e-4e68252551c3591845d0c50e021.pdf
35. Lepakshi VA. Machine learning and deep learning based AI tools for development of diagnostic tools. In: Comput. Approaches Nov. Ther. Diagnostic des. To mitigate SARS-CoV2 infect. Revolut. Strateg. To combat pandemics. INC; 2022.
36. Chang CL, Chen CH. Applying decision tree and neural network to increase quality of dermatologic diagnosis. Expert Syst Appl. 2009;36:4035–41.
37. Ilker Ali Ozkan MK. Skin lesion classification using machine learning algorithm; 2017. p. 285–9.
38. Chrimes D. Using decision trees as an expert system for clinical decision support for COVID-19 corresponding author, vol. 12; 2023. p. 1–12.
39. Ramezankhani R, Sajjadi N, Esmaeilzadeh RN, Jozi SA, Shirzadi MR. Application of decision tree for prediction of cutaneous leishmaniasis incidence based on environmental and topographic factors in Isfahan Province, Iran. Geospat Health [Internet]. 2018 [cited 2024 Feb 19];13:172–8. Available from: https://pubmed.ncbi.nlm.nih.gov/29772890/
40. Dhivyaa CR, Sangeetha K, Balamurugan M, Amaran S, Vetriselvi T, Johnpaul P. Skin lesion classification using decision trees and random forest algorithms. J Ambient Intell Humaniz Comput. 2020:1–13.
41. Kolkur MS, Kalbande DR, Kharkar V. Machine learning approaches to multi-class human skin disease detection. Int J Comput Intell Res. 2018;14:29–39.
42. Shinde A, Ingle PM. Hybrid approach for skin disease classification : integrating machine learning and. 2023;
43. Saha DK. Computer vision based skin disease recognition. Res Sq. 2023:1–14.
44. Kalaivani A. Detection and classification of skin diseases with ensembles of deep learning networks in medical imaging. 2022;6:13624–37.
45. Syahab AS, Pamungkas GL, Akmal S. The ultraviolet index classification using logistic regression and random Forest methods for predicting extreme conditions [internet]. Atlantis Press International BV; 2023. Available from: https://2fyahootechpulse.easychair.org/publications/preprint_download/TfZq

46. Brinker TJ, Hekler A, Enk AH, Berking C, Haferkamp S, Hauschild A, et al. Deep neural networks are superior to dermatologists in melanoma image classification. Eur J Cancer. 2019;119:11–7.
47. Esteva A, Kuprel B, Novoa RA, Ko J, Swetter SM, Blau HM, et al. Dermatologist-level classification of skin cancer with deep neural networks. Nat Publ Gr [Internet] 2017;542:115–118. Available from: https://doi.org/10.1038/nature21056.
48. Seeja RD, Suresh A. Deep learning based skin lesion segmentation and classification of melanoma using support vector machine. 2019;20:1555–61.
49. Huynh QT, Nguyen PH, Le HX, Ngo LT, Trinh N, Tsuji K, et al. Automatic acne object detection and acne severity grading using smartphone images and artificial intelligence 2022;1–13.
50. Yotsu R, Ding Z, Hamm J, Blanton R. Deep learning for AI-based diagnosis of skin-related neglected tropical diseases: a pilot study. PMC Artic. 2023:1–26.
51. Nigat TD, Sitote TM, Gedefaw BM. Fungal skin disease classification using the convolutional neural network. J Healthc Eng. 2023;2023:1–9.
52. Achararit P. Artificial intelligence-based diagnosis of Oral lichen planus using deep convolutional neural networks. Thieme. 2023:1–8.
53. Collenne J, Monnier J, Iguernaissi R, Nawaf M, Grob J, Gaudy-marqueste C, et al. Jo l P of. J Invest Dermatol [Internet]. 2023.; Available from: https://doi.org/10.1016/j.jid.2023.09.289.
54. Ba W, Wu H, Chen WW, Wang SH, Zhang ZY, Wei XJ, et al. Convolutional neural network assistance significantly improves dermatologists diagnosis of cutaneous tumours using clinical images. Eur J Cancer. 2022;169:156–65.
55. Li S, Li W, Cook C, Zhu C, Gao Y. Independently recurrent neural network (IndRNN): building a longer and deeper RNN. Proc IEEE Comput Soc Conf Comput Vis Pattern Recognit. 2018:5457–66.
56. Abbas Q, Celebi ME. DermoDeep-A classification of melanoma-nevus skin lesions using multi-feature fusion of visual features and deep neural network. Multimed Tools Appl. 2019;78:23559–80.
57. Zhang L, Zhang J, Gao W, Bai F, Li N, Ghadimi N. A deep learning outline aimed at prompt skin cancer detection utilizing gated recurrent unit networks and improved orca predation algorithm. Biomed Signal Process Control. 2024;90:105858.
58. Bu J, Lin Y, Qing LQ, Hu G, Jiang P, Hu HF, et al. Prediction of skin disease using a new cytological taxonomy based on cytology and pathology with deep residual learning method. Sci Rep [Internet] 2021;11:1–9. Available from: https://doi.org/10.1038/s41598-021-92848-y.
59. Kumar RPR, Varun R, Sreeshwan J, Kumar KA, Rana U, Rajyalakshmi A. Feasible skin lesion detection using CNN and RNN. E3S Web Conf. 2023;430:1–8.
60. Zhang B, Zhou X, Luo Y, Zhang H, Yang H, Ma J, et al. Opportunities and challenges : classification of skin disease based on deep learning. Chinese. J Mech Eng. 2021:1–14.
61. Li H, Pan Y, Zhao J, Zhang L. Neurocomputing skin disease diagnosis with deep learning : a review. Neurocomputing [Internet]. 2021;464:364–93. Available from: https://doi.org/10.1016/j.neucom.2021.08.096
62. Sun J, Yao K, Huang G, Zhang C, Leach M, Huang K, et al. Machine learning methods in skin disease recognition: a systematic review. PRO. 2023;11:1–16.
63. Ilter N, Guvenir HA. Dermatology Data Set. UCI Mach Learn Repos [Internet]. 1998;90:2–4. Available from: https://archive.ics.uci.edu/ml/datasets/Dermatology
64. Cassidy B, Kendrick C, Brodzicki A, Jaworek-korjakowska J, Hoon M. Analysis of the ISIC image datasets : usage, benchmarks and recommendations. Med Image Anal [Internet]. 2022;75:102305. Available from: https://doi.org/10.1016/j.media.2021.102305
65. Tschandl P, Rosendahl C, Kittler H. Data descriptor: the HAM10000 dataset, a large collection of multi-source dermatoscopic images of common pigmented skin lesions. Sci Data [Internet] 2018;5:1–9. Available from: https://doi.org/10.1038/sdata.2018.161.
66. Li Y, Shen L. Skin lesion analysis towards melanoma detection using deep learning network. Sensors (Switzerland). 2018;18:1–16.

67. Shinde RK, Alam MS, Hossain MB, Md Imtiaz S, Kim JH, Padwal AA, et al. Squeeze-MNet: precise skin cancer detection model for low computing IoT devices using transfer learning. Cancers (Basel). 2023;15:1–13.
68. Bajwa MN, Muta K, Malik MI, Siddiqui SA, Braun SA, Homey B, et al. Computer-aided diagnosis of skin diseases using deep neural networks. Appl Sci. 2020;10:1–13.
69. Balaji VR, Suganthi ST, Rajadevi R, Krishna Kumar V, Saravana Balaji B, Pandiyan S. Skin disease detection and segmentation using dynamic graph cut algorithm and classification through naive Bayes classifier. Meas J Int Meas Confed [Internet]. 2020;163:107922. Available from: https://doi.org/10.1016/j.measurement.2020.107922
70. Codella NCF, Lin CC, Halpern A, Hind M, Feris R, Smith JR. Collaborative human-AI (CHAI): Evidence-based interpretable melanoma classification in dermoscopic images. Lect Notes Comput Sci (including Subser Lect Notes Artif Intell Lect Notes Bioinformatics). 2018;11038 LNCS:97–105.
71. Alam TM, Shaukat K, Khan WA, Hameed IA, Almuqren LA, Raza MA, et al. An efficient deep learning-based skin cancer classifier for an imbalanced dataset. Diagnostics 2022, Vol 12, Page 2115 [Internet] [cited 2023 Sep 18]. 2022;12:2115. Available from: https://www.mdpi.com/2075-4418/12/9/2115/htm
72. Alwakid G, Gouda W, Humayun M, Sama NU. Melanoma detection using deep learning-based classifications. Healthc. 2022:10.
73. Yang K, Qinami K, Fei-Fei L, Deng J, Russakovsky O. Towards fairer datasets: filtering and balancing the distribution of the people subtree in the ImageNet hierarchy. FAT* 2020. Proc 2020 Conf fairness, accountability, Transpar. 2020:547–58.
74. Shorten C, Khoshgoftaar TM. A survey on image data augmentation for deep learning. J Big Data [Internet]. 2019:6. Available from: https://doi.org/10.1186/s40537-019-0197-0
75. Aggarwal LP. Data augmentation in dermatology image recognition using machine learning. Skin Res Technol. 2019;25:815–20.
76. Wan L, Ai Z, Chen J, Jiang Q, Chen H, Li Q, et al. Detection algorithm for pigmented skin disease based on classifier-level and feature-level fusion. Front Public Heal. 2022;10 https://doi.org/10.3389/fpubh.2022.1034772.
77. Saad I, Abdelhalim A, Mohamed MF, Mahdy YB. Data augmentation for skin lesion using self-attention based progressive generative adversarial network. Expert Syst Appl [Internet]. 2021;165:113922. Available from: https://doi.org/10.1016/j.eswa.2020.113922
78. Al-abayechi AAA. Enhancement of dermoscopic images and feature extraction for classification of skin lesions. Baghdad, Iraq: University of Baghdad; 2015.
79. Akyel C. LinkNet-B7 : Noise removal and lesion segmentation in images of skin cancer; 2022. p. 1–15.
80. Wei L, Pan SX, Nanehkaran YA, Rajinikanth V. An optimized method for skin cancer diagnosis using modified thermal exchange optimization algorithm. Comput Math Methods Med. 2021:2021.
81. Gupta A, Bhatnagar M, Issac A, Travieso CM, Dutta MK. Imaging method for noise removal and segmentation of Skin Lesions from Dermoscopic Images. 2019:1–5.
82. Gütlein M, Helma C, Karwath A, Kramer S. A large-scale empirical evaluation of cross-validation and external test set validation in (Q). SAR. 2013:516–28.
83. Srinivasu PN, Sivasai JG, Ijaz MF, Bhoi AK, Kim W, Kang JJ. Networks with MobileNet V2 and LSTM. MDPI. 2021;21:1–27.
84. Thieme AH, Zheng Y, Machiraju G, Sadee C, Mittermaier M, Gertler M, et al. A deep-learning algorithm to classify skin lesions from mpox virus infection. 2023;29
85. Sihore S, Tijare K, Paradhi R, Meshram P, Chandankhede S, Shelke S. Skin disease recognition using texture analysis. Int J Innov Eng Sci. 2023;8:410–3.
86. Wei L, Gan Q, Ji T. Skin disease recognition method based on image color and texture features. Comput Math Methods Med. 2018;2018:8145713.
87. Islam N, Member S, Gallardo-alvarado J, Abu M, Salman NA. Skin disease recognition using texture. Analysis. 2017;1:4–5.

88. Kolkur S, Kalbande D, Shimpi P, Bapat C, Jatakia J. Human skin detection using RGB. HSV and YCbCr Color Models. 2017;137:324–32.
89. Al-Zyoud W, Abu Helou A, Alqasem E, Rawashdeh NA. Visual feature extraction from dermoscopic colour images for classification of melanocytic skin lesions. EurAsian J Biosci. 2020;14:1299–307.
90. Arunkumar TR, Jayanna HS. A Novel Approach for Automatic Identification of Psoriasis Affected Skin Area. 2nd Int Conf Emerg Comput Inf Technol. 2017:1–8.
91. Chakravorty R, Liang S, Abedini M, Garnavi R. Dermatologist-like feature extraction from skin lesion for improved asymmetry classification in PH2 database. Proc Annu Int Conf IEEE Eng Med Biol Soc EMBS. 2016;2016:3855–8.
92. Bhati P, Singhal M. Early stage detection and classification of melanoma. Int Conf Commun Control Intell Syst CCIS. 2015;2016:181–5.
93. Varoquaux G, Cheplygina V. Machine learning for medical imaging: methodological failures and recommendations for the future. npj Digit Med. 2022:5.
94. Nguyen VD, Bui ND, Do HK. Skin lesion classification on imbalanced data using deep learning with soft attention. Sensors. 2022;22:1–24.
95. Hesamian MH, Jia W, He X, Kennedy P. Deep learning techniques for medical image segmentation: achievements and challenges. J Digit Imaging. 2019;32:582–96.
96. Militello M, Yang RA, Anderson JB, Szeto MD, Presley CL, Laughter MR. Social media and ethical challenges for the dermatologist. Curr Dermatol Rep [Internet] 2021;10:120–7. Available from: https://doi.org/10.1007/s13671-021-00340-7.
97. Mehrabi N, Morstatter F, Saxena N, Lerman K, Galstyan A. A survey on bias and fairness in machine learning. ACM Comput Surv. 2021;54
98. Yaqoob A, Musheer R, Navneet A. Applications and techniques of machine learning in cancer classification : a systematic review. Human-Centric Intell Syst. 2023:1–28.
99. Collins E, Rozanov N, Zhang B. Evolutionary data measures: understanding the difficulty of text classification tasks. CoNLL. 22nd Conf Comput Nat Lang Learn Proc. 2018, 2018:380–91.
100. Yang G, Ye Q, Xia J. Unbox the black-box for the medical explainable AI via multi-modal and multi-Centre data fusion: a mini-review, two showcases and beyond. Inf Fusion [Internet]. 2022;77:29–52. Available from: https://doi.org/10.1016/j.inffus.2021.07.016
101. Zhao Y, Wong ZSY, Tsui KL. A framework of rebalancing imbalanced healthcare data for rare event's classification: a case of look-alike sound-alike mix-up incident detection. J Healthc Eng. 2018;2018:1–11.
102. Deniz-Garcia A, Fabelo H, Rodriguez-Almeida AJ, Zamora-Zamorano G, Castro-Fernandez M, del Pino Alberiche Ruano M, et al. Quality, usability, and effectiveness of mHealth apps and the role of artificial intelligence: current scenario and challenges. J Med Internet Res [Internet] 2023. [cited 2023 Sep 24];25. Available from: /pmc/articles/PMC10196903/.
103. Guan H, Liu M. Domain adaptation for medical image analysis: a survey. IEEE Trans Biomed Eng. 2022;69:1173–85.
104. Mohta A, Mohta A. Ethical considerations of artificial intelligence-how sacrosanct is health care? Int J Clin Exp Dermatology Int J Clin Expl Dermatol [Internet]. 2022;2022:80–1. Available from: https://opastpublishers.com/
105. Li CX, Shen CB, Xue K, Shen X, Jing Y, Wang ZY, et al. Artificial intelligence in dermatology: past, present, and future. Chin Med J. 2019;132:2017–20.
106. Ramezankhani R, Sajjadi N, Esmaeilzadeh RN, Jozi SA, Shirzadi MR. Application of decision tree for prediction of cutaneous leishmaniasis incidence based on environmental and topographic factors in Isfahan province. Iran Geospat Health. 2018;13:172–8.

Chapter 13
A Comparative Study of Different Machine Learning Techniques for Skin Disease Detection

Deba Prasad Dash, Maheshkumar H. Kolekar, and Eva Mishra

13.1 Introduction

A certain kind of cancer that starts in the skin is called skin cancer. It is the most normal kind of cancer, and it mostly occurs in the outermost layer of skin, known as the epidermis. Although there are different types of skin cancer, the following are the most prevalent ones:

Basal Cell Carcinoma (BCC): BCC is the most widespread skin cancer. It generally shows as a shiny bump, small, or a red, scaly patch. Early detection usually results in a cure for BCC, which rarely contaminates the other areas of the human body.

Squamous Cell Carcinoma (SCC): The other most widespread kind of skin cancer is SCC. It often looks like a red, scaly patch or a sore that will not heal. Compared to certain other cancers, SCC is less prone to spread, but it can still be more aggressive than BCC.

Melanoma: The most serious kind of skin cancer that can contaminate other areas of the human body is melanoma. It frequently manifests as a new mole or changes to the shape, size, or color of an already-existing mole. Early diagnosis and solution are important for enhancing the prognosis of melanoma.

Other, less prevalent types of skin cancer are known as dermatofibrosarcoma protuberans, Merkel cell carcinoma, and various types of cutaneous lymphoma.

D. P. Dash (✉)
Thapar Institute of Engineering and Technology, Patiala, Punjab, India
e-mail: deba.dash@thapar.edu

M. H. Kolekar
Indian institute of Technology, Patna, Bihar, India
e-mail: mahesh@iitp.ac.in

E. Mishra
Chitkara University, Rajpura, Punjab, India

T. Choudhury, A. Katal (eds.), *Machine Learning for Disease Detection, Prediction, and Diagnosis*, https://doi.org/10.1007/978-981-96-4241-0_13

Some of the risk factors linked to skin cancer are exposure to ultraviolet radiation from artificial sources (like tanning beds) or from the sun, a history of skin cancer in the family, fair skin, and a history of severe sunburns. Preventive measures include the use of sunscreen, avoiding high sun exposure, and wearing protective clothing.

Dermoscopy is a non-invasive examination method that uses oil immersion and incident light to enable visual inspection of the skin's underlying components. Dermatoscopy can detect melanoma more often than unassisted observation-based detection; however, the efficiency of the diagnosis relies on the dermatologist's training. Particularly in the initial stages, the detection of melanoma from melanocytic nevi is rarely simple. For this reason, an automatic diagnosis tool is crucial for doctors.

Diagnosis typically involves a physical examination and may include a biopsy to confirm the kind and stage of the cancer. Treatment options for skin cancer differ based on the location, stage, and variety of the cancer but may comprise radiation therapy, surgical removal, chemotherapy, targeted therapy, or immunotherapy. The choice of treatment depends on the individual patient's situation.

Early detection and regular skin self-examinations are crucial for identifying skin cancer at an early, more treatable stage. If the individual has concerns about skin changes or risk factors for skin cancer, it is important to consult a healthcare professional or a dermatologist for a thorough evaluation and guidance on prevention and early detection.

DL, a subset of ML [1], has demonstrated high efficiency in assisting with the classification of skin cancer. Out of many applications of deep learning, one of the applications is the development of computer-aided diagnosis systems for dermatologists and healthcare professionals. These systems can analyze images of skin lesions and provide diagnostic assistance. The following steps explain how deep learning be used for skin cancer detection:

Data Collection: Deep learning models require a high and varied set of images of skin lesions. These datasets often contain various types of skin conditions, including benign and malignant lesions. High-quality, labeled data is essential for training an accurate model.

Image Preprocessing: Before feeding the data into a deep learning model, images are often preprocessed. This may involve resizing, normalizing pixel values, and augmenting the dataset by applying transformations like flipping, rotation, or adjusting contrast and brightness.

Convolutional neural networks (CNNs): CNNs are largely employed for image classification processes, such as the diagnosis of skin cancer. These networks can automatically learn relevant features from the images. They comprise multiple layers of convolutional and pooling operations to capture different levels of details from the images.

Training: The DL model learns the pattern utilizing the dataset having a label. In the process of training, the model learns to classify features and patterns in the images that are indicatory of skin cancer. This typically involves minimizing a loss function that calculates the difference between the ground truth labels and the model's predictions.

Validation and Testing: After training, the model accuracy is verified using a different dataset not used at the time of training to assess its performance. Hyperparameters, such as learning rates and network architecture, may be fine-tuned during this phase. The model's recall, precision, accuracy, and F1 score are often calculated.

Deployment: After training and validating the DL model, it can be used to solve real-world problems. Dermatologists and healthcare professionals can use the model as a tool to aid in their clinical decisions. This may involve providing a probability score or diagnosis based on the input image.

Continuous Improvement: Deep learning models can be continuously improved by updating them with new data and fine-tuning the model based on feedback and actual-world performance. The ongoing process supports the model's adaptation to novel variations.

It is essential to note that, while DL models can be powerful tools in skin cancer detection, they should be used as aids to medical professionals rather than as standalone diagnostic tools. They are not a replacement for the expertise of dermatologists and healthcare providers.

Research in this field continues to advance, and deep learning models are becoming increasingly accurate in their ability to detect skin cancer. However, regulatory and ethical considerations are essential when integrating AI into medical practice, and the development of such systems should follow established medical and ethical guidelines.

Jain et al [2] presented the melanoma skin cancer classification method using an image analysis technique. Geometry-based features such as circularity index, greatest diameter, area, perimeter, and irregularity index were collected from the segmented image. The threshold approach was used for the classification of images. In another approach, the Squeeznet model was used for skin cancer detection [3]. The model's cancer detection accuracy was 99.77%. Discrete wavelet filters along with a support vector machine (SVM) classifier detected skin cancer efficiently as proposed by Ragab et al [4]. Apart from CNN and SVM [5], other classifiers such as VGG16 and ResNet have found good applications in skin cancer classification [6]. Different online and clinical databases were used to design and validate the model. ISIC dataset was largely used to test the model. Different machine learning algorithms were explored for classifying skin cancer and achieved good accuracy. Modified K-nearest neighbor (KNN) classified cancerous and noncancerous cells with 98.3% accuracy [7]. Self-organizing map and feed-forward neural network applied on 50 skin lesion images achieve accuracy between 96% and 98% [8]. The generative adverse neural network applied to the International Skin Imaging Collaboration (ISIC) 2018 dataset achieved 86.1% accuracy [9]. Self-attention dependent Progressive generative adversarial network classified vascular/BCC/pigmented benign keratosis/pigmented Bowen's/nevus/dermatofibroma skin cancer with 70% accuracy [10]. Transfer learning is a good option for training on large datasets instead of going for training the DL network from scratch. Kalouche fine-tuned the VGG16 CNN approach for the automatic detection of skin cancer [11].

13.2 Methodology

13.2.1 Database Used

The standard HAM10000 dataset was used to train the model [12], which has 10,015 skin lesion images categorized into seven groups. Dermatoscopic images from different populations captured and saved using various modalities are included in the dataset. The various forms of skin cancer include BCC, actinic keratosis, dermatofibroma, melanoma, benign keratosis-like lesions, melanocytic nevi, and vascular lesions. Figure 13.1 shows an example of skin cancer images from the HAM10000 database. Figure 13.2 displays many forms of skin cancer.

13.3 Dataset Division

Figure 13.3 shows the distribution of the HAM100000 dataset. The JPEG formats of the dataset images are saved at a resolution of 600 × 450 pixels. For the preprocessing part, first, the three-channel RGB images in the dataset are resized to 224 × 224 × 3, and then they are scaled to 0 mean and unit variance.

The medical field frequently deals with imbalanced datasets that have large class disparities, and this dataset was no exception. This turned out to be a major obstacle that needed to be overcome in the suggested investigation. The performance of the model is typically impacted by the biased or skewed prediction that results from this imbalanced dataset.

Data augmentation can create a balanced dataset by expanding the sample size for certain unbalanced classes. The method of generating new training data out of the preexisting dataset is known as data augmentation, and it is a method applied in computer vision and ML to make datasets larger. The goal of data augmentation is to enhance the accuracy and generalization of a machine learning model. Data augmentation involves applying transformations to the existing data to create new examples that are similar to the original data but not identical. The transformations can include cropping, rotating, scaling, flipping, adding noise, or changing the brightness and contrast of images.

The HAM10000 dataset has 10,015 photos, of which 70% are in training sets, and 30% are in testing sets. A training set has a total of 90% of the complete dataset and the remaining 10% of the entire dataset is used for the validation of the model.

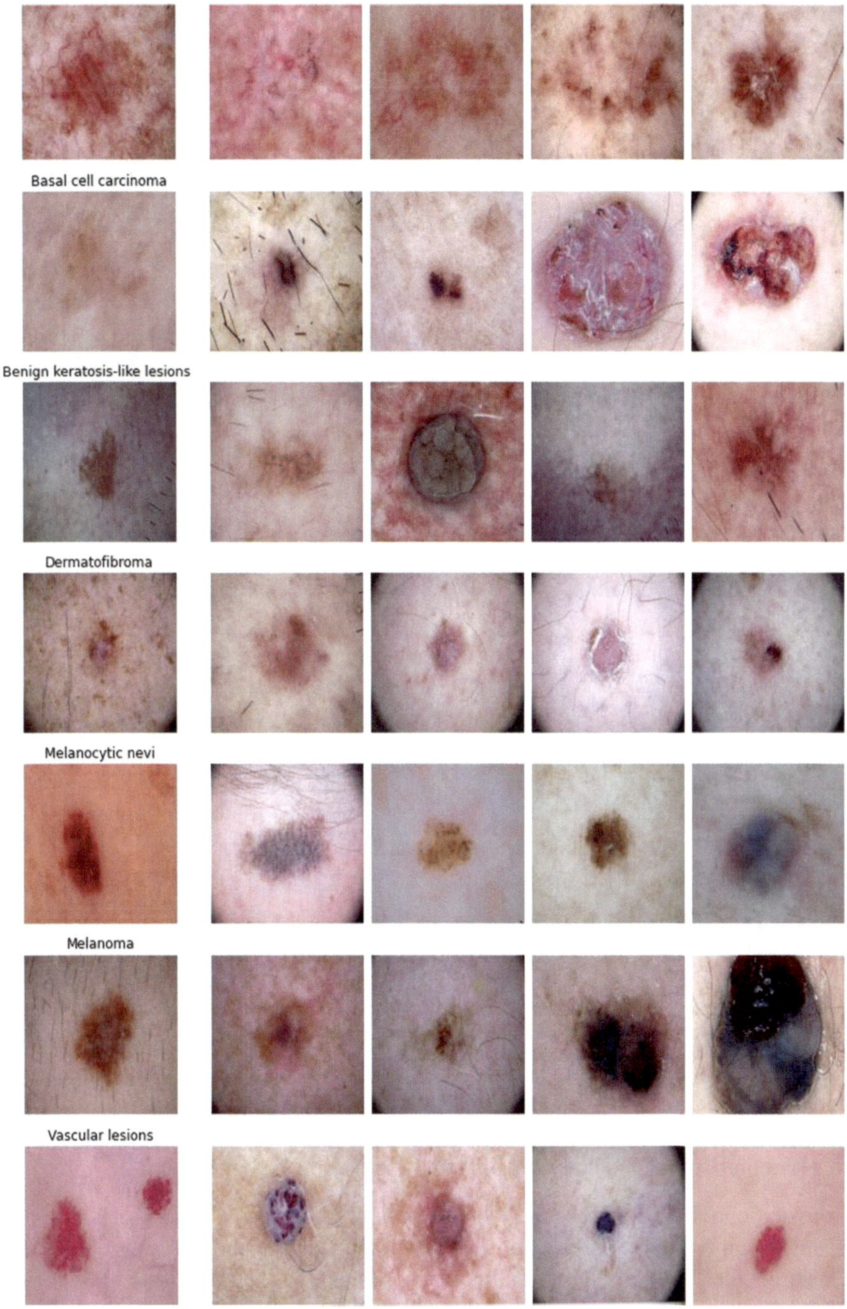

Fig. 13.1 Different types of skin cancer figures

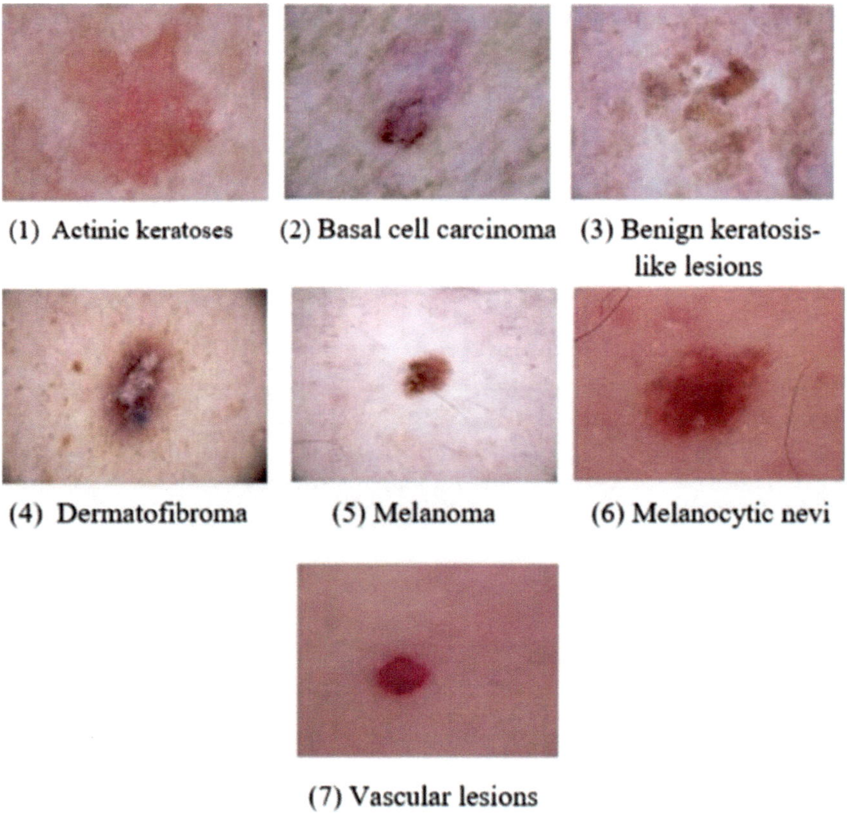

Fig. 13.2 Seven types of skin lesion categories from the HAM10000 database

13.4 Convolutional Neural Network

The DL approach utilized for a variety of computer vision applications is the CNN [13]. This algorithm is used for tasks including object detection, image segmentation, and classification. For jobs involving grid-like data, including photos and movies, CNNs are especially well-suited. Their widespread usage in various applications has transformed the domain of computer vision.

Key components and concepts of CNNs include the following:

Convolutional Layers: These are the core building blocks of CNNs. Convolutional layers apply filters (also called kernels) to input data, scanning across the data to extract local patterns. The result is a feature map that highlights the presence of particular features in the input. Convolutional layers are responsible for capturing hierarchical features in an image, starting from edges and shapes and progressing to more complex features.

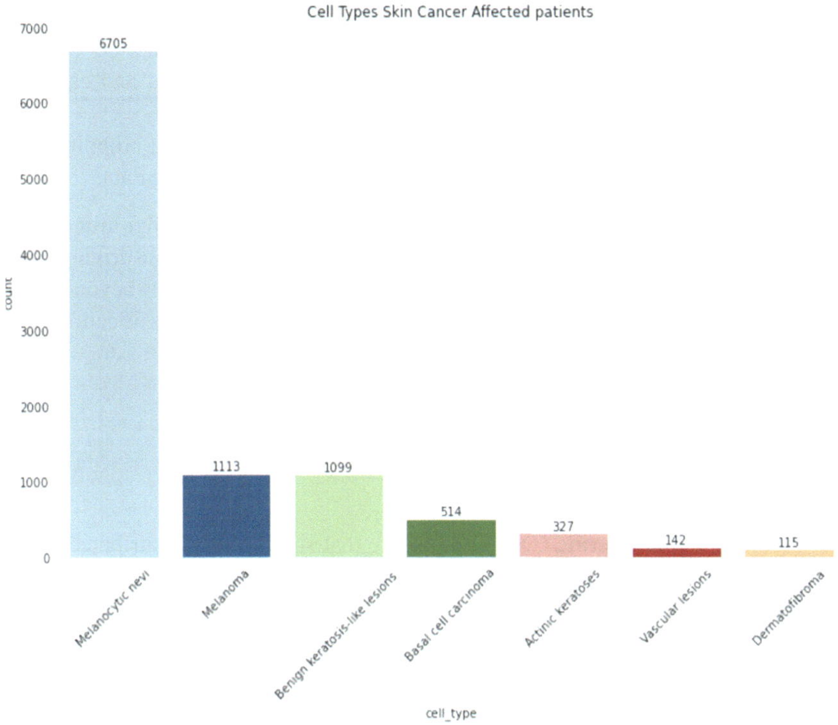

Fig. 13.3 Class distribution of the HAM10000 dataset

Pooling Layers: Pooling layers aid in lowering computation and managing overfitting by minimizing the feature maps' spatial dimension. A popular kind of pooling layer called max pooling keeps the maximum value in a certain area.

Fully Connected Layers: After many convolutional layers and pooling layers, another layer is added to the model named fully connected layers which are typically added to make predictions as per the learned features. These layers join every node in one layer to every other neuron using the output from the preceding layers that have been flattened.

Activation Functions: Rectified linear unit (ReLU) is a type of activation function used to bring non-linearity into the model. ReLU, for example, replaces all negative data with zero, facilitating the network to learn complex, non-linear patterns.

CNN Architecture: CNN architectures can vary in complexity and depth. Common CNN architectures include LeNet-5, VGGNet, GoogLeNet (Inception), AlexNet, and ResNet, each with its unique design and performance characteristics.

Training: CNNs are trained through a process called backpropagation, where they learn to optimize a specified objective, such as image classification. This process involves adjusting the network's parameters (weights and biases) using gradient descent and a loss function that quantifies the model's error.

Transfer Learning: Transfer learning is a famous method used to fine-tune pre-trained CNN models for particular tasks (typically on huge datasets such as ImageNet). This method frequently produces positive outcomes and can save time and resources.

Data Augmentation: To generalize the model, flipping, rotation, and cropping data augmentation methods are used to increase the variety of training data.

CNNs have demonstrated remarkable success in image recognition tasks, enabling applications in fields such as medical image analysis, self-driving cars, facial recognition, and many others. They are also being used in tasks beyond computer vision, such as natural language processing (e.g., in text classification) and reinforcement learning for game playing.

13.5 Alexnet

AlexNet [14] is a pioneering CNN architecture that significantly contributed to the advancement of computer vision and DL. It was proposed by Ilya Sutskever, Geoffrey Hinton, and Alex Krizhevsky, and the model got the ImageNet Large Scale Visual Recognition Challenge in 2012, demonstrating a dramatic improvement in image classification accuracy.

Key features of AlexNet include the following:

Architecture: AlexNet is made up of five convolutional layers, three fully connected layers, and a total of eight NN layers. It was one of the first CNN architectures to be deep and wide, with a large number of learnable parameters.

ReLU: AlexNet utilized the ReLU activation function in the hidden layers. ReLU helps introduce non-linearity and accelerates the training of deep networks.

Local Response Normalization (LRN): In the original AlexNet architecture, LRN was applied after the activation function. This normalization helps in creating competition between neuron activations, which can enhance the network's generalization.

Dropout: Dropout is a regularization method where random neurons are omitted while training to prevent overfitting. AlexNet used dropout in the fully connected layers.

Overlapping Max-Pooling: AlexNet used max-pooling layers with a window size of 3×3 and a stride of 2, which led to overlapping pooling regions. This was different from previous architectures and helped in improving the network's performance.

Large Number of Parameters: AlexNet had approximately 60 million parameters, which was quite large for its time. This made it computationally expensive but also contributed to its remarkable performance.

Data Augmentation: The model was generalized using horizontal flipping and random cropping data augmentation methods on the training images.

AlexNet's success in the ImageNet competition marked a turning point in deep learning and spurred the development of deeper and more sophisticated CNN architectures. Its design principles, including the use of deep layers, ReLU activations, and dropout, have become standard in many subsequent CNN architectures. Today, deeper models such as GoogLeNet, ResNet, and VGGNet have built upon the lessons learned from AlexNet and continue to advance the field of computer vision.

13.6 VGG16

VGG16 [11] is a deep CNN model that belongs to the VGG family of models. It was designed by the VGG at the Oxford University. VGG16 is simple and has a uniform architecture, which makes it easy to understand and implement.

Key features of the VGG16 architecture include the following:

Architecture Depth: VGG16 is relatively deep, comprising 16 weight layers, hence the name "VGG16." It includes three fully connected layers and 13 convolutional layers. The use of a deep architecture allows it to capture complex hierarchical features in images.

Convolutional Layers: The convolutional layers in VGG16 have a small receptive field (3×3) with a stride of 1, which helps preserve spatial information and detailed features in the input image.

Max-Pooling: Max-pooling layers are applied after every set of convolutional layers, with a window size of 2×2 and a stride of length 2, reducing the spatial dimensions of the feature set.

Filter Size: VGG16 uses 3×3 filters in all of its convolutional layers, which contributes to its uniform and simple design.

Fully Connected Layers: The final three layers of the network are fully connected layers, which are used for making class predictions. The architecture gradually reduces the spatial dimensions, resulting in a flattened feature map before these fully connected layers.

Activation Function: The VGG16 model applies the ReLU activation function in all its layers, contributing to the non-linearity and making training deep networks more efficient.

Parameter Count: While VGG16's uniform design makes it easy to understand and implement, it has a relatively large number of parameters (about 138 million), making it computationally expensive.

VGG16 is famous for its good efficiency in image classification problems, particularly on the ImageNet Large Scale Visual Recognition Challenge dataset. Its simplicity, in comparison to more complex architectures such as GoogLeNet and ResNet, has made it a popular choice for transfer learning. Researchers and developers often use pre-trained VGG16 models as a starting point for various computer vision applications, fine-tuning them on specific datasets.

While VGG16 was a significant milestone in the development of deep learning for computer vision, it has been succeeded by more efficient and powerful architectures in subsequent years. However, it remains an essential reference point in the history of CNNs and deep learning in computer vision.

13.7 Pre-trained InceptionV3

InceptionV3 [11] is a CNN architecture proposed by Google's Inception team. It is designed for image classification and object detection tasks and is part of the Inception family of models, which are known for their efficiency and accuracy. The "V3" in InceptionV3 indicates that it is the third version in the Inception series, succeeding InceptionV1 and InceptionV2.

A pre-trained InceptionV3 model refers to a trained InceptionV3 network that is trained on a large dataset, typically ImageNet, which has millions of labeled figures. This pre-training is done to learn features and patterns from a diverse set of figures, and these learned features can then be fine-tuned or used as feature extractors for various computer vision tasks.

The key features of InceptionV3 include the following:

Inception Modules: InceptionV3 makes use of "Inception modules," which are composed of convolutional layers with different filter sizes and depths. These modules are made to capture features at multiple scales and provide a trade-off between computational complexity and representational power.

Global Average Pooling (GAP): GAP is used at the end of the network by InceptionV3 instead of fully connected layers. GAP reduces the spatial dimensions and the number of parameters, making the network more efficient and less prone to overfitting.

Auxiliary Classifiers: During training, InceptionV3 uses auxiliary classifiers, which are additional output layers placed at intermediate stages in the network. This technique is used to combat the vanishing gradient problem and helps train the network more effectively.

Batch Normalization: Batch normalization is utilized in the network to speed up training and improve the generalization of the model.

Pre-trained InceptionV3 models are readily available in various DL frameworks, such as TensorFlow and Keras. The pre-trained models can be downloaded and used for a different range of computer vision problems, object recognition, image detection, and feature extraction. Fine-tuning a pre-trained InceptionV3 model on a specific dataset is a common practice to achieve high performance with minimal training time and computational resources, especially for transfer learning scenarios.

By leveraging a pre-trained InceptionV3 model, the user can quickly harness the power of deep learning for computer vision tasks, even if the researcher has a small amount of labeled data for specific applications.

13.8 Resnet50

ResNet-50 [15], short for Residual Network 50, is a CNN architecture within the ResNet family. ResNet is known for introducing a groundbreaking concept called residual learning, which greatly improved the training of very deep networks. ResNet-50, specifically, is a variant with 50 layers, making it moderately deep, and it is commonly applied for image classification and feature extraction tasks.

Key features of the ResNet-50 architecture include the following:

Residual Blocks: The fundamental innovation in ResNet is the residual block. In traditional CNNs, each layer learns a transformation from the input data to the output feature maps. However, in residual blocks, the layer learns a residual transformation. In other words, it learns the difference between the input and the desired output. This enables the network to learn both shallow and deep features effectively.

Bottleneck Architecture: ResNet-50, like other ResNet models, employs a bottleneck architecture in its residual blocks. This means that, within each block, there is a sequence of 1×1, 3×3, and 1×1 convolutions, which helps reduce the computational cost while preserving the network's representational power.

Skip Connections: Another essential feature of ResNet is the skip connections or "shortcut connections." These connections allow gradients to flow directly back through the network without passing through the multiple non-linear activation functions. This mitigates the gradient vanishing problem and makes deep network training more manageable.

GAP: Like many modern CNN architectures, ResNet-50 utilizes GAP as a way to minimize the spatial size of the feature maps before the final classification layer. GAP calculates the mean output of each feature set, resulting in a fixed-size vector for each image.

Multiple Stacks of Residual Blocks: ResNet-50 contains multiple residual blocks stacked on top of each other, with increasing numbers of filters in each block as the researcher moves deeper into the network. The network architecture has four convolutional stages, with varying numbers of residual blocks in each stage.

Pre-Trained Models: Pre-trained ResNet-50 models are available in popular deep learning frameworks such as TensorFlow and PyTorch, trained on large datasets like ImageNet. These pre-trained models are often used for transfer learning in various computer vision tasks.

ResNet-50 and other ResNet variants have achieved state-of-the-art performance in different computer vision problems, for example, object detection, image classification, and image segmentation. Their deep and residual architecture makes them suitable for a wide range of applications, and fine-tuning pre-trained ResNet-50 models can be an effective approach to achieving excellent performance on specific tasks with limited training data.

13.9 Results

Different deep learning models such as CNN, AlexNet, VGG16, InceptionV3, and Rsnet50 were trained on the skin cancer images and the trained model was tested for accuracy.

As can be observed from Table 13.1, the Pre-trained ResNet50 model achieved the highest accuracy of 92.05% compared to other models implemented by the authors such as ResNet50, Pre-trained inceptionV3, InceptionV3, VGG16, Pre-trained VGG16, AlexNet, and CNN. The pre-trained model implemented by the authors achieved better accuracy compared to models such as CNN and AlexNet. The deep learning model CNN achieved the lowest accuracy.

The layers of the RsNet50 models are as follows:

1. Input layer (224*244*3)
2. GAP layer
3. Dense layer
4. Output layer

Softmax is used as an activation function. By default, the ResNet50 has 48 convolutional layers, one max pooling layer, and one average pooling layer. Simple ResNet50 implementation achieved 76.77% accuracy.

The pre-trained inceptionV3 model achieved a good accuracy of 91.35%.

Table 13.2 gives the accuracy comparison of different existing methods with that of the proposed approach. The accuracy difference is there because of the total number of figures used for training and testing purposes. For example, Sajid et al (2018) used 500 lesion images for training and testing of data, whereas, in our model, 10,015 images were used for training the model and achieved 92.05% accuracy. Kadampur et al [3] used 80% data for training and 10% for testing and 10% for validation of the model. In our approach, we have used 70% data for training and 30% for testing purposes.

Table 13.1 Accuracy of various DL models in skin cancer detection

DL model	Accuracy	F1 score
CNN	76.74	0.65
AlexNet	83.12	0.73
VGG16	79.36	0.68
Pre-trained VGG16	89.31	0.82
InceptionV3	77.14	0.49
Pre-trained InceptionV3	91.35	0.85
ResNet50	76.77	0.51
Pre-trained ResNet50	92.05	0.84

Table 13.2 Comparison table of state-of-the-art approaches

Author	Model	Accuracy
Kadampur et al (2020) [3]	Squeeznet model	99.77
Rashid et al (2019) [9]	GAN	86.1
Abdelhalim et al (2021) [10]	Self-attention-based PGAN	70.1
Sajid et al (2018) [7]	Modified KNN	98.3
Proposed approach	Pre-trained ResNet50	92.05

13.10 Conclusion and Future Scope

Different types of deep learning models were trained and tested for their efficiency on the skin cancer image data. The pre-trained model achieved good accuracy in skin cancer detection compared to other state-of-the-art models. Maximum accuracy was obtained by the Pre-trained ResNet50 model. In the future, the method will be tested on the clinical dataset collected from Indian patients and the software will be validated in a clinical setup for real-life use.

References

1. Dash DP, Kolekar M, Chakraborty C, Khosravi MR. Review of machine and deep learning techniques in epileptic seizure detection using physiological signals and sentiment analysis. ACM Trans Asian Low-Resour Lang Inf Process. 2024;23(1):1–29.
2. Jain S, Pise N, et al. Computer aided melanoma skin cancer detection using image processing. Procedia Comput Sci. 2015;48:735–40.
3. Kadampur MA, Al Riyaee S. Skin cancer detection: applying a deep learning based model driven architecture in the cloud for classifying dermal cell images. Inf Med Unlocked. 2020;18:100282.
4. Ragab M, Choudhry H, Al-Rabia MW, Binyamin SS, Aldarmahi AA, Mansour RF. Early and accurate detection of melanoma skin cancer using hybrid level set approach. Front Physiol. 2022;13:965630.
5. Kolekar MH, Dash DP. A nonlinear feature based epileptic seizure detection using least square support vector machine classifier. In: TENCON 2015-2015 IEEE region 10 conference. IEEE; 2015. p. 1–6.
6. Mahbod A, Schaefer G, Wang C, Ecker R, Ellinge I. Skin lesion classification using hybrid deep neural networks. In: ICASSP 2019-2019 IEEE international conference on acoustics, speech and signal processing (ICASSP); 2019. p. 1229–33.
7. Sajid PM, Rajesh DA. Performance evaluation of classifiers for automatic early detection of skin cancer. J Adv Res Dyn Control Syst. 2018;10:454–61.
8. Acosta MFJ, Tovar LYC, Garcia-Zapirain MB, Percybrooks WS. Melanoma diagnosis using deep learning techniques on dermatoscopic images. BMC Med Imaging. 2021;21:1–11.
9. Rashid H, Tanveer MA, Khan HA. Skin lesion classification using GAN based data augmentation. In: 2019 41St annual international conference of the IEEE engineering in medicine and biology society (EMBC); 2019. p. 916–9.
10. Abdelhalim ISA, Mohamed MF, Mahdy YB. Data augmentation for skin lesion using self-attention based progressive generative adversarial network. Expert Syst Appl. 2021;165:113922.

11. Kalouche S, Ng A, Duchi J. Vision-based classification of skin cancer using deep learning. In: 2015, conducted on Stanfords Machine Learning course (CS 229) taught; 2016.
12. Tschandl P, Rosendahl C, Kittler H. The HAM10000 dataset, a large collection of multi-source dermatoscopic images of common pigmented skin lesions. Sci Data. 2018;5(1):1–9.
13. Amirreza Rezvantalab, Habib Safigholi, and Somayeh Karimijeshni. Dermatologist level dermoscopy skin cancer classification using different deep learning convolutional neural networks algorithms. arXiv preprint arXiv:1810.10348, 2018.
14. Dorj U-O, Lee K-K, Choi J-Y, Lee M. The skin cancer classification using deep convolutional neural network. Multimed Tools Appl. 2018;77:9909–24.
15. Alavikunhu Panthakkan SM, Anzar SJ, Mansoor W. Concatenated xception-resnet50—a novel hybrid approach for accurate skin cancer prediction. Comput Biol Med. 2022;150:106170.

Chapter 14
Leveraging MLP Mixer for Improved Melanoma Diagnosis Using Skin Lesion Images

Jyoti Dabass, Bhupender Singh Dabass, Prasad Pramod Maharana, and Manju Dabass

14.1 Introduction

Melanoma eventuating from melanocytes or color-bearing cells is a type of skin tumor. In the UK, its mean figure of cases from 2016 to 2018 was 16,744, while the average number of deaths from 2017 to 2019 was 2341. Notably, 87% of cases of melanoma survive, and 86% are prevented [1]. Numerous methods, including invasive ones like biopsy and histological examination, can be used to diagnose melanoma. Thermography, dermatoscopy, and computer-aided analysis are a few non-invasive methods [2]. The BCN_20000 dataset [3], the HAM10000 ("Human Against Machine with 10000 training images") dataset [4], and the MSK dataset [5] are a few of the data repositories that combine a variety of skin cancer images. Because of their dependability and accuracy in feature identification for classification applications, convolutional neural networks (CNNs) have been widely favored. The MNIST database showed an error rate of 0.23% in 2012. "Afterward, AlexNet attained a top-5 error rate of 18.2% to win the 2012 ImageNet Large Scale Visual Recognition Challenge [6]." Vision Transformers (ViTs) used self-attention learning to enhance ViT's training. As a result, ViT was able to outperform CNNs that were trained only on ImageNet. The best technique achieved an accuracy of 94.55% on

J. Dabass (✉)
Yardi School of Artificial Intelligence, Indian Institute of Technology, Hauz Khas, Delhi, India

B. S. Dabass
Department of Law, Institute of Law and Research, Faridabad, Haryana, India

P. P. Maharana
Artificial Intelligence and Machine Learning Programme, Liverpool John Moores University, Liverpool, UK

M. Dabass
EECE Department, The NorthCap University, Gurugram, India

T. Choudhury, A. Katal (eds.), *Machine Learning for Disease Detection, Prediction, and Diagnosis*, https://doi.org/10.1007/978-981-96-4241-0_14

CIFAR-100, and 88.58% on ImageNet, outpacing the existing methods on numerous image identification functions [7]. Transformers and CNNs have demonstrated impressive performance; however, these methods typically come with significant processing overhead. In 2021, a straightforward and effective model made up of multilayer perceptron (MLP) was presented in light of the high computational expenses connected with CNNs and attention modules in Transformers [8]. Instead of using convolution, this new design made use of methods such as matrix multiplication, token mixing, and channel mixing. Our curiosity was piqued by this novel architecture's assertion that, despite convolution and attention providing good performance, neither is required. When trained on larger datasets using regularization approaches, the MLP mixer delivers good results on image classification benchmarks, with inference costs and pre-training in contrast to those of existing models. MLP can work efficaciously with small model sizes, but over-fitting severely degrades the model as the size grows. This research work would use an MLP mixer to address some of the shortcomings of CNN and Transformers in the image classification task—that is, the identification of melanoma. To give the model a variety of images to learn from, the three datasets have been mixed. This proposed effort will concentrate on developing a model without complex computing modules, such as convolution or attention, and aim to achieve performance comparable to that of cutting-edge CNN and transformer models. Here is a brief overview of the principal findings:

1. To categorize images of pigmented skin lesions, an MLP mixer architecture-based classification model has been proposed.
2. To focus on a less computationally demanding lightweight model that trains more quickly than cutting-edge models without sacrificing functionality.

14.2 Problem Statement and Related Works

CNN versions have shown remarkable effectiveness in image recognition and classification. Using novel designs and techniques, CNNs have been surpassing their performance. Nevertheless, to identify features in images and reach conclusions, these models rely on convolution. Convolution is useful for identifying features, but it is not required. The complexity and depth of CNN models have increased since they have more layers than AlexNet, VGGNet, ResNet, and Inception. Recent models (networks) have input layers with receptive fields that span the whole input image. This implies that all of the input image pixels are included in the context that the features used in the final output feature map [9]. Convolution is still the core technique of deep neural networks, despite the quick development of neural network designs. Convolution kernels, which are based on the well-known image filtering method, are incredibly resilient due to two remarkable characteristics: they are channel-specific and spatially-agnostic [10]. According to [11], their proposed study using CNN to diagnose skin cancer produced a training accuracy of 83.04%, recall of 0.80, and precision of 0.81. According to several ViT research, self-attention alone—without the use of convolution—is sufficient for a wide category of computer vision applications [12]. In the primary experiment, images were directly transformed using a traditional transformer with the least amount of editing. To accomplish this, the image was divided into patches, which were then supplied to the transformer as a series of linear embeddings. Similar to how words are handled in an NLP application, image patches were also processed in this manner. Subsequently, the model underwent supervised training to become an image classifier. This architecture produced low-key accuracies analogous to ResNet of similar dimensions when trained on medium-sized datasets such as ImageNet shorn of any substantial regularization [7].

A new transformer structure called the ViT made its debut at International Conference on Learning Representations (ICLR) in 2021. AUC (area under the receiver operating characteristic (ROC) curve) of 0.771 ± 0.018, which postulates upgrading in skin cancer analysis errands, was obtained from testing ViT on skin cancer detection [13]. Researchers looked for ways to improve ViT training after ViTs gave computer vision a self-attention-based foundation. This further led to the suggestion of a knowledge distillation technique that enabled ViT to outperform CNNs trained only on ImageNet [14]. Transformers and CNNs have demonstrated impressive performance; however, these methods typically come with significant processing overhead. When developing models, resources such as training time, RAM, and graphics processing unit (GPU) are crucial. Even if a model is proficient in classifying images at a high computational cost, it will not produce an effective result. Given the high computational costs of Transformers' attention modules and CNNs' convolution modules, a straightforward and effective model made of MLP was suggested. To epitomize the association between tokens and channels, correspondingly, this novel architecture used methods including token besides channel mixing [15]. MLP mixer is the term given to this novel design. MLPs have long been used in machine learning and statistical modeling but were eventually abandoned because they produced unsatisfactory results. However, the introduction of the MLP mixer opened up new possibilities for computer vision. This new approach showed that, while neither convolution nor attentiveness is necessary for optimal performance, they are both sufficient for it. Using ImageNet-21 K, the authors achieved an accuracy of 84.15% [8]. In the MLP mixer, a completely linked layer was suggested as a projection layer follower. The images were supplied via the projection layer after being divided into equal patches. "Table X" was indicated as the projection layer's output. Upon assessing whole techniques, the MLP mixer needed the minimum proportion of training period despite the slow pace of the model's convergence [16]. When trained from scratch, the Tolstikhin-developed MLP mixer obtained a Canadian Institute For Advanced Research (CIFAR)10 accuracy of almost 80%. Neyshabur gained 85.19% accuracy on the CIFAR10 dataset by compounding an MLP model with a regularizer [17].

MLP can work efficaciously with small-sized models, but over-fitting severely degrades the model as the size grows. The author thinks that overfitting is the main hindrance keeping MLP from accomplishing State Of The Art (SOTA) efficacy [18]. MLPs form the base of the MLP mixer's structure. MLP mixer is fit to extract the main information from the image by thoroughly combining data from several patches and channels. Overfitting of base models can be dodged by upsetting this architecture's information mixing mechanism, which can supplementarily enhance the transferability of adversarial instances across models [19]. After investigating several design possibilities for this architecture, the research team found that multiplicative gating in conjunction with linear spatial projections yields the best results. Since the model is made up of fundamental MLP layers with gating, it was given the moniker gMLP [20]. To enhance communications between spatial places, gMLP [21] created a gating operation. An affine transform layer was proposed by ResMLP to help stack a lot of MLP blocks. On CIFAR-10 and CIFAR-100, ResMLP managed to get top-1 accuracy of 98.1% and 87.0%, respectively [22].

When it comes to MLP-based models, there are some contemporary models such as ResMLP and MLP_mixer that flatten the spatial proportions of the input data and then apply a linear projection beside the spatial dimension [8]. However, there is another variant called "Vision Permutator" that conserves the original spatial sizes of the input tokens [23]. This chapter aims to develop a lightweight, MLP-based architecture capable of identifying features in images of lesions with pigmented skin. The HAM10000, BCN_20000, and MSK datasets together will be used to train the model. Tables 14.1 and 14.2 summarize the existing literature on skin cancer detection.

HAM 10000 dataset as well as the ISIC dataset, being dependable public datasets for skin cancer and their precise labels (diagnostic), have been the subjects of the majority of research and innovative models [4]. An additional discernible trend is the extensive use of computational resources throughout the model training process. Numerous models in the aforementioned studies were trained on GPUs such as the

Table 14.1 CNN models' performance on the skin lesion categorization task

References	Classification	Dataset	Methodology	Results
[24]	Binary (malign/benign)	ISIC 2017	VGGNet	Sensitivity:78.6 6%
[25]	Binary (non-cancer/cancer)	ISIC 2017	ResNet	AUC is 0.915
[26]	Binary (non-cancer/cancer)	Combination of data from the Stanford Hospital and ISIC, Dermofit Library22 data	Inception V3 with pre-trained weights (ImageNet)	Accuracy: 72.1%, AUC: 0.91
[27]	Binary (Melanoma/Melanocytic nevi)	ISIC2017	Ensemble of CNN models	Specificity and sensitivity are 88% and 58% correspondingly
[28]	Binary (Melanoma/Melanocytic nevi)	ISIC2017	Google's inception v4 CNN architecture	Sensitivity: 95%, AUC: 0.95, and Specificity: 80%

Table 14.2 Transformer models' performance on the skin categorization task

References	Classification	Dataset	Methodology	Results
[29]	Two-class (Melanoma/ Non- Melanoma)	ISIC 2017	Parameter-free attention module SimAM and the pre-trained swin transformer	Precision: 74.0% and AUC:0.900
[30]	Two-class (Melanoma Vs se borrheic keratosis)	ISIC 2017	Convolution layers and the ViT model are used to categorize and segment images.	Sensitivity: 68.0, accuracy: 89.2%
[31]	Multiclass	HAM10000	Mutual attention transformer	Accuracy: 92.5%
[32]	Multiclass	HAM10000 dataset	Pre-trained ViT	Accuracy:73.7%
[33]	Multiclass	ISIC 2018	Spatial pyramid transformer and a full transformer network	Sensitivity: 85.7, accuracy: 89.6%
[34]	Multiclass classification	The Human Against Machine (HAM 10000 dataset)	ViT model	Accuracy is 94.1%

V100 and A100, with a smaller number using Giga Texel Shader eXtreme (GTX) Titan. The computing cost of these GPUs is high for training these models. For the model to be generalized, the training time—which is dependent on the hardware and data pre-processing—also matters a lot. Even while CNNs and Transformers perform remarkably well on classification tasks, training and inference come at a very high cost. The majority of studies have failed to produce models that can contend with existing techniques. Therefore, we need a model that can attain the same level of performance with less computing overhead and training time. We anticipate using more recent image classification algorithms in our research, which are quicker to deploy and require less computing power to train.

14.3 Methodology

Earlier research focused on modeling transformer and CNN models for skin cancer images. These techniques have demonstrated impressive results and their predictive potential. However, there are drawbacks to these models, including high computing costs, significant data augmentations, and high data consumption. Therefore, our goal is to present a lightweight model that requires less computing power and trains more quickly, all without significantly sacrificing prediction accuracy or performance. This methodology has the potential to yield good outcomes while cutting down on overtraining and deployment time.

We established the following criteria as we investigated our possibilities for selecting the ideal model:

1. Lightweight architecture and design that does not rely on attention or convolution.
2. Relatively quicker to train and having good generalization.
3. Makes reduced use of GPU and memory for processing.
4. Performance comparable to CNN and transformer models.

JAX/Flax model pre-trained on the dual public data repositories of ImageNet and ImageNet-21 k is utilized in [35], as the MLP mixer model was relatively new. JFT-300 M, an exclusive dataset of 300 M samples and 18 k categories is used for training in [36] for assessing efficiency on a bigger scale. As we had limited choices for converting weights and existing models written in Flax/Jax to TensorFlow in the tractable model, we preferred to train the model from scratch in TensorFlow. Training from scratch was simpler and less complicated thanks to the model's features, which include faster training and convergence.

14.3.1 Description of Dataset

The HAM10000 dataset has been selected for our investigation [4]. These cases include all vital diagnostic criteria linked to pigmented lesions provided in Table 14.3. This dataset is altered to convert a multi-class dataset into a binary classification task. As several CNN models have already used this dataset to demonstrate high predictive qualities and accuracy, the dataset is enlarged to 100 × 100 for providing an unbiased comparison with transformer and CNN models. Notably, 70:10:20 is the fraction for the training, validation, and testing set.

Table 14.3 Image distribution as per classes in the dataset

No.	Skin lesion category	Count of images
1	Vascular lesions (pyogenic granulomas angiomas, angiokeratomas and hemorrhage, **vasc**).	142
2	Basal cell carcinoma (**bcc**)	514
3	Actinic keratoses and intraepithelial carcinoma/Bowen's disease (**akiec**)	327
4	Benign keratosis-like lesions (seborrheic keratoses/solar lentigines and lichen-planus-like keratoses, **bkl**)	1099
5	Melanoma (**mel**)	1113
6	Dermatofibroma (**df**)	115
7	Melanocytic nevi (**nv**)	6705
All images		10,015

14.3.2 Pre-processing of Data

Data is pre-processed to boost image attributes and to increase the focus of model training on the targeted region.

14.3.2.1 Data Grouping

There are 10,015 images in the original dataset, separated into seven classifications. We reorganized the data into two classes: non-melanoma and malignant (a binary classification). Melanoma (mel) class image can be found in the category "Melanoma." The strategy for classifying the remaining groupings as non-cancer was different. Since they are not malignant in character, we regarded dermatofibroma (df), vascular lesions, solar lentigines, seborrheic keratoses, and lichen-planus-like keratoses (bkl) as benign keratosis-like lesions and dermatocytic nevi (nv) as non-cancer [37]. The remaining two categories, which include Bowen's disease (akiec) and actinic keratoses and intraepithelial carcinoma (bcc), are either precancerous or malignant and have not been taken into consideration [38]. Table 14.4 provides the image count.

14.3.2.2 Image Normalization

In image processing, normalization is done to fine-tune the value of pixel intensity. Normalization also known as contrast widening or histogram stretching provides intensity smoothing. To move the original image's intensities into the range [0,255], we used a normalization logic. The "Min-max normalization" norm is applied, and it operates by linearly transforming the initial data. Using this method, the scaled data falls within the intended range. In terms of math, it may be stated as follows:

$$New\left(I_N\right)=\left(\frac{I-I_{MIN}}{I_{MAX}-I_{MIN}}\right)*I_{NEW\_MAX} \tag{14.1}$$

where I_N denotes the Nth pixel intensity of the image; I_{MIN} and I_{MAX} indicate the minimum and maximum intensity values for the entire image; I_{NEW_MAX} shows the maximum intensity value of the desired range; and $New(I_N)$ signifies the new intensity of the Nth pixel of the image. Figure 14.1 shows the normalized and original skin image.

Table 14.4 Image distribution after grouping data

Category	Image count
Non-melanoma (class 0)	8061
Melanoma (class 1)	1113

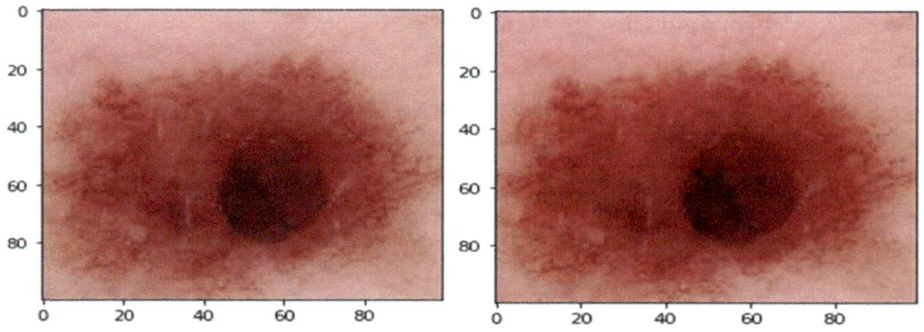

Fig. 14.1 Before and after normalization (left and right) of the skin image

14.3.2.3 Balance of Classes and Augmentation of Data

When one class has noticeably fewer observations than the others, this is referred to as a class imbalance. This problem is common in scenarios where anomaly identification plays a crucial role, such as power theft, financial fraud, diagnosing uncommon illnesses, and so on. In these cases, machine learning techniques can be unreliable and biased. Therefore, to make sure that our model is devoid of bias, we must rebalance the classes. Accordingly, we must perform "over-sampling" in which sample seven images for every image depicts "melanoma" [39]. Data augmentation involves employing deep learning models or image processing tools to create new data points or make small changes to the existing data. We have applied augmentations that vary the hue-saturation, color, intensity, and shift of the pixels in the image. In addition, we applied random rotation (limited to 70°), equalization, sharpening, and Gaussian noise to the images.

14.3.3 Model Structure

The model is implemented in TensorFlow using the existing Keras documentation. We created a new model, trained it, and then deduced the outcomes. Additionally, we adjusted the hyperparameters to maximize the model's performance. The proposed design uses patches as input, feeding it to a linear classification head, fully linked layers with a mixer layer, and Gaussian error linear unit (GELU) activations. Also, it uses regularization techniques such as layer norm and skip connections. It has three major parts as given below:

1. Linear embedding per patch.
2. Blender strata.
3. Head of classification.

14.3.3.1 Linear Embedding Per-Patch

For this, the input images having proportions (H, W) are segregated into a set of S mutually distinct and dissimilar image darns where each patch has two dimensions. The projection layer linked with output dimension C projects these patches linearly. Here network grows linearly instead of quadratically.

14.3.3.2 Mixer Layers

The mixer layer contains two distinct MLP layer types. It is important to know that the idea for these two layers came from CNN.

1. Channel Mixing.
   Multiple channels of communication are made possible by this MLP. It operates independently on each location using distinct rows from the table as inputs. After processing one patch of the image, it just processes the channels.
2. Mixing Tokens.
   MLP assists the communication between various spatial locations (or tokens) within an image. It uses distinct columns from the table as inputs and runs on each channel independently. Patches of images are provided as input to the network. Mathematically, the mixer layers can be written as follows:

$$U_{*,i} = X_{*,i} + W_2\sigma\left(W_1 LAYERNORM\left(X\right)_{*,i}\right), for\, i = 1,\ldots\ldots\ldots c \tag{14.2}$$

$$Y_{*,i} = U_{j,*} + W_4\sigma\left(W_3 LAYERNORM\left(U\right)_{j,*}\right), for\ j = 1,\ldots\ldots\ldots s \tag{14.3}$$

   where X and s denote the linear protrusion table's columns and input chronology of mutually distinct image darns correspondingly. W_n, c, *and* σ specify the weights of layer n, hidden dimension, and pixel-wise GELU activation correspondingly. Additionally, the layer norm utilized in transformer architectures is used in the architecture. The GELU activation includes skip connections for regularization and dropout for non-linearity.

14.3.3.3 Classification Head

The classification head is the MLP-mixer architecture's final layer which is made up of a linear classifier, a standard categorization head, and a global average pooling layer. Additionally, a dropout layer in the head addresses overfitting.

14.3.3.4 Loss Function

During training, we work on minimizing the output gap between the desired and expected results utilizing binary cross-entropy loss. It can be computed using probability where $p(i)$ = probability that the class is 1 and $1–p(i)$ = probability that the class is 0 for measuring loss between actual and predicted numerals.

Then binary cross entropy can be denoted as

$$BCE = \frac{1}{n}\sum_{i=1}^{N}\left(y_i * \log\left(P_i\right)\right) + \left(1 - y_i\right) * \log\left(1 - P_i\right)) \quad (14.4)$$

14.3.3.5 Optimizer

Every epoch of the deep learning model's training requires us to adjust the weights and decrease the loss function. An optimizer helps in improving accuracy and decreasing total loss.

AdamW Optimizer

Weight decay is only applied in AdamW, an improved version of Adam. Because AdamW has established superior training loss and models trained with it generalize significantly finer than models trained with Adam, this improved variant can contend with stochastic gradient descent [40].

Algorithm Showing AdamW Optimizer Algorithm Stages [40]

Adam with L_2 regularization and Adam with weight decay (AdamW).

1. "**assume** α =0.001, $\beta_1 = 0.9$, $\varepsilon = 10^{-8}$, $\beta_2 = 0.999, \omega \epsilon \mathfrak{R}$.
2. **Set** time step $t \leftarrow 0$, parameter vector $x_{t=0} \in \mathcal{R}^n$, first moment vector $m_{t=0} \leftarrow 0$, second moment vector $v_{t=0} \leftarrow 0$, schedule multiplier $\eta_{t=0}\ \varepsilon\ \mathfrak{R}$.
3. reiterate,
4. $t \leftarrow t+1$
5. $\nabla f_t(x_{t-1}) \leftarrow$ SelectBatch(x_t − 1) Δ select batch and reciprocate the respective gradient.
6. $g_t \leftarrow \nabla f_t(x_{t-1}) + \omega x_{t-1}$
7. $m_t \leftarrow \beta_1 m_{t-1} + (1 - \beta_1)g_t$ Δ here and below all actions are element-wise.
8. $v_t \leftarrow \beta_2 v_{t-1} + (1 - \beta_2)g_t^2$
9. $\widehat{m_t} \leftarrow m_t / (1 - \beta_1^t)$ Δ β_1 is taken to the power of t.
10. $\widehat{v_t} \leftarrow v_t / (1 - \beta_2^t)$ Δ β_2 is taken to the power of t.
11. $\eta_t \leftarrow$SetScheduleMultiplier (t) Δ can be fixed, decay, or also be used for warm restarts.
12. $x_t \leftarrow x_{t-1} - \eta_t \left(\alpha \widehat{m_t} / \left(\sqrt{\widehat{v_t}} + \varepsilon\right) + \omega x_{t-1}\right)$
13. **until** stopping criterion is achieved.
14. **return** optimized parameter $\boldsymbol{x_t}$".

14.3.4 Applying the MLP-Mixer Paradigm

We examined the author's JAX/Flax MLP mixer code before we put the model into TensorFlow. This served as the foundation for our review and validation of our execution. The model design is divided into three pieces.

1. MLP Block.

   MLP blocks include GELU activation and dense layers. The Flax/JAX application of the MLP block is shown below.

```
"class MlpBlock (nn. Module):
 mlp_dim:  int
 @nn. compact
 def _ _call_ _ (self, x):
    y = nn. Dense (self. Mlp_dim) (x)
     y= nn. gelu(y)
      return nn. Dense (shape [-1] (y)"
```

2. Mixer Block.

   The input obtained from each MLP is combined by the mixer block before being sent back to the classifier head. Channel mixing and Token mixing are the two MLP blocks that make up the mixer layers. Flax/JAX implementation of the Mixer block is as follows.

```
"class MixerBlock (nn. Module):
  tokens_mlp_dim: int
  channels_mlp_dim: int
  @nn. compact
  def _ _call_ _ (self, x):
   y=nn. LayerNorm () (x)
   y=jnp. swapaxes (y, 1, 2)
   y=MlpBlock (self. tokens_mlp_dim, name= 'token_
     mixing') (y)
   y=jnp. swapaxes (y, 1, 2)
   x= x+ y
   y= nn. LayerNorm () (x)
   return x+ MlpBlock (self. Channels_mlp_dim, name=
     'channel_mixing') (y)."
```

3. MLP-mixer model assembling with classifier head.

   It is the last block in which the classifier head tailored to the classification task (Binary/Multiclass) is merged after the mixer and MLP blocks.

14.3.5 Improving MLP-Mixer Model Performance

Subsequent actions are used to refine the model provided below.

1. ***Build** the Patch extraction to convert the images to patches.*
2. ***Build** the MLP mixer module with the two MLP blocks.*
3. *Add a Global Average Pooling to flatten the output into 1-dimension from the MLP mixer layer in Step 2.*
4. *Add a Dropout Layer to the network to apply regularization.*
5. *Add a Dense neuron with Sigmoid Layer activation for our Binary classification task.*
6. *Train the layers from scratch along with callbacks (**Early Stopping and Learning Rate Scheduler)** along with a low learning rate.*

14.3.6 Experimentation Methodologies

The patch size, hidden units/embedding size for the MLPs, number of blocks, Keras Tuner (Hyperband algorithm), dropout rate, and learning rate are tuned to get better performance. In a non-stochastic, infinite-armed bandit issue, the Hyperband technique distributes a pre-decided reserve, such as iterations to randomly selected configurations [41].

14.3.6.1 Keras Tuner

Keras Tuner is a user-friendly framework for hyperparameter optimization which is premeditated to be effortlessly enhanced by scholars to test out innovative search tactics, in addition to having built-in systems for random search, Hyperband, and Bayesian optimization [42].

14.3.7 Evaluation

The confusion matrix provided in Table 14.5 helps in evaluating the model's efficiency on unknown data.

1. Accuracy (A).

   It keeps track of how many positive and negative observations are accurately identified:

$$A = \frac{T_p + T_n}{T_p + T_n + F_p + F_n} \tag{14.5}$$

Table 14.5 Confusion matrix

Predicted	Actual negative (0)	Actual positive (1)
Positive (1)	False positive (F_p)	True positive (T_p)
Negative (0)	True negative (T_n)	False negative (F_n)

2. Precision *(P)*.
   It estimates the number of observations agreeing with the positive outcomes:

$$P = \frac{T_p}{T_p + F_p} \tag{14.6}$$

3. Recall and Sensitivity (R).
   It quantifies the fraction of categorized positive observations out of all positive observations:

$$R = \frac{T_p}{T_p + F_n} \tag{14.7}$$

4. $F1$-Score ($F1$).
   It is the recall and precision harmonic mean:

$$F1 = 2 * \left(\frac{P * R}{P + R} \right) \tag{14.8}$$

5. ROC AUC Score.
   The area under the ROC curve or ROC AUC computes the area to show how good our curve is.

14.4 Results

This part will go over the outcomes of putting our proposed model into practice, tweak it, and present some test set predictions. Additionally, we talked about the data's relevance as compared to previous research.

14.4.1 Performance of MLP Mixer Model on Testing Set

Using the available Keras and TensorFlow documentation, we constructed the MLP mixer having batch sizes of 32 and 3323 testing set images. We have mentioned our base model's performance in Table 14.6.

Table 14.6 MLP mixer base model performance

	Parameters				Result	Weighted avg	Weighted avg	
Architecture	Size of patch	Dropout rate	Embedding dim	Num blocks	Accuracy of test data	Precision	Recall	F1-score
Base MLP mixer	8	0.1	512	4	84.04%	0.84	0.84	0.84

14.4.2 Outcomes of the Fine-Tuned Model

14.4.2.1 Performance

The best-performing model parameters were figured out using the Keras tuner with the Hyperband technique.

Findings

We can see from Table 14.7 that performance tends to decrease as patch size and embedding dimensions rise. As the image size is fixed at 100 × 100, the performance is negatively impacted when patch sizes are increased beyond a certain point. According to the original design [8], the model has also been applied to 224 × 224 quality images. Our goal is to concentrate on lightweight models that require less computational power and train more quickly.

14.4.2.2 Additional Performance Adjustments

We choose the optimal hyperparameters from the tuner search to boost the model's efficacy. Apart from those hyperparameters, we modified the optimizer AdamW's weight decay from 0.0001 to 0.00001 and used a learning rate of 0.6.

14.4.2.3 Lightweight

The model is lightweight and compact with better performance. The weights and trained model require 25 MB of disc space. The size comparison of techniques is shown in Table 14.8.

14.4.2.4 Training Time

The model trains more quickly on inexpensive computing resources since it depends less on convolution or attention methods. The Kaggle Notebook's Nvidia Tesla T4 16 GB graphics card allowed for a relatively short training period. In less than 60 minutes, the complete model was trained.

Table 14.7 Model assessment for hyperparameter choice

Model	Optimizer	Dropout rate	Patch size	Embedding dim	Num blocks	Test accuracy (%)
MLP mixer	AdamW	0.1	8	256	4	84.0
		0.1	8	256	8	85.2
		0.2	16	384	16	77.2
		0.2	16	384	32	74.8
		0.1	32	512	8	73.2
		0.3	32	512	32	65.4
		0.1	**8**	**512**	**8**	**87.3**

Table 14.8 Comparison of size of MLP mixer with CNN architectures

Model name	Size (in MB)	Size (in MB)	Difference
ResNet50V2	25	98	−73.0
Xception	25	88	−63.0
ResNet50	25	98	−73.0
VGG19	25	549	−524.0
VGG16	25	528	−503.0

14.4.3 Best Model

We get the highest 87.3% accuracy on the testing set using the optimal model having dropped out (0.1), embedding dim (512), and patch size (8). Without any pre-training, the proposed MLP mixer demonstrated good dataset generalization. Table 14.9 illustrates the guidelines of the best-performing model.

Table 14.10 provides the results of the fine-tuned MLP mixer model. From Table 14.10, it can be observed that the performance of the fine-tuned MLP mixer is better compared to the base MLP mixer.

14.4.3.1 Confusion Matrix

Table 14.11 shows the results of confusion metrics. From Table 14.11, it can be noted that we got good results for all metrics.

14.4.3.2 Area Under Curve—Receiver Operating Characteristic Curve

The AUC-ROC curve embodies the model's capability to discriminate between classes. Greater AUC advances a model's capacity to differentiate between patients with the illness and those without. An AUC that is near to one designates a high degree of separability in an excellent model. An AUC that is around zero indicates that a model is weak since it has the lowest degree of separability. The AUC for the

Table 14.9 Parameters of the best-performing MLP mixer model

Name of parameters	Value
Metric	Accuracy
Input shape/input_size	100 × 100, 3 channels
Epochs/epochs	100
Loss/loss	Binary cross-entropy
Weight decay/weight_decay	0.0001
Optimizer/optimizer	AdamW
Learning rate/learning rate	0.001
Layers/num_blocks	8
Hidden units/embedding_dim	512
Patch size/patch_size	8

Table 14.10 Results of the fine-tuned MLP mixer model

Model	Hyper-Parameters				Outcomes		Weighted avg	Weighted avg
Name	Size of patch	Dropout rate	Num blocks	Embedding dim	Test accuracy	F1	Precision	Recall
Base MLP mixer	8	0.1	4	512	84.04%	0.84	0.84	0.84
MLP mixer (fine-tuned)	**8**	**0.1**	**8**	**512**	**87.33%**	**0.873**	**0.876**	**0.872**

Table 14.11 Metric assessment based on confusion matrix

Metrics	Value (for Melanoma)	Value (for Non-Melanoma)
Misclassification rate	12.0%	12.7%
Accuracy	87.0%	87.3%
AUC score	0.87	0.87
Precision	91.0%	89.7%
F1-score	88.0%	86.3%
Recall	85.%	83.2%

random classifier is 0.5. For a classifier to be considered good, its AUC must be higher than 0.5.

14.4.4 Assessment with Research on Transformer and CNN Models

Existing models used the HAM10000 dataset to identify skin cancer by focusing on multiclass or binary classification methods for melanoma detection. Our model has outperformed the other models in this classification task. The comparison summary is included in Table 14.12.

14.5 Discussion

The proposed model is a quick-to-train, light-weight, broadly applicable model which makes it better compared to attention or convolution-based models. It provides 87.3% accuracy and a Dice coefficient (F1-score) of 87% for binary categorization. We used Albumentations to have class balance and Keras documentation to develop the MLP-mixing model. Table 14.7 shows the results obtained during preliminary research for hyperparameter selection. It helped us in selecting the hyperparameters for training the MLP mixer model to get better performance. Test data accessed the model's performance. Tables 14.9 and 14.11 display the optimal model parameters and performance, respectively. Table 14.11 shows the detailed results as per the confusion matrix. It is observed that the model is somewhat overfitting and that the validation accuracy fluctuates. Our attempt to raise the dropout rate resulted in worse performance. Therefore, we did not alter the optimized parameters. It is also significant to note that, in comparison to CNNs, the training is a little more unstable due to the large fluctuations in losses. The AUC-ROC outline given in Fig. 14.2 illustrates how well the classifier can reliably categorize the samples.

As we know an ideal classifier has a curve leaning toward the upper left corner with area = 1 under the curve. As per the resulting image displayed in Fig. 14.2, our architecture has an AUC of 0.87 and a curvature tilting toward the y-axis. Our proposed model getting an accuracy of 87.3%, a precision of 87.6%, a recall of 87.2%, and an AUC of 0.87 falls contentedly between the CNN and transformer architectures as provided in Table 14.12.

Significance of the Study Melanoma, the deadliest cancer claiming lives every year, can be early diagnosed and treated with the help of non-surgical diagnosis methods such as imaging. These imaging methods necessitate the purchase of costly imaging machinery and correspondingly qualified doctors. Deep learning helps in detecting melanoma more quickly. The most advanced deep learning techniques available today, such as CNN and Transformers, have extremely high accuracy rates, but they are also quite complex, need large datasets, require lengthy training periods, and incur significant processing costs. CNN models require more time to train because of features such as Max Pooling and a higher layer count (Depth). With all these limitations, training and applying a model to a particular application

Table 14.12 Comparison of the MLP mixer model with existing works

References	Technique	Precision	Accuracy	Recall	AUC	F1-score
[25]	ResNet	–	–	–	0.915	–
[24]	VGGNet	–	–	78.66	–	–
[26]	Inception V3	–	72.1	–	0.91	–
[27]	CNN ensemble	–	–	58	–	–
[28]	Inception V4	–	–	95	0.95	–
[43]	MobileNetV2	–	91.33	–	–	–
[44]	Inception V3	–	–	–	0.943	–
[29]	Swin transformer	74	–	–	0.9	–
[30]	ViT	–	89.2	68	–	–
[45]	SqueezeNet	–	92.18	80.77	–	80.84
Proposed work	**MLP-mixer model**	**87.6**	**87.33%**	**87.2**	**0.87**	**87.3**

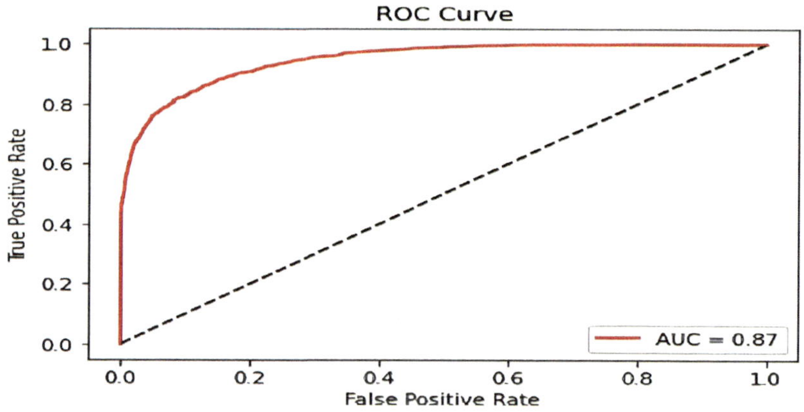

Fig. 14.2 Curvature of the ROC for the MLP-mixer model

is difficult. The proposed work provides an effective MLP-mixer-based architecture where features are frequently combined at one place or between several locations in today's image processing networks. For example, CNNs use convolutions, kernels, and pooling to execute both mixes, while ViTs use self-attention. The model also takes advantage of skip connections and regularization. This work aimed to develop a competitive substitute for CNNs that can achieve comparable performance while utilizing lighter architectures and quicker training. This will help the medical diagnostics sector develop reliable and affordable imaging systems that can accurately and cost-effectively identify melanoma without the need for pricey imaging equipment. To shorten the time, it takes to start therapy, this might be taken into consideration for primary diagnosis before consulting. Additionally, we hope that this will encourage additional interest in and development of MLP for computer vision applications.

14.6 Conclusion

The proposed chapter focuses on creating a lightweight and dependable model to categorize skin lesion images into non-melanoma and melanoma images. The presented work provides 87.33% accuracy, 87.6% precision, and 87.2% recall on the HAM10000 dataset, thus outperforming all previous techniques developed for this dataset. The results of the proposed model are verified by expert doctors making it a trustworthy and useful method for melanoma detection. With the use of this portable model, medical personnel may more precisely distinguish skin lesions from melanoma.

With an accuracy of 87.33% for binary classification and a Dice coefficient (F1-score) of 87.3%, we have created a model that generalizes well and requires fewer resources to train. Our study was limited to the identification of melanoma using images of skin lesions. This can be expanded upon to include multiclass skin lesion classification. With sophisticated augmentations such as CutMix, Rand Augment, and Mixup, the model can outperform CNN models on huge datasets and generalize effectively. Overfitting in the model can be avoided via regularization approaches such as gradient clipping and early halting. Since the model was trained for binary categorization of skin lesion images, it may be evaluated for multiclass classification in the future.

References

1. Cancer Research UK. Melanoma skin cancer statistics. [online]. 2022. https://www.cancerresearchuk.org/health-professional/cancer-statistics/statistics-by-cancer-type/melanoma-skin-cancer#heading-Four. Accessed 25 Jul 2022.
2. Narayanamurthy V, Padmapriya P, Noorasafrin A, Pooja B, Hema K, Firus Khan AY, Nithyakalyani K, Samsuri F. Skin cancer detection using non-invasive techniques. RSC Adv. 2018;849:28095–130. [online]. https://doi.org/10.1039/C8RA04164D.
3. Combalia M, Codella NCF, Rotemberg V, Helba B, Vilaplana V, Reiter O, Carrera C, Barreiro A, Halpern AC, Puig S, Malvehy J. BCN20000: dermoscopic Lesions in the Wild. [online]. 2019. http://arxiv.org/abs/1908.02288.
4. Tschandl P, Rosendahl C, Kittler H. The HAM10000 dataset, a large collection of multi-sources dermatoscopic images of common pigmented skin lesions. Sci Data. 2018;51:180161. [online]. http://www.nature.com/articles/sdata2018161.
5. Codella NCF, Gutman D, Celebi ME, Helba B, Marchetti MA, Dusza SW, Kalloo A, Liopyris K, Mishra N, Kittler H, Halpern A., Skin lesion analysis toward melanoma detection: a challenge at the 2017 International symposium on Biomedical Imaging (ISBI), Hosted by the International Skin Imaging Collaboration (ISIC). [online]. 2017. http://arxiv.org/abs/1710.05006.
6. Krizhevsky A, Sutskever I, Hinton GE. ImageNet classification with deep convolutional neural networks. Commun ACM. 2017;606:84–90. [online]. https://dl.acm.org/doi/10.1145/3065386.
7. Dosovitskiy A, Beyer L, Kolesnikov A, Weissenborn D, Zhai X, Unterthiner T, Dehghani M, Minderer M, Heigold G, Gelly S, Uszkoreit J, Houlsby N. An image is worth 16x16 words: transformers for image recognition at scale. [online]. 2020. http://arxiv.org/abs/2010.11929.

8. Tolstikhin I, Houlsby N, Kolesnikov A, Beyer L, Zhai X, Unterthiner T, Yung J, Steiner A, Keysers D, Uszkoreit J, Lucic M, Dosovitskiy A. MLP-mixer: an all-MLP architecture for vision. [online]. 2021. http://arxiv.org/abs/2105.01601.
9. Araujo A, Norris W, Sim J. Computing receptive fields of convolutional neural networks. Distill 411. 2019. [online]. https://distill.pub/2019/computing-receptive-fields. Accessed 25 Jul 2022.
10. Li D, Hu J, Wang C, Li X, She Q, Zhu L, Zhang T, Chen Q. Involution: inverting the inherence of convolution for visual recognition. [online]. 2021. http://arxiv.org/abs/2103.06255.
11. Subramanian RR, Achuth D, Kumar PS, Naveen Kumar Reddy K, Amara S, Chowdary AS. Skin cancer classification using convolutional neural networks. In: 2021 11th International conference on cloud computing, data science & engineering (confluence). [online] IEEE, 2021, pp. 13–9. https://ieeexplore.ieee.org/document/9377155/.
12. Tatsunami Y, Taki M. Sequencer: deep LSTM for image classification. [online]. 2022. http://arxiv.org/abs/2205.01972.
13. de Lima LM, Krohling RA. Exploring advances in transformers and CNN for skin lesion diagnosis on small datasets. [online]. 2022. http://arxiv.org/abs/2205.15442.
14. Wei G, Zhang Z, Lan C, Lu Y, Chen Z. ActiveMLP: an MLP-like architecture with active token mixer. [online]. 2022. http://arxiv.org/abs/2203.06108.
15. Guo J, Tang Y, Han K, Chen X, Wu H, Xu C, Xu C, Wang Y. Hire-MLP: vision MLP via hierarchical rearrangement. [online]. 2021. http://arxiv.org/abs/2108.13341.
16. Kumar NS, Ramaswamy Karthikeyan B. Diabetic retinopathy detection using CNN, transformer and MLP based architectures. In: ISPACS 2021 – International symposium on intelligent signal processing and communication systems: 5G dream to reality, proceeding. Institute of Electrical and Electronics Engineers; 2021.
17. Lv T, Bai C, Wang C. MDMLP: image classification from scratch on small datasets with MLP. [online]. 2022. http://arxiv.org/abs/2205.14477.
18. Zhai X, Kolesnikov A, Houlsby N,Beyer L. Scaling vision transformers. In: Zhao Y, Wang G, Tang C, Luo C, Zeng W, Zha Z-J, editors. A battle of network structures: an empirical study of CNN, transformer, and MLP. [online] 2021. http://arxiv.org/abs/2108.13002.
19. Lyu H, Wang Y, Tan Y, Zhou H, Zhao Y, Zhang Q. Boosting adversarial transferability of MLP-mixer. [online]. 2022. http://arxiv.org/abs/2204.12204.
20. Liu H, Dai Z, So DR, Le QV. Pay attention to MLPs. [online]. 2021. http://arxiv.org/abs/2105.08050.
21. Tang C, Zhao Y, Wang G, Luo C, Xie W, Zeng W. Sparse MLP for image recognition: is self-attention really necessary? [online].2022. www.aaai.org.
22. Touvron H, Bojanowski P, Caron M, Cord M, El-Nouby A, Grave E, Izacard G, Joulin A, Synnaeve G, Verbeek J, Jégou H. ResMLP: feedforward networks for image classification with data-efficient training. [online]. 2021. http://arxiv.org/abs/2105.03404.
23. Hou Q, Jiang Z, Yuan L, Cheng M-M, Yan S, Feng, J., Vision permutator: a permutable MLP-like architecture for visual recognition. [online]. 2021. http://arxiv.org/abs/2106.12368.
24. Romero-Lopez A, Giro-i-Nieto X, Burdick J, Marques O. Skin lesion classification from dermoscopic images using deep learning techniques. In: Biomedical engineering. Calgary: ACTA Press; 2017.
25. Bi L, Kim J, Ahn E, Feng D. Automatic skin lesion analysis using large-scale dermoscopy images and deep residual networks. 2017.
26. Esteva A, Kuprel B, Novoa RA, Ko J, Swetter SM, Blau HM, Thrun S. Dermatologist-level classification of skin cancer with deep neural networks. Nature. 2017;5427639:115–8. [online]. http://www.nature.com/articles/nature21056.
27. Marchetti MA, Codella NCF, Dusza SW, Gutman DA, Helba B, Kalloo A, Mishra N, Carrera C, Celebi ME, DeFazio JL, Jaimes N, Marghoob AA, Quigley E, Scope A, Yélamos O, Halpern AC. Results of the 2016 international skin imaging collaboration international symposium on biomedical imaging challenge: comparison of the accuracy of computer algorithms to dermatologists for the diagnosis of melanoma from dermoscopic images. J Am Acad Dermatol. 2018;782:270–7.

28. Haenssle HA, Fink C, Schneiderbauer R, Toberer F, Buhl T, Blum A, Kalloo A, Hassen ABH, Thomas L, Enk A, Uhlmann L, Alt C, Arenbergerova M, Bakos R, Baltzer A, Bertlich I, Blum A, Bokor-Billmann T, Bowling J, Braghiroli N, Braun R, Buder-Bakhaya K, Buhl T, Cabo H, Cabrijan L, Cevic N, Classen A, Deltgen D, Fink C, Georgieva I, Hakim-Meibodi L-E, Hanner S, Hartmann F, Hartmann J, Haus G, Hoxha E, Karls R, Koga H, Kreusch J, Lallas A, Majenka P, Marghoob A, Massone C, Mekokishvili L, Mestel D, Meyer V, Neuberger A, Nielsen K, Oliviero M, Pampena R, Paoli J, Pawlik E, Rao B, Rendon A, Russo T, Sadek A, Samhaber K, Schneiderbauer R, Schweizer A, Toberer F, Trennheuser L, Vlahova L, Wald A, Winkler J, Wölbing P, Zalaudek I. Man against machine: diagnostic performance of a deep learning convolutional neural network for dermoscopic melanoma recognition in comparison to 58 dermatologists. Ann Oncol. 2018;298:1836–42.
29. Xie J, Wu Z, Zhu R, Zhu H. Melanoma detection based on Swin transformer and SimAM. In: 2021 IEEE 5th information technology, networking, electronic and automation control conference (ITNEC). IEEE; 2021. p. 1517–21.
30. Chen J, Chen J, Zhou Z, Li B, Yuille A, Lu Y. MT-TransUNet: mediating multi-task tokens in transformers for skin lesion segmentation and classification. 2021.
31. Zhou L, Luo Y. Deep features fusion with mutual attention transformer for skin lesion diagnosis. In: 2021 IEEE international conference on image processing (ICIP). IEEE. 2021, pp. 3797–801.
32. Pedro R, Oliveira AL. Assessing the impact of attention and self-attention mechanisms on the classification of skin lesions. In: 2022 International joint conference on neural networks (IJCNN). IEEE. 2022, pp. 1–8.
33. He X, Tan E-L, Bi H, Zhang X, Zhao S, Lei B. Fully transformer network for skin lesion analysis. Med Image Anal. 2022;77:102357.
34. Xin C, Liu Z, Zhao K, Miao L, Ma Y, Zhu X, Zhou Q, Wang S, Li L, Yang F, Xu S, Chen H. An improved transformer network for skin cancer classification. Comput Biol Med. 2022;149:105939.
35. Deng J, Dong W, Socher R, Li L-J, Li K, Fei-Fei L. ImageNet: a large-scale hierarchical image database. In: 2009 IEEE conference on computer vision and pattern recognition. IEEE; 2009. p. 248–55.
36. Sun C, Shrivastava A, Singh S, Gupta A. Revisiting unreasonable effectiveness of data in deep learning era. In: 2017 IEEE international conference on computer vision (ICCV). IEEE; 2017. p. 843–52.
37. Luba MC, Bangs SA, Mohler AM, Stulberg DL. Common benign skin tumors. Am Fam Physician. 2003;674:729–38.
38. National Health Service for England, Skin cancer (non-melanoma). [online]. 2020. https://www.nhs.uk/conditions/non-melanoma-skin-cancer/. Accessed 23 Nov. 2022.
39. Ando S, Huang CY. Deep over-sampling framework for classifying imbalanced data. 2017, pp. 770–85.
40. Loshchilov I, Hutter F. Decoupled weight decay regularization. arXiv preprint arXiv:1711.05101. 2017.
41. Li L, Jamieson K, DeSalvo G, Rostamizadeh A, Talwalkar A. Hyperband: a novel bandit-based approach to hyperparameter optimization. 2016.
42. O'Malley T, Bursztein E, Long J, Chollet F Jin H, Invernizzi L, et al. KerasTuner. 2019. https://github.com/keras-team/keras-tuner. Accessed 25 Nov. 2022.
43. Ech-Cherif A, Misbhauddin M, Ech-Cherif M. Deep neural network based mobile dermoscopy application for triaging skin cancer detection. In: 2019 2nd international conference on computer applications & information security (ICCAIS). IEEE; 2019. p. 1–6.
44. Pham TC, Hoang VD, Tran CT, Luu MSK, Mai DA, Doucet A, Luong CM. Improving binary skin cancer classification based on best model selection method combined with optimizing full connected layers of deep CNN. In: 2020 international conference on multimedia analysis and pattern recognition (MAPR). IEEE; 2020. p. 1–6.
45. Abayomi-Alli OO, Damaševičius R, Misra S, Maskeliūnas R, Abayomi-Alli A. Malignant skin melanoma detection using image augmentation by oversampling in nonlinear lower-dimensional embedding manifold. Turk J Electr Eng Comput Sci. 2021;29SI-1:2600–14.

Chapter 15
Application of AI to Detect Brain Tumors

Karabi Kalita, Prachi Mall, and Deepika Raina

15.1 Introduction

Having a malignant brain has a 5-year life expectancy of 34% for men and 36% for women, making it the tenth significant cause of death, along with other neurological cancers [61]. Additionally, the World Health Organization states that there were around 120,000 fatalities from brain tumors in the years prior, and an estimated 400,000 persons worldwide are affected by them. The brain tumor is diagnosed in approximately 250,000 people per year and 2% of those instances are identified as cancerous tumors. Between 2004 and 2020, the global rate of brain tumors grew from roughly 10% to 15% [68]. In India, almost 28,000 cases of brain tumors are documented and over 24,000 individuals die every year were estimated by the International Agency for Research on Cancer. The brain is a vital and complex organ of an individual's body, governing both purposeful and involuntary processes [53]. It is among the most sophisticated and vulnerable organs, controlling a wide range of vital processes such as cognition, motor skills, emotional processes, hearing, vision, and bulk of bodily processes including analyzing, integrating, organizing, determining, and proving instructions that affect the rest of human body [58]. The anatomical structure of the human brain is exceedingly complex. Because their presenting symptoms, which include headaches, cognitive impairments, and seizures, are sometimes mistaken for other, more benign ailments, brain tumors are considered to be among the most difficult illnesses to diagnose. The diagnosis, analysis, and development of an effective treatment for illnesses such as stroke, confusion, brain tumors, and headaches in central neurocytoma are incredibly challenging [70]. A brain tumor is the formation of an irregular cluster of cells in the brain. There are four common classifications for the existence of a brain tumor, namely,

K. Kalita · P. Mall · D. Raina (✉)
School of Pharmacy, Graphic Era Hill University, Dehradun, Uttarakhand, India
e-mail: deepikaraina@gehu.ac.in

T. Choudhury, A. Katal (eds.), *Machine Learning for Disease Detection, Prediction, and Diagnosis*, https://doi.org/10.1007/978-981-96-4241-0_15

metastatic, glioblastoma, meningioma tumor, and astrocytoma [56]. Benign (non-cancerous) and malignant (cancerous) brain tumors are two main categories, and brain-based benign tumors are non-progressive and less invasive and do not spread throughout a person's body. On the contrary, cancerous tumors are malignant and propagate throughout the body quickly [42]. Though benign brain tumors can cause several serious problems, they are not malignant, grow slowly, and seldom spread to other tissues. They also have more well-defined borders, making surgical excision easier, and they do not usually reappear following removal. Malignant brain tumors are cancerous, develop quickly, and can propagate to other regions of the brain or central nervous system, resulting in potentially fatal consequences. They can have an impact on various areas of the brain that control various activities. This implies that the adverse effect of any brain tumor is determined by the tumor's location, size, and aggressiveness [41]. Studies showed that higher levels of depression symptoms in survivors of brain tumors were connected with poor life quality, self-assurance, and lack of confidence. According to qualitative studies, brain tumors and treatment impacted many elements of everyday life [19].

Artificial intelligence (AI), which is extensively utilized in many fields nowadays, can perform effectively and rapidly. This is because machine learning (ML) algorithms are continuously being created and improved to address a variety of problems, including those in the medical field, making the application of AI in medical imaging one of the most significant scientific concerns [46]. AI aims to create machines that behave and work like humans. In recent years, improvements in AI have occurred, with the technology growing increasingly complicated, autonomous, and smart. As the range of data increases exponentially, so does the computer power advances which significantly accelerates AI development. The lives of humans are currently being impacted by AI's practical achievements in a wide range of domains. AI has the power to enhance people's health and well-being by enhancing the diagnostic work of experts, placing a strong emphasis on prevention, and providing individualized treatment recommendations via electronic health records (EHRs) [15, 26]. AI is being used to improve automatic diagnosis in medical imaging, which is dependent on image interpretation. Radiology diagnosis has grown more accurate as a result of the incorporation of AI. Object recognition and segmentation using ML in medical visuals is a crucial use of AI. An important and challenging issue in pattern recognition, computer vision, and image processing is image segmentation. It is employed for many different tasks, such as medical imaging analysis and target segmentation [40]. A lot of speculation is going on about how AI could result in job automation and significant workforce displacement. Deep learning (DL) models are trained at labs and start-ups for specific image recognition tasks. Distinct imaging equipment manufacturers and DL algorithms have independent focal points for the likelihood of a lesion, cancer, characteristic, or location of a nodule (Fig. 15.1) [13].

Due to the high resolution of magnetic resonance imaging (MRI), which offers extensive information on the framework of the brain, MRI scans have a significant impact on automated medical image processing. MRI techniques are primarily used by researchers to detect and analyze brain tumors. Researchers recently created

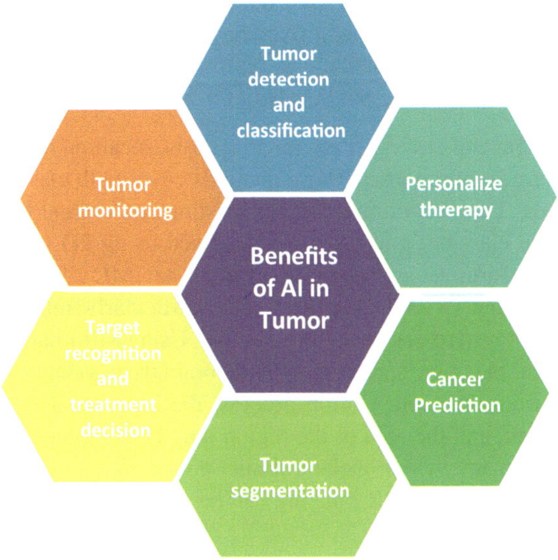

Fig. 15.1 Benefits of AI in tumor

several kinds of automated brain tumor identification and categorization systems using MRI scans [37].

15.2 Current Methods of Brain Tumor Detection

Any brain cancer that keeps growing raises the patient's risk of progressive neurological damage from compression or invasion, persistently high intracranial pressure due to tumor growth, and secondary hydrocephalus. When brain tumors are detected early, there is a chance of the following:

- When suspected cases are diagnosed sooner, it lowers the chance of developing a neurological impairment as a result of brain damage caused by tumors before, during, or after radiation therapy. This could lessen the burden of a residual big cerebral tumor and improve tolerance to and responsiveness from radiation therapy and chemotherapy [66].
- It may have an effect on the severity of symptoms at diagnosis, provide different surgical options (such as resection of the tumor instead of just a biopsy), and affect the decision about additional oncology treatment [24].
- Early diagnosis may eventually increase the prognosis for patients with brain tumors. Reducing delays in the diagnostic process might help lessen service consumers' discontent with the healthcare system and their lack of confidence in primary care [24].

- Minimize patients' and families' concerns about the repercussions of preventable delays in diagnosis, whether those delays are caused by physicians or systemic delays [39].
- Strengthen the public's belief in the healthcare system.

Several strategies for the diagnosis, location, and classification of brain tumors have been reported. These methods cover both classic ML methods and DL methods. This section investigates existing and current brain tumor detection methods. The classifier's features were derived using 2D Gabor filter and discrete wavelet transform techniques [39]. Traditional ML (TML) approaches, support vector machine (SVM) classifiers, and particularly multi-layer perceptron were often employed to diagnose brain tumors. Using MR imaging to detect and segment brain tumors is a critical topic. Normal brain tissues can be separated from damaged ones during the segmentation process, the location of the tumor region can be recognized, and the area of the region may be computed by using numerous features of the image, which include tissue, boundaries, brightness, and contrast. Tumor differentiation in MR imaging can be challenging due to tumor size, form, and location [50].

It is challenging to manually evaluate the numerous MRI images generated frequently by the healthcare sector. TML algorithms have grown in popularity for picture classification, but they are limited by the need to define "feature vectors" for extraction from raw data [35]. Tumor segmentation in MR imaging can be challenging due to parameters such as tumor location, shape, and size [33]. Because there are so many MRI images produced during medical practice, visually segmenting tumors or lesions is a tedious and arduous process. Since it frequently needs a lot of data, segmenting brain tumors from MRIs is a critical endeavor in image processing. Additionally, a tumor's soft tissue margins may not be clearly defined [30]. Traditional MR imaging is strongly reliant on the tumor's fundamental anatomy, position, composition, mass effect, and frequency, and some shortcomings still pose difficulties, such as differentiating brain tumors from simulated lesions, identifying and evaluating tumors, and separating recurrent tumors from tissue necrosis. For classification purposes, conventional ML techniques are applied to tackle an issue demand for the creation of a custom feature. While getting accurate classification results, DL methods can be created so that no manual feature extraction is necessary [32].

Early tumor detection can be crucial in determining the possibility of recovery, treatment, and a better chance of survival. Tumor segmentation is essential before any therapy to sustain healthy tissues and eliminate tumor cells. By enabling the quickest possible diagnosis of brain tumors, AI being applied in medical imaging, resulting in reasoning of images and providing relevant information on the shape, width, position, and metabolic process of brain tumors, aids in auto diagnosis. Even though these techniques are employed to provide the most reliable information about brain tumors, tumor tissues must be detected, properly recognized, and separated to segment a tumor [7]. Malignant from benign tumors must be distinguished by pathologists from other tumors since these distinctions directly affect the range of therapy choices that can be used. The preliminary identification of minor brain metastasis in the patient over some time is difficult as MRI susceptibility fluctuates and various features of MRI acquisition can affect performance. As stereotactic radiosurgery procedures as well as therapeutic decisions rely on the quantity and location of even minor metastases, neuroradiologists are concerned about early detection, considering the major influence on the patient's eventual outcome [51]. As a result, the bulk of computer-aided detection technologies available in neuro-oncology are primarily focused on the computerized diagnosis of brain metastases. Early-stage tumor diagnosis is critical because it aids in the prevention of tumor growth through medical interventions. Over time, ML, image processing, and pattern recognition algorithms aid this process, and new techniques emerge periodically [29]. Brain tumor patients endure long visits to specialized hospitals, more time at home, rising expenses, and costly home renovations while experiencing life-limiting mental and physical difficulties [14].

15.3 Application of AI in Brain Tumor Detection

DL is an AI approach that replicates how the human brain processes information and creates prototypes that might be used to make sound decisions. Enormous amounts of data can be examined using ML. They may also require additional effort and resources to properly train, despite providing faster and more accurate results for identifying lucrative possibilities or significant risks. Combining AI with ML and cognitive approaches may assist it handle enormous volumes of data more efficiently [63]. The fundamental goal of automated tumor detection is to compile essential diagnostic information on the nature, existence, and location of tumors. The information collected from clinical imaging can influence and control upcoming treatments, resulting in the right tumor identification and therapy [3]. Segmentation is a vital procedure in a variety of programs, which comprises brain cancer diagnosis and detection, precise and unified quantification of disease burden, and large-scale assessment for further investigation. An accurate characterization of tumor borders is critical for brain tumor management planning [18]. AI is a type of invention that models human cognitive functions such as analyzing, learning, thinking, and problem-solving. From its beginning as a branch of computer science, AI has expanded to become a broad field of study that invites contributions from physics, statistics, and information technology (Fig. 15.2) [9].

- Deep Learning

DL is a method that uses advanced and deep networks to give highly predictive performance, whereas ML is a broad concept that refers to a computer's ability to learn and hence improve procedures and frameworks of analysis. It is a type of artificial brain network that was built specifically to interpret pixel input, and it involves a mathematical action known as convolution, which is a specialized type of linear operation. Simple neural networks having at least one layer that employs

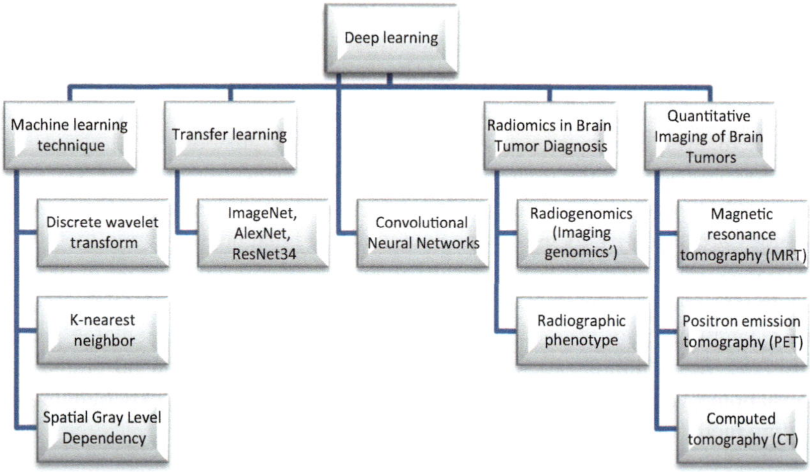

Fig. 15.2 Application of AI in tumor imaging

convolution rather than standard matrix multiplication are known as convolutional networks. The structure of convolutional neural network (CNN) neurons is more akin to the frontal lobe of the human brain, which is crucial for processing visual stimuli. The standard neural network's challenge with piecemeal image processing is overcome since different kinds of neurons have been programmed to embrace the whole field of vision [10]. To quickly achieve high classification accuracy for glioma, meningioma, and pituitary brain illnesses, a deep convolution neural network design is proposed. The consolidated strategy, which includes the hidden layers and consists of different stages, produces the most visible effect on tumor apprehension [43]. By reducing the extent of damage and its interactions with brain regions, AI-driven technologies have already started to play an important function in surgical preparation. This allows for precise brain surgery that is as extreme as is logically required to maintain the standard of life. Ultimately, AI-assisted models can anticipate problems, recurrences, and therapeutic responses, recommending the best course of action. AI-powered algorithms are anticipated to eventually combine biochemical and physiological characteristics in order to divide susceptible and influence patients to individualized screening techniques [16]. Data mining is based on ML, a type of computational intelligence that attempts to automatically identify significant patterns in data. Initial data must be provided in a specified format for ML algorithms to be trained to perform a specific task connected to medical visuals, depending on whether or not the prediction algorithm runs supervised or unsupervised. The next step involves using the computing capacity of contemporary computers to carry out these functions automatically or partially automatically to take the place of or enhance the efficacy of an individual decision-maker [17, 64].

Brain tumors may be swiftly and accurately classified or segmented using an automated multi-layered deep neural network technique utilized by DL. DL enables the application of a constructed CNN approach for brain tumor classification. This model was created for several programs, including AlexNet, GoogleNet, and ResNet-34 [71]. It presents a complicated 3D model-based upgraded CNN-based brain tumor classification model. The experimental results reveal that the classification produced through this model shows the best results since the CNN optimization hyperparameters improved the CNN's performance. Adaptive Dynamic Sine Cosine Function Grey Wolf Optimizer Algorithm (ADSCFGWO) method combines the gray wolf optimizer and sine cosine algorithms in a flexible framework that benefits from both algorithms' advantages (Table 15.1) [1].

Table 15.1 Ongoing AI algorithms with rate of accuracy for detection of brain tumor

Sl.	AI Algorithm	Accuracy	References
1.	Diverse capsule networks (DCNet++) employ a hierarchical learning architecture and advanced capsule algorithms networks (DCNet) effectively add a deeper convolutional network.	Both the DCNet++ and DCNet algorithms had accuracy values of 95.03% and 93.04%, respectively.	Phaye et al. [49]
2.	Three CNN algorithms, VGG-16, VGG-19, and AlexNet, were used to distinguish between healthy and tumor brain visuals.	AlexNet performed the best, with 99.55% accuracy.	Abd-Ellah et al. [2]
3.	CNN was used to identify pituitary, meningioma, and glioma tumors by extracting hidden characteristics from images and selecting features.	The accuracy of the proposed method is 93.68%.	Pashaei et al. [48]
4.	Three automated processes were used: Brain image processing, brain categorization, and extraction of traits. It was decided to use the Principal Component Analysis – Normalized GIST descriptor (PCA-NGIST) technique, which combines PCA and the normalized GIST descriptor.	This method has an accuracy level of 93.68%.	Gumaei et al. [25]
5.	In the learning process, a CNN, or "Adam" optimizer, which is a stochastic optimization technique based on the random gradient descent principle, was used.	98.51% and 84.19%, respectively, were the training and validation accuracy percentages.	Abiwinanda et al. [5]
6.	Different techniques were compared to the CNN model to increase performance in classifying brain tumors from MRI scans.	The average rate of accuracy was 87.14, 93.74, 95.97, 96.65, and 100 by CNN-based AlexNet.	Tandel et al. [59]
7.	Glioma Isocitrate Dehydrogenase (IDH) mutations and multiparametric imaging features utilizing 3D-CNN	It shows effectiveness with a specificity of 97% and AUC of 99%.	Yogananda et al. [69]
8.	High-density network and a softmax layer were included in the deep inception residual network.	The model's accuracy outperforms cutting-edge methods by 99.69%.	Kokkalla et al. [34]
9.	Transfer learning is utilized by the VGG16, VGG19, dense net, and ResNet50 networks to detect brain tumors.	Performance is improved by 99.02%.	Özlem and Güngen [47]
10.	CNN algorithms such as Inceptionv3, AlexNet, ResNet-50, GoogleNet, and VGG-16 were used.	The first CNN design has a 99.33% accuracy.	Irmak [31]

15.4 Benefits of AI in Brain Tumor Detection

Brain tumor patient outcomes may be improved by AI-driven computer-aided diagnosis (CAD). Otsu's segmentation technique is among the thresholding methods that have had an immense impact on medical imaging and imaging for nanomedicine. Precision, sensitivity, Dice coefficient, and specificity are some of the evaluation metrics standards of assessment employed for evaluating the effectiveness of tumor segmentation [8]. The harmony search algorithm, which initially was developed for photonic color images, was converted to work well on MRI grayscale data for brain segmentation. The suggested approach holds promise for helping clinicians make an immediate and precise diagnosis of the location and size of an MRI brain tumor rather than having to wait hours [6]. The sole fixed parameter is the total quantity of levels necessary for segmentation. It is also semi-automatic and does not demand training. Studies that evaluated the effectiveness of methods for determining the existence of a Tumor in an image were referred to as detection studies. AI-assisted technologies are the link between precision diagnosis and precision therapies [54]. Another advantage of employing AI for medical purposes in an era of massive data is that sophisticated structures and relationships within data can be detected algorithmically, eliminating the need for more healthcare personnel to analyze data. AI reduces costs and saves time for the healthcare sector in several ways [12].

By considering interaction, confounders, and increased occurrence of multimorbidity, the use of AI in the medical field gives physicians the chance to provide patient care that is customized to each patient's needs. In healthcare settings, increased caseloads and inadequate documentation of medical data may increase the likelihood of human mistakes. Unlike doctors, AI systems can discover and diagnose ailments faster and with less inaccuracy [23].

The development of CAD tools can increase diagnostic accuracy in the detection of small metastatic brain lesions, to enable early and correct treatment planning, especially stereotactic radiosurgery [16].

A study suggests using an AI-based framework (CNN-long short-term memory (LSTM)) in the Internet of Things healthcare sector for diagnosing and classifying brain tumors. The CNN model extracts features from medical MRI data, while the LSTM model learns about dependencies and predicts tumor class. The model was evaluated using brain MRI datasets, and data augmentation techniques improved its performance during training [28]. The CNN model uses TL techniques to improve predictive accuracy, using pre-trained models such as ResNet-50, VGG-16, Inception V3, DenseNet201, Xception, and MobileNet. The weights from these models are then transferred to the CNN model for effective training. Data augmentation techniques are used to increase the dataset size for better training. The proposed integrated diagnosis framework ResNet-CNN achieves high accuracy compared to baseline models [27].

CNN has been demonstrated to be highly accurate for classification, despite the usage of many ML and DL techniques. CNN is frequently used to distinguish between two types of brain tumors: abnormal and normal. The creation of an autonomous system for detecting brain tumors needs to take computation time, accuracy, and dependability into account [11]. Since the intensity of proteinaceous fluid in MR images is higher, the intensity of the image is analyzed using watershed methods. MRI scan pictures are analyzed using the YOLOv7 real-time object identification model to detect brain tumors. There are 10,288 photos in the dataset from different subsets. Notably, 20% is set up for testing, and the remaining 80% is used for training. The training method is enhanced by data augmentation approaches, yielding 51,448 photos for tumor identification. This method locates tumors precisely and detects them [4]. Combining various MRI sequence types, a new LSTM model has been built for the classification of brain tumors. More information is provided by this fused sequence than by a single sequence. If the 14-layer model is trained on six datasets for glioblastoma and stroke lesions, the 23 CNN model is trained using Brat's series for glioma detection [11].

15.5 Potential Challenges and Limitations

DL-based AI is opaque, keeping clinicians in the dark regarding diagnostic signs. The crucial query therefore becomes how to offer solid proof for the assertions made in the responses. There remains some discrepancy between AI models and human cognition, referred to as "black-box" transparency [65]. However, positive results from randomized trials or other forms of testing might be sufficient to demonstrate the effectiveness and safety of AIs, at least in some circumstances, negating the need to open the "black box." Present brain tumor detection technologies and systems centered around CNNs often have the following issues: (1) Because the procedure is solitary and the degree of precision is low, it cannot provide useful information to clinicians; (2) heavy reliance on human assistance along with data pre-processing; and (3) the majority of them are processed using one technique, and MRI data from many modalities is not used properly [36]. It might be difficult to hold AI systems accountable if they make mistakes in patient diagnosis and treatment. There is also a chance that, in some situations, people will be able to receive medical information from AI systems that they would prefer to get from a physician. Healthcare algorithms that use ML could be biased, predicting a greater probability of ailment based on factors such as gender or ethnicity even when these variables are not the true causes of the disease [20]. Consequently, at every stage of the product development process, AI developers must be aware of this danger and act to reduce any potential biases. To be more precise, they should examine these stages when deciding (1) which ML techniques and technologies to employ to train the algorithms and (2) which datasets to use, taking into consideration factors such as quality and variety. The most difficult task in many healthcare industries is ensuring that AI technologies are integrated into regular clinical practices, not if or when they will be mature enough to be useful. Before AI systems can be widely deployed, they must be certified by regulatory bodies, sufficiently standardized so that related products perform similarly, integrated with EHR systems, explained to physicians, sold by either private or private payer organizations, and refined over time in the field. As AI is being used more and more, there is still a knowledge gap that needs to be filled to help healthcare professionals, researchers, and the general public understand how AI functions. Furthermore, concerns about data access, collaboration, and encryption require being considered to guarantee full patient and public participation [21]. Stakeholders that guarantee the validity and trustworthiness of the datasets, update the software often, and are forthright and honest about any systemic issues, such as biased data, can help to successfully integrate AI in clinical practice. Employing ML methods and histogram-based analysis of CBV maps, we can distinguish between Glioblastoma Multiforme (GBMs) and brain metastatic tumors in a binary manner. The discovery is related to the KNN-based technique with 0.95 and Area Under the Receiver Operating Characteristic curve (AUROC) with 0.94 accuracies, and data from peritumoral edema was insufficient for distinguishing between these two units [52]. For separation among GBMs and brain metastases, ML models produced noticeably higher AUROC values (AUROC, 0.92; SVM-based approach) [60]. Additionally, when examined alongside radiologists (AUROC, 0.72–0.86), GBMs and Primary Central Nervous System Lymphoma (PCNSL) may be distinguished (AUROC, 0.98; severe gradient boosting-based model) [45]. Recent breakthroughs in ML, notably DL, have made it possible to recognize and classify patterns in medical imaging. Currently, CNNs are the most effective image processing techniques due to their many layers and excellent diagnostic accuracy when dealing with a huge number of input images. Studies have also demonstrated the excellent diagnostic accuracy of performance ratings [67]. To oversee AI-based healthcare innovations, specifically Software as a Medical Device, the United States Food and Drug Administration formed a digital health section [38].

15.6 Ethical Considerations

AI must be used in the healthcare system in a way that is both morally and legally acceptable because this growth will present many challenges. Traditional testing methods may hinder innovation, and the regulatory framework in many countries is not up-to-date with AI developments. With proper oversight, AI/ML-based medical devices can improve care quality by delivering safe and effective software functionality. To fully realize the potential of AI advancements in healthcare, the law must change. Medical professional responsibility and data privacy laws are not well defined under current regulations, which could be used for profit. The creation and regulation of e-health platforms necessitates the consideration of ethical, legal, and cultural aspects. The Indian Supreme Court has ruled that the right to privacy is a basic right, and the Ministry of Health and Family Planning is drafting rules tailored to this particular industry (Fig. 15.3) [11].

Fig. 15.3 Certain ethical and legal challenges connected with the implementation of AI in healthcare

- Safety and Transparency

AI in healthcare raises safety concerns. For instance, based on made-up cancer instances, IBM Watson for Oncology has suggested dangerous therapies. The software was trained using synthetic data, which prompted concerns about its validity and reliability. Transparency is vital for patient safety and confidence. Recommendations for "black-box" systems may need complex modifications that are outside the purview of professionals and patients. Positive results from randomized trials may demonstrate AI's effectiveness and safety. To improve AI accountability and support informed decision-making and address ethical and legal concerns, it is possible to develop interpretable algorithms, decision-making visualizations, and comprehensible explanations for AI forecasts [62].

- Algorithmic Fairness and Biases

AI presents the potential for bias and discrimination, even if it can improve healthcare in affluent environments and democratize expertise. AI product development must reduce potential biases at all stages since algorithms might exhibit biases that result in unjust outcomes, such as inaccurate diagnoses and treatments that are ineffective for some subpopulations. Some biases may be eliminated with the use of increased data availability and data gathering from minority communities. Meanwhile, some companies may argue that their work is a trade secret and object to publication. Furthermore, in situations when resources are scarce, AI designed for highly qualified experts may not always recommend safe, secure, and appropriate behavior. Additional deliberation over monetary backing and regulatory obligations is required to ensure that AI improves life in both high-income and low- and middle-income countries [22].

• Data Privacy

It is not necessary to sacrifice fundamental privacy rights to achieve innovation; the processing of patient data was not sufficiently disclosed to the parties involved in data sharing. Google DeepMind and the Royal Free NHS Foundation Trust agreed to provide patient data for Streams testing in exchange for the Trust being able to use the app for free for 5 years. Strict antidiscrimination laws and thoughtful patient data withdrawal rights are required to protect patients from uses of their data that go beyond the doctor-patient relationship, like impacts on health or insurance premiums, job opportunities, or interpersonal relationships. From an ethical and legal standpoint, data privacy is crucial for AI equity since it significantly affects patient autonomy, AI trust, and legal compliance. Strong security measures, such as anonymization and encryption, are necessary to prevent unauthorized access to patient data. To maintain accountability, prevent data misuse, and monitor and track data use, strict access controls and audit procedures must be implemented [44].

AI-based health research employs EHRs to enhance clinical treatment and conduct scientific investigations; nevertheless, a certain degree of transparency needs to be guaranteed in the interest of patient confidence and safety. On the other hand, the effectiveness depends on the caliber of the IT infrastructure. AI is capable of pattern recognition, trend analysis, and the creation of cutting-edge healthcare delivery strategies. Its capacity to manage massive volumes of data enables a deeper understanding of clinical processes and patient outcomes, which drives the creation of more beneficial clinical guidelines and support materials, elevates the standard of care, and impacts evidence-based practices [57]. Ethical questions are raised by the expanding use of AI in healthcare. Objective real-time data and diverse programming teams are required to guarantee ethical use and prevent data bias. AI can help doctors make better judgments, but it is important to hold developers and users responsible. The healthcare sector upholds moral principles, although worries regarding the potential effects of unproven AI systems on safety have been raised [55].

15.7 Conclusion

AI-enabled detection can increase precision and effectiveness, boost patient outcomes, and save healthcare costs. Early brain tumor discovery determines the possibility of recuperation, medical treatment, and a greater probability of survival. The study, classification, and segmentation of cancer imaging data, especially that related to brain tumors, as well as the precise and error-free identification of tumors so that they are differentiated from other diseases that present similarly, can all benefit greatly from the application of DL and other ML techniques. In this chapter, we aimed to explain the limitations of earlier studies as well as how AI is now used in brain tumor identification. DL enables the utilization of a previously generated CNN model for analyzing medical images, notably for brain tumor classification.

AI equity in healthcare requires robust data privacy measures to safeguard patient autonomy, trust, and legal compliance. Ethical concerns arise from AI's trend analysis and data bias prevention. This AI framework enables better communication with health-related information scientists while showing physicians and researchers the appropriate time to use AI algorithms for various types of healthcare data.

References

1. Abdelhamid AA, El-Kenawy ESM, Alotaibi B, Amer GM, Abdelkader MY, Ibrahim A, Eid MM. Robust speech emotion recognition using CNN+LSTM based on stochastic fractal search optimization algorithm. IEEE Access. 2022;10:49265–84. https://doi.org/10.1109/ACCESS.2022.3172954.
2. Abd-Ellah MK, Awad AI, Khalaf AAM, Hamed HFA. Two-phase multi-model automatic brain tumor diagnosis system from magnetic resonance images using convolutional neural networks. Eurasip J Image Video Process. 2018;1(97) https://doi.org/10.1186/s13640-018-0332-4.
3. Abd-Ellaha MK, Awad AI, Khalaf AAM, Hamed HFA. A review on brain tumor diagnosis from MRI images: practical implications, key achievements, and lessons learned. Magn Reson Imaging. 2019;61:300–18. https://doi.org/10.1016/j.mri.2019.05.028.
4. Abdusalomov AB, Mukhiddinov M, Whangbo TK. Brain tumor detection based on deep learning approaches and magnetic resonance imaging. Cancers. 2023, August 18;15(16):4172. https://doi.org/10.3390/cancers15164172.
5. Abiwinanda N, Hanif M, Hesaputra ST, Handayani A, Mengko TR. Brain tumor classification using convolutional neural network. IFMBE Proc. 2019:183–9. https://doi.org/10.1007/978-981-10-9035-6_33.
6. Aboussaleh I, Riffi J, Mahraz AM, Tairi H. Brain tumor segmentation based on deep learning's feature representation. J Imaging. 2021;7(12):269. https://doi.org/10.3390/jimaging7120269.
7. Ahmadi M, Sharifi A, Fard MJ, Soleimani N. Detection of brain lesion location in MRI images using convolutional neural network and robust PCA. Int J Neurosci. 2021; https://doi.org/10.1080/00207454.2021.1883602.
8. Aleid A, Alhussaini K, Alanazi R, Altwaimi M, Altwijri O, Saad AS. Artificial intelligence approach for early detection of brain tumors using MRI images. Appl Sci. 2023;13:3808. https://doi.org/10.3390/app13063808.
9. Alghoul A, Al Ajrami S, Al Jarousha G, Harb G, Abu-Naser SS. Email classification using artificial neural network. IJAER. 2018;2(11):8–14. https://www.researchgate.net/publication/329307944_Email_Classification_Using_Artificial_Neural_Network
10. Almadhoun HR, Abu Naser SS. Detection of brain tumor using deep learning. IJAER. 2022;6(3):29–47. ISSN: 2643-9085
11. Amin J, Sharif M, Haldorai A, Yasmin M, Nayak RS. Brain tumor detection and classification using machine learning: a comprehensive survey. Complex Intell Syst. 2021, November 8;8(4):3161–83. https://doi.org/10.1007/s40747-021-00563-y.
12. Apell P, Eriksson H. Artificial intelligence (AI) healthcare technology innovations: the current state and challenges from a life science industry perspective. Technol Anal Strateg Manag. 2023;35(2):179–93. https://doi.org/10.1080/09537325.2021.1971188.
13. Arabahmadi M, Farahbakhsh R, Rezazadeh J. Deep learning for smart healthcare—a survey on brain tumor detection from medical imaging. Sensors. 2022;22:1960. https://doi.org/10.3390/s22051960.
14. Bharati S, Podder P, Mondal MRH. Artificial neural network based breast cancer screening: a comprehensive review. IJCISIM. 2020a;2:125–37. https://www.researchgate.net/publication/342109095_Artificial_Neural_Network_Based_Breast_Cancer_Screening_A_Comprehensive_Review

15. Bharati S, Podder P, Mondal MRH. Hybrid deep learning for detecting lung diseases from X-ray images. Inf Med Unlocked. 2020b;20:100391. https://doi.org/10.1016/j.imu.2020.100391.
16. Cè M, Irmici G, Foschini C, Danesini GM, Falsitta LV, Serio ML, Fontana A, Martinenghi C, Oliva G, Cellina M. Artificial intelligence in brain tumor imaging: a step toward personalized medicine. Curr Oncol. 2023;30(3):2673–701. https://doi.org/10.3390/curroncol30030203.
17. Cellina M, Cè M, Irmici G, Ascenti V, Caloro E, Bianchi L, Pellegrino G, D'Amico N, Papa S, Carrafiello G. Artificial intelligence in emergency radiology: where are we going? Diagnostics (Basel). 2022a;12(12):3223. https://doi.org/10.3390/diagnostics12123223.
18. Cellina M, Cè M, Irmici G, Ascenti V, Khenkina N, Toto-Brocchi M, Martinenghi C, Papa S, Carrafiello G. Artificial intelligence in lung cancer imaging: unfolding the future. Diagnostics. 2022b;12(11):2644. https://doi.org/10.3390/diagnostics12112644.
19. Cheung AT, Li WHC, Ho LLK, Ho KY, Chiu SY, Chan CG, Chung OK. Impact of brain tumor and its treatment on the physical and psychological well-being, and quality of life amongst pediatric brain tumor survivors. Eur J Oncol Nurs. 2019;41:104–9. https://doi.org/10.1016/j.ejon.2019.06.003.
20. Davenport TH, Dreyer K. AI will change radiology, but it won't replace radiologists. Harv Bus Rev. 2018;27. https://hbr.org/2018/03/ai-will-change-radiology-but-it-wont-replace-radiologists
21. Davenport T, Kalakota R. The potential for artificial intelligence in healthcare. Future Healthc J. 2019;6(2):94–8. https://doi.org/10.7861/futurehosp.6-2-94.
22. Gerke S, Minssen T, Cohen G. Ethical and legal challenges of artificial intelligence-driven healthcare. Artif Intell Healthcare. 2020:295–336. https://doi.org/10.1016/b978-0-12-818438-7.00012-5.
23. Gill SK, Karwath A, Uh H, Cardoso VR, Gu Z, Barsky A, Slater L, Acharjee A, Dall'Olio JDL, Bouhaddani S, Chernbumroong S, Stanbury M, Haynes S, Asselbergs FW, Grobbee DE, Eijkemans MJC, Gkoutos GV, Kotecha D, Consortium B, The cardaic Group. Artificial intelligence to enhance clinical value across the spectrum of cardiovascular healthcare. Eur Heart J. 2023;44(9):713–25. https://doi.org/10.1093/eurheartj/ehac758.
24. Grant R, Dowswell T, Tomlinson E, Brennan PM, Walter FM, Ben-Shlomo Y, Hunt DW, Bulbeck H, Kernohan A, Robinson T, Lawrie TA. Interventions to reduce the time to diagnosis of brain tumors. Cochrane Database Syst Rev. 2020, September 4;2020(9) https://doi.org/10.1002/14651858.cd013564.pub2.
25. Gumaei A, Hassan MM, Hassan MR, Alelaiwi A, Fortino G. A hybrid feature extraction method with regularized extreme learning machine for brain tumor classification. IEEE Access. 2019;7:36266–73. https://doi.org/10.1109/ACCESS.2019.2904145.
26. Gupta A, Anpalagan A, Guan L, Khwaja AS. Deep learning for object detection and scene perception in self-driving cars: survey, challenges, and open issues. Array. 2021;10(10):100057. https://doi.org/10.1016/j.array.2021.100057.
27. Haq AU, Li JP, Agbley BLY, Khan A, Khan I, Uddin MI, Khan S. IIMFCBM: intelligent integrated model for feature extraction and classification of brain tumors using mri clinical imaging data in IoT-healthcare. IEEE J Biomed Health Inform. 2022a;26(10):5004–12. https://doi.org/10.1109/jbhi.2022.3171663.
28. Haq AU, Li JP, Khan S, Alshara MA, Alotaibi RM, Mawuli C. DACBT: deep learning approach for classification of brain tumors using MRI data in IoT healthcare environment. Sci Rep. 2022b;12(1) https://doi.org/10.1038/s41598-022-19465-1.
29. Hephzipah JJ, Thirumurugan P. Performance analysis of meningioma brain tumor detection system using feature learning optimization and ANFIS classification method. IETE J Res. 2022;68(2):1542–50. https://doi.org/10.1080/03772063.2020.1844079.
30. Hossain T, Shishir FS, Ashraf M, Al Nasim MA, Shah FM. Brain tumor detection using convolutional neural network. ICASERT; 2019. https://doi.org/10.1109/ICASERT.2019.8934561.
31. Irmak E. Multi-classification of brain tumor MRI images using deep convolutional neural network with fully optimized framework. Iran J Sci Technol Trans Electr Eng. 2021;45(1) https://doi.org/10.1007/s40998-021-00426-9.

32. Kaka H, Zhang E, Khan N. Artificial intelligence and deep learning in neuroradiology: exploring the new frontier. Can Assoc Radiol J. 2021;72(11) https://doi.org/10.1177/0846537120954293.
33. Kareem SW, Abdulrahman BF, Hawezi RS, Khoshaba FS, Askar S, Muheden KM, Abdulkhaleq IS. Operationalising ethics in artifcial intelligence for healthcare: a framework for AI developers. Comparative evaluation for detection of brain tumor using machine learning algorithms. IAES (IJ-AI). 2023;12(1):469–77. https://doi.org/10.11591/ijai.v12.i1.pp469-477.
34. Kokkalla S, Kakarla J, Venkateswarlu IB, Singh M. Three-class brain tumor classification using deep dense inception residual network. Soft Comput. 2021;25:8721–9. https://doi.org/10.1007/s00500-021-05748-8.
35. LeCun Y, Bengio Y, Hinton G. Deep learning. Nature. 2015;521(7553):436–44. https://doi.org/10.1038/nature14539.
36. Li M, Kuang L, Xu S, Sha Z. Brain tumor detection based on multimodal information fusion and convolutional neural network. IEEE Access. 2019;7:180134–46. https://doi.org/10.1109/ACCESS.2019.2958370.
37. Litjens G, Kooi T, Bejnordi BE, Setio AAA, Ciompi F, Ghafoorian M, van der Laak JAWM, van Ginneken B, Sánchez CI. A survey on deep learning in medical image analysis. Med Image Anal. 2017;42:60–88. https://doi.org/10.1016/j.media.2017.07.005.
38. Malathi M, Sinthia P. Brain tumor segmentation using convolutional neural network with tensor flow. Asian Pac J Cancer Prev. 2019;20(7):20952101. https://doi.org/10.31557/APJCP.2019.20.7.2095.
39. Mohsen H, El-Dahshan ESA, El-Horbaty ESM, Salem ABM. Classification using deep learning neural networks for brain tumors. Future Comput Inform J. 2018;3(1):68–71. https://doi.org/10.1016/j.fcij.2017.12.001.
40. Mondal MRH, Bharati S, Podder P. CO-IRv2: optimized inceptionResNetV2 for COVID-19 detection from chest CT images. PLoS One. 2021;16(10):e0259179. https://doi.org/10.1371/journal.pone.025917.
41. Mousavi SMH, Ilanloo A. Seven Staged Identity Recognition System Using Kinect V. 2 Sensor. In: 2022 9th Iranian Joint Congress on Fuzzy and Intelligent Systems (CFIS), IEEE; 2022. p. 1–7. https://doi.org/10.1109/CFIS54774.2022.9756435.
42. Mousavi SMH, Mirinezhad SY. Iranian Kinect face database (IKFDB): a color-depth based face database collected by Kinect v. 2 sensor. SN Appl Sci. 2021;3(19):1–17. https://doi.org/10.1007/s42452-020-03999-y.
43. Musallam AS, Sherif AS, Hussein MK. A new convolutional neural network architecture for automatic detection of brain tumors in magnetic resonance imaging images. IEEE Access. 2022;10:2775–82. https://doi.org/10.1109/ACCESS.2022.3140289.
44. Naik N, Hameed BMZ, Shetty DK, Swain D, Shah M, Paul R, Aggarwal K, Ibrahim S, Patil V, Smriti K, Shetty S, Rai BP, Chlosta P, Somani BK. Legal and ethical consideration in artificial intelligence in healthcare: who takes responsibility? Front Surg. 2022, March 14;9 https://doi.org/10.3389/fsurg.2022.862322.
45. Nakagawa M, Nakaura T, Namimoto T, Kitajima M, Uetani H, Tateishi M, Oda S, Utsunomiya D, Makino K, Nakamura H, Mukasa A, Hirai T, Yamashita Y. Machine learning based on multi-parametric magnetic resonance imaging to differentiate glioblastoma multiforme from primary cerebral nervous system lymphoma. Eur J Radiol. 2018;108:147–54. https://doi.org/10.1016/j.ejrad.2018.09.017.
46. Nazar M, Alam MM, Yafi E, Mazliham M. A systematic review of human-computer interaction and explainable artificial intelligence in healthcare with artificial intelligence techniques. IEEE Access. 2021;9:3127881. https://doi.org/10.1109/ACCESS.2021.3127881.
47. Özlem P, Güngen C. Classification of brain tumors from MR images using deep transfer learning. J Supercomput. 2021;77:7236–52. https://doi.org/10.1007/s11227-020-03572-9.
48. Pashaei A, Sajedi H, Jazayeri N. Brain tumor classification via convolutional neural network and extreme learning machines. In: 2018 8th international conference on computer and knowledge engineering (ICCKE). IEEE. 2018, p. 314–9, https://doi.org/10.1109/ICCKE.2018.8566571.

49. Phaye SS, Sikka A, Dhall A, Bathula D. Dense and diverse capsule networks: making the capsules learn better. 2018. arXiv:1805.04001
50. Philip AK, Samuel BA, Bhatia S, Khalifa SAM, El-Seedi HR. Artificial intelligence and precision medicine: a new frontier for the treatment of brain tumors. Life (Basel). 2022;13(1):24. https://doi.org/10.3390/life13010024.
51. Pope WB. Brain Metastases: neuroimaging. Handb Clin Neurol. 2018;149:89–112. https://doi.org/10.1016/B978-0-12-811161-1.00007-4.
52. Qin J, Li Y, Liang D, Zhang Y, Yao W. Histogram analysis of absolute cerebral blood volume map can distinguish glioblastoma from solitary brain metastasis. Med (Baltim). 2019;98(42):e17515. https://doi.org/10.1097/MD.0000000000017515.
53. Quader S, Kataoka K, Cabral H. Nanomedicine for brain cancer. Adv Drug Deliv Rev. 2022;182:114115. https://doi.org/10.1016/j.addr.2022.114115.
54. Rudie JD, Rauschecker AM, Bryan RN, Davatzikos C, Mohan S. Emerging applications of artificial intelligence in neuro-oncology. Radiology. 2019;290(3):607–18. https://doi.org/10.1148/radiol.2018181928.
55. Smith H. Clinical AI: opacity, accountability, responsibility and liability. AI & Soc. 2020, July 25;36(2):535–45. https://doi.org/10.1007/s00146-020-01019-6.
56. Soomro TA, Zheng L, Afifi AJ, Ali A, Soomro S, Yin M, Gao J. Image segmentation for MR brain tumor detection using machine learning: a review. IEEE Rev Biomed Eng. 2022;16:70–90. https://doi.org/10.1109/RBME.2022.3185292.
57. Stephenson J. WHO offers guidance on use of artificial intelligence in medicine. JAMA Health Forum. 2021, July 13;2(7):e212467. https://doi.org/10.1001/jamahealthforum.2021.2467.
58. Talukder MA, Islam MM, Uddin MA, Akhter A, Pramanik MAJ, Aryal S, Almoyad MAA, Hasan KF, Moni MA. An efficient deep learning model to categorize brain tumor using reconstruction and fine-tuning. Expert Syst Appl. 2023;230:120534. https://doi.org/10.1016/j.eswa.2023.120534.
59. Tandel GS, Balestrieri A, Jujaray T, Khanna NN, Saba L, Suri JS. Multiclass magnetic resonance imaging brain tumor classification using artificial intelligence paradigm. Comput Biol Med. 2020;122:103804. https://doi.org/10.1016/j.compbiomed.2020.103804.
60. Tateishi M, Nakaura T, Kitajima M, Uetani H, Nakagawa M, Inoue T, Kuroda JI, Mukasa A, Yamashita Y. An initial experience of machine learning based on multi-sequence texture parameters in magnetic resonance imaging to differentiate glioblastoma from brain metastases. J Neurol Sci. 2020;410:116514. https://doi.org/10.1016/j.jns.2019.116514.
61. Tiwari P, Pant B, Elarabawy MM, Abd-Elnaby M, Mohd N, Dhiman G, Sharma S. CNN based multiclass brain tumor detection using medical imaging. Comput Intell Neurosci. 2022;1830010 https://doi.org/10.1155/2022/1830010.
62. Ueda D, Kakinuma T, Fujita S, Kamagata K, Fushimi Y, Ito R, Matsui Y, Nozaki T, Nakaura T, Fujima N, Tatsugami F, Yanagawa M, Hirata K, Yamada A, Tsuboyama T, Kawamura M, Fujioka T, Naganawa S. Fairness of artificial intelligence in healthcare: review and recommendations. Japanese journal of. Radiology. 2023, August 4;42(1):3–15. https://doi.org/10.1007/s11604-023-01474-3.
63. Van der Velden BH, Kuijf HJ, Gilhuijs KG, Viergever MA. Explainable artificial intelligence (XAI) in deep learning-based medical image analysis. Med Image Anal. 2022;79:102470. https://doi.org/10.1016/j.media.2022.102470.
64. Vicini S, Bortolotto C, Rengo M, Ballerini D, Bellini D, Carbone I, Preda L, Laghi A, Coppola F, Faggioni L. A narrative review on current imaging applications of artificial intelligence and radiomics in oncology: focus on the three most common cancers. Radiol Med. 2022;127(8):819–36. https://doi.org/10.1007/s11547-022-01512-6.
65. von Eschenbach WJ. Transparency and the black box problem: why we do not trust AI. Philos Technol. 2021;34:1607–22. https://doi.org/10.1007/s13347-021-00477-0.
66. Walker D, Hamilton W, Walter FM, Watts C. Strategies to accelerate diagnosis of primary brain tumors at the primary-secondary care interface in children and adults. CNS Oncol. 2013, September;2(5):447–62. https://doi.org/10.2217/cns.13.36.

67. Wong KKL, Fortino G, Abbott D. Deep learning-based cardiovascular image diagnosis: a promising challenge. Futur Gener Comput Syst. 2020;110:802–11. https://doi.org/10.1016/j.future.2019.09.047.
68. Xie Y, Zaccagna F, Rundo L, Testa C, Agati R, Lodi R, Manners DN, Tonon C. Convolutional neural network techniques for brain tumor classification (from 2015 to 2022): review, challenges, and future perspectives. Diagnostics. 2022;12(8):1850. https://doi.org/10.3390/diagnostics12081850.
69. Yogananda CGB, Shah BR, Vejdani-Jahromi M, Nalawade SS, Murugesan GK, Yu FF, Pinho MC, Wagner BC, Mickey B, Patel TR, Fei B, Madhuranthakam AJ, Maldjian JA. A novel fully automated MRI-based deep-learning method for classification of IDH mutation status in brain gliomas. Neuro-Oncology. 2020;22(3):402–11. https://doi.org/10.1093/neuonc/noz199.
70. Zahoor MM, Qureshi SA, Bibi S, Khan SH, Khan A, Ghafoor U, Bhutta MR. A new deep hybrid boosted and ensemble learning-based brain tumor analysis using MRI. Sensors. 2022;22(7):2726. https://doi.org/10.3390/s22072726.
71. ZainEldin H, Gamel SA, El-Kenawy ESM, Alharbi AH, Khafaga DS, Ibrahim A, Talaat FM. Brain tumor detection and classification using deep learning and sine-cosine fitness grey wolf optimization. Bioengineering. 2023;10(18) https://doi.org/10.3390/bioengineering10010018.

Chapter 16
Revolutionizing Brain Tumor Detection: Unleashing the Power of Artificial Intelligence

Nikita Shinde, Mrunal Patil, Rakesh Kumar Sharma, Sachin Kadam, and Shivaji Kashte

16.1 Introduction

It is estimated that one million people worldwide receive a brain tumor diagnosis. Of these, 81% are gliomas that are malignant, which are highly concerning because of their rapid neurological decline and tendency to impact younger patients [1]. According to cancer statistics, 24,810 adults in the USA were found to have primary brain and spinal cord tumors in 2023 [2]. "Brain tumors, characterized by abnormal cell growth, are a common and often fatal brain disease affecting individuals of every age" [3]. Two main subdivisions may be used to classify brain tumors broadly. Malignant refers to cancer with varying degrees of malignancy, whereas benign refers to non-cancerous conditions. These tumors are typically associated with specific brain areas, such as gliomas in glial tissues, meningiomas in the brain's protective membrane, and pituitary tumors within the gland [4]. "Gliomas are the most prevalent type of brain tumor arising from glial cells" [5]. They exhibit significant diversity in clinical and biological characteristics, ranging from slow-growing, relatively benign types like pilocytic astrocytoma to highly aggressive ones like glioblastoma multiforme (GBM) with poor prognosis. Hence, prompt and precise diagnosis is crucial for effective treatment and improved long-term survival [5].

N. Shinde · M. Patil · S. Kashte (✉)
Department of Stem Cell and Regenerative Medicine, Centre for Interdisciplinary Research, D. Y. Patil Education Society (Deemed to be University), Kolhapur, MS, India
e-mail: ShivajiKashte.cir@dypgroup.edu.in

R. K. Sharma
D.Y. Patil Medical College, D. Y. Patil Education Society (Deemed to be University), Kolhapur, India

S. Kadam
Sophisticated Analytical and Technical Help Institute, Indian Institute of Technology, Delhi, India

T. Choudhury, A. Katal (eds.), *Machine Learning for Disease Detection, Prediction, and Diagnosis*, https://doi.org/10.1007/978-981-96-4241-0_16

Diagnosing and effectively treating brain tumors pose significant challenges in the medical field. Conventional diagnostic methods, largely dependent on human skills, encounter inherent drawbacks, including subjectivity and variations in interpretation. These obstacles can result in delayed diagnoses and less precise evaluations of the tumor's attributes and advancement, ultimately impacting patient outcomes [6].

In neuroimaging, notably brain tumor diagnosis, the incorporation of artificial intelligence (AI) has emerged as a revolutionary factor. The advancement of AI technology, utilizing computer vision, deep learning (DL), and machine learning (ML), has demonstrated great promise in enhancing the precision and effectiveness of brain tumor diagnosis [7]. AI can potentially reduce human errors, increase diagnostic precision and accuracy, and optimize treatment in brain tumor detection. The demand for more dependable and impartial techniques in brain tumor identification has spurred the investigation of AI in medical imaging. AI algorithms have demonstrated the ability to scrutinize images with accuracy and reliability, surpassing that of humans. These algorithms can navigate extensive datasets, uncovering intricate patterns and anomalies that might escape even the most seasoned radiologists [8].

This chapter will introduce fundamental AI concepts in medical imaging, encompassing computer vision, ML, and DL techniques. These concepts will lay the groundwork for comprehending AI's application in detecting brain tumors. The chapter will also include case studies (Table 16.1) demonstrating the successful implementation of AI and its positive influence on patient outcomes.

16.1.1 Review of Literature

Detecting a brain tumor is a lengthy procedure that greatly depends on the skills and knowledge of radiologists. As the number of patients increases, the amount of data that needs to be analyzed also rises significantly. This has made traditional approaches costly and inefficient. The brain's magnetic resonance imaging (MRI) is typically used to identify and outline brain tumors, as it is a non-invasive technique. Despite technological progress, a constant obstacle persists: technological advancement surpasses the pace at which healthcare professionals can be educated to employ these advancements proficiently [9].

Over the past couple of years, the significance of AI has increased significantly, particularly in the healthcare industry. AI demonstrates significant potential as a decision-making aid for precisely identifying and categorizing brain tumors. ML and DL methods are increasingly used to quickly and accurately categorize brain tumors [10].

Traditional methods in image processing typically involve extracting features within lower layers, but this approach may not be optimal for analyzing medical images. Lakshmi and Rao conducted a study to enhance the early detection of brain tumors by employing a DL approach and optimizing hyperparameters considering the Inception-v3 CNN model. Their research revealed that the Inception-v3

Table 16.1 Case studies of AI for brain tumor detection

Sr. No.	AI tool	Brain tumor detection	Accuracy/ finding	References
1.	DS-detect dataset	Glioma	86.3%	[55]
2.	VGG16, inception V3,and ResNet-50, MobileNet	Brain tumor detection	98%	[34]
3.	Neural network, KNN, decision tree, naive Bayes, random forest	Identifying and classifying brain tumors	97.305%	[35]
4.	DeepMedic and 3D U-net	Small lesion detection and tumor segmentation	91.3%	[42]
5.	CNN	Brain tumor classification	98.51% and 84.19%	[43]
6.	T1-weighted	Meningioma, glioma, pituitary tumor	84.19%	[43]
7.	Xception, MobileNet, VGG-16, inception V3, DenseNet201, and ResNet-50	Meningioma, glioma, and pituitary	99.90%	[45]
8.	OGSA and RNN	Classification of tumor and non-tumor	96%	[50]
9.	CNN	Glioblastoma analysis	89.9%	[51]
10.	NVIDIA K80 GPU with 25 GB of RAM	Meningioma and glioblastoma	97.77%	[71]
11.	OASIS and the BRATS2018 MATLAB 2018b	Tumors and without tumors	95.78%	[63]
12.	Logistic regression model	Tumor classification	89%	[72]
13.	ML model	Early tumor detection	98.95%	[31]
14.	POS-SVM model	Tumor classification	95.23%	[38]
15.	CNN model	Identification of meningioma, glioma, and pituitary tumors	93.68%	[43]
16.	Bayesian FCM algorithm	Distinguish between tumor and non-tumor	98.5%	[48]
17.	CNN model (ResNet34)	Distinguish between normal and abnormal brain	100%	[49]
18.	CNN model	Types of brain tumors	84.19%	[44]
19.	Recurrent neural network (RNN)	Classification of meningioma, glioma, and pituitary	98%	[52]
20.	RNN	Classification of tumor	98.41%	[53]
21.	FRNN	Classification of tumor	87.8%	[54]
22.	Decision trees, naive Bayes, SVM, ANN, KNN, and PSO	Distinguish between normal and abnormal brain	100%	[56]
23.	FCM	Distinguishing between benign and malignant brain tumors	98.9%	[57]

(continued)

Table 16.1 (continued)

Sr. No.	AI tool	Brain tumor detection	Accuracy/ finding	References
24.	local, BRATS 2013, and BRATS 2015	Detecting brain tumors	90%, 93%, and 97%	[35]
25.	Kaggle dataset	Detecting brain tumors	98.42%	[64]
26.	Inception-v3 CNN model	Early detection of brain tumor	99.34%	[11]
27.	Fuzzy logic framework	Types of brain tumor	93%	[12]
28.	ELM-LRF algorithm	Tumor prediction	97.18%	[13]
29.	Fuzzy sure entropy (NS-EMFSE)	Benign or malignant	95.62%	[14]
30.	FCM	Tumor detection	87.50%	[15]
31.	DeepTumorNet	Types of brain tumor	99.67%	[16]

algorithm shows an accuracy of 99.34% in trained data and 89% in validated data, showcasing its effectiveness in accurately detecting brain tumors [11].

Gordillo and his team developed a fuzzy logic technique to segment brain tumor images. They combined features extracted from MR images with expert knowledge to create this framework. The system is entirely automated and operates unsupervised. They tested it using GBM and meningioma and achieved satisfactory results. The accuracy varied between 71% and 93%, proving the effectiveness of the proposed methodology [12].

Ari and colleagues used a three-stage "Extreme Learning Machine Local Receptive Fields (ELM-LRF)" technique to show the prognosis of the tumor. Initially, the method focuses on removing image noise using local and nonlocal techniques. Following this, the ELM-LRF algorithm forecasts whether a tumor is benign or malignant. Finally, segmentation of the tumor is carried out. The approach particularly relies on cranial MR images due to their suitability for thorough analysis. The proposed method has shown effectiveness and 97% accuracy in predicting malignant tumors [13].

A group of researchers led by Ozyurt has developed a new detection system capable of accurately classifying MRI images as benign or malignant. This is a significant breakthrough in medical imaging, as it could lead to earlier and more accurate cancer diagnoses. Initially, a team of researchers employed a fusion of neutrosophic set-expert maximum fuzzy sure entropy (NS-EMFSE) techniques for segmenting images. Subsequently, they implemented DL approaches, particularly CNN, to extract pertinent attributes from the segmented MRI images. Support vector machine (SVM) and k-nearest neighbor (KNN) classifiers were used to classify these extracted traits into benign or malignant tumors. It was determined through thorough comparison and analysis that the SVM classifier, in conjunction with features extracted by CNN, achieved the highest accuracy of 95.62% [14].

Praveen and his colleagues developed a method that uses artificial neural networks (ANNs) and fuzzy c-means (FCM) to detect tumors. Initially, an FCM algorithm was employed to segment images for localizing the tumor. Subsequently, significant attributes were extracted from the MRI images that had undergone segmentation using the gray-level run-length matrix methodology. Finally, the team used ANN to classify the MRI images. The experiment involved over a hundred MRI images and 87% accuracy [15].

Raza and colleagues developed a hybrid deep-learning method called "DeepTumorNet" to categorize meningioma, glioma, and pituitary tumors. Instead of solely relying on existing models, they enhanced the GoogLeNet architecture by incorporating an additional 15 layers. Their approach demonstrated outstanding performance metrics, achieving a 100% Precision, Accuracy, Recall, and F1 score. Their assessment compared DeepTumorNet results against various pre-trained models, including MobileNetv2, AlexNet, ShuffleNet, Darknet-53, ResNet101, Excption-Net, and SqueezeNet [16].

In summary, incorporating AI methodologies, notably ML and DL algorithms, has exhibited encouraging outcomes in boosting the accuracy and effectiveness of brain tumor identification and categorization from MRI images. However, it is essential to acknowledge the limitations of this review, including potential biases in the selection and interpretation of studies and the evolving nature of AI technologies in healthcare. Furthermore, this research has been warranted to address these limitations and validate the effectiveness of these methodologies in real-world clinical settings.

16.2 Traditional Brain Tumor Detection Methods

16.2.1 Neuroimaging Techniques

Neuroimaging methods have been the foundation for diagnosing brain tumors for a significant time. These approaches offer valuable information about the brain's structure and functioning, allowing anomalies like tumors to be detected. MRI and computed tomography (CT) scans are the most common neuroimaging methods. CT scans, for example, utilize X-rays to produce in-depth cross-sectional brain images. While CT scans are effective at identifying dense structures such as tumors, they might not possess the precision necessary to differentiate between benign and malignant tumors [17].

On the other hand, MRI uses potent radio waves and magnetic fields to create intricately specified pictures of the brain's soft tissues. This technique differentiates between various brain tumor types and assesses their characteristics, such as size, location, and vasculature [18]. However, traditional neuroimaging techniques have limitations. The heavy reliance on human interpretation in these methods can introduce subjectivity and variability into diagnoses, potentially resulting in delayed or missed detections of small or subtle tumors [19]. Integrating AI into neuroimaging aims to mitigate these challenges by providing objective and consistent analyses of imaging data. AI algorithms can swiftly and precisely evaluate enormous volumes of visual data, which helps detect early brain tumors and differentiate benign from malignant growths [20].

16.2.2 Challenges in Conventional Diagnosis

Conventional brain tumor diagnosis faces several challenges, including the following:

Interobserver Variability: Because human judgment is subjective, there can be variation in interpretations among radiologists, leading to contradictory diagnoses [21].

Limited Sensitivity: Conventional neuroimaging methods may not always detect small or hidden tumors, especially in their early stages [22].

Diagnostic Delays: Delays in diagnosis and treatment initiation, which can impact patient outcomes, are often caused by the time-consuming process of manual image analysis [22].

Conventional brain tumor diagnosis confronts numerous challenges that can have substantial implications for patient care and outcomes. Due to these challenges, increasing enthusiasm has been shown for using AI to address these issues effectively.

16.3 Artificial Intelligence in Brain Tumor Detection

AI can be described as technology replicating brainpower functions such as acquiring knowledge, logical thinking, and addressing complex problems. Originating from computer science, contemporary AI has evolved into a multidisciplinary field, incorporating insights from various areas, including statistics, informatics, and physics [5].

16.3.1 Types of Artificial Intelligence

Categories of AI applications:

16.3.1.1 Predictive Analysis (ML)

ML empowers computers to acquire knowledge from data and make predictions, playing a vital role in data-driven decision-making, especially in finance and healthcare.

16.3.1.2 Deep Learning

DL has gained popularity as a predominantly utilized method in supervised ML. It is commonly employed for tackling intricate ML assignments involving text, speech, recommender systems, images, and videos [23].

16.3.1.3 Image Processing (Computer Vision)

Computer vision enables machines to interpret visual data, with practical uses encompassing facial recognition, autonomous vehicles, and quality assurance in manufacturing [24].

16.3.2 Artificial Intelligence: Solution for Brain Tumor Detection

The AI industry has experienced rapid growth, surpassing expectations, thanks to significant investments from technology firms. In addition, the brain tumor diagnostics and detection industry is anticipated to grow significantly over the next several years. The industry's was estimated to be worth $844.63 million in 2021. Forecasts indicate that the market is expected to grow significantly, with a target of $2476.14 million by 2028. Between 2021 and 2028, there will be a significant 16% annual growth rate [25]. AI's contributions are instrumental in helping healthcare professionals make informed decisions through complex computations and reasoning processes. Employing AI-based approaches in cancer imaging offers significant benefits, including improved tumor detection and characterization, continuous monitoring of treatment responses, and early detection of potential cancerous indications in various body regions. The effectiveness of AI-driven imaging models, including the CXR-vision, LIDCIDRI, LUNA16, and volumetric analysis using CT scans, has been proven in brain tumor detection [26]. Integrating AI into neuroimaging and brain tumor diagnostics shows great potential for addressing multiple challenges within the field (Table 16.1). AI algorithms play a crucial role by delivering objective and consistent image analysis, effectively mitigating Interobserver variability. Moreover, these algorithms enhance sensitivity, enabling the early detection of even minute or concealed tumors. AI also speeds up image analysis, which lowers diagnostic delays and improves medical treatment overall [27, 28].

16.4 Machine Learning in Brain Tumor Detection

ML involves creating and evaluating algorithms that empower computer systems to enhance the functioning for assignment through trial-and-error learning from data without explicit programming. There are three primary classifications of ML algorithms: supervised learning (SL), unsupervised learning (UL), and reinforcement learning (RL) [29].

Supervised ML relies on human supervision, usually involving labeled data that indicates the desired output or correct answers. This approach is suitable when there is ample high-quality data with known targets. Models are trained and evaluated using known input–output pairs, enabling predictions on new data. In contrast, unsupervised categories of AI applications:

16.4.1 Machine Learning (ML)

ML empowers computers to acquire knowledge from data and make predictions, playing a vital role in data-driven decision-making, especially in finance and healthcare.

16.4.2 Deep Learning

DL has gained popularity as a predominantly utilized method in supervised ML. It is commonly employed for tackling intricate ML assignments involving text, speech, recommender systems, images, and videos [23].

16.4.3 Image Processing (Computer Vision)

Computer vision enables machines to interpret visual data, with practical uses encompassing facial recognition, autonomous vehicles, and quality assurance in manufacturing [24].

Supervised ML relies on human supervision, usually involving labeled data that indicates the desired output or correct answers. This approach is suitable when there is ample high-quality data with known targets. Models are trained and evaluated using known input–output pairs, enabling prediction learning to operate without labeled data or predefined targets and aiming to extract valuable insights from unstructured information. It uncovers hidden patterns or inherent structures within the input data. RL takes a distinct approach to ML, prioritizing attaining maximum rewards rather than predicting specific output data based on input [23].

16.4.4 Key Terms in the Mechanics of Machine Learning

Classification Labeling a set of pixels with a specific class or category, like identifying a "tumor" using a segmentation algorithm. If a segment marks an area as an "abnormal brain," the classifier assesses if it is harmless or cancerous tissue.

Model The ML system acquires factors or pivotal choices, and once it has learned them, the model can make predictions about the class of an unfamiliar sample.

Algorithm The process of building a model that accurately predicts classes based on training example features.

Labeled Data It is a compilation of instances, such as images, each accompanied by an accurate answer. The answer could indicate the correct tumor boundary, the presence of cancer, or even the exact kind of malignancies the lesion may indicate.

Training The stage where labeled sample data, such as details about tumor types or lesion boundaries, are used to train the ML algorithm. Model weights or decision points are adjusted iteratively until significant performance improvement stops.

Testing The collection of examples utilized in training, also known as the training set. Additionally, a third set of examples, often called a test set, may be employed for real-world testing. As the algorithm improves with the validation set, it may become biased towards the training set's unique characteristics. Evidencing strong performance on a novel test dataset increases trust in the algorithm's real-world precision. The terminology used for validation and testing may differ, so it is essential to clarify their meanings, which one's engineering or statistical background can sometimes influence.

Nodes This neural network element features a function to activate multiple inputs. This function usually applies a minimum or maximum value, performs a specific operation, and integrates the inputs to create an output.

Layers A group of nodes that compute outputs based on any number of inputs, usually from the input layer to the next layer (or from earlier layers to the output layer).

Weights During the training process, each input feature is assigned a weight called "weighting." These weights are then modified to obtain the best possible model. Based on their training approaches, ML methods are categorized into SL, UL, and RL. For example, in brain image analysis, these training styles can be understood by distinguishing tumor (malignant or benign) regions from normal (non-diseased) tissue regions [30].

16.4.5 Machine Learning Models

Transfer learning, a commonly used technique in ML, often involves leveraging pre-trained neural networks. In image categorization and object detection, commonly utilized transfer learning architectures, sayResNet-50, VGG16, and Inception V3, are regularly used. These transfer learning techniques offer significant advantages, especially in their cost-effectiveness and time-saving capabilities [31]. Using 50 deep layers and a dataset of over a million illustrations generated by the ImageNet database, ResNet-50 is a variation based on the ResNet models. A series of convolutional units with typical pools make up ResNet-50 [32].

In 2014, Zisserman et al. introduced VGG16, part of the Visual Geometry Group Network (VGG-NET) family of network structures. VGG16 is a deep network like AlexNet, designed for image detection and classification, and can be trained using the ImageNet database [33].

Inception V3, predominantly employed for image categorization and detection, is a demanding model to train, especially on less powerful computer setups, often necessitating multiple training days. It is an improvement over Inception V1 (GoogLeNet, 2014). On the other hand, when handling vast datasets and intricate context recognition tasks, VGG16 demonstrates higher performance and accuracy [31]. Using an MRI dataset, Almadhoun et al. unveiled a sophisticated educational model created especially for identifying malignancies in the brain. Together with this technique, they employed "Inception V3, VGG16, ResNet-50, and MobileNet" as supplementary transfer learning methods. They used a dataset of 10,000 MR scans to assess the performance of these algorithms. With 5000 photos piece, this dataset was divided into two distinct groups: one for non-malignant brain tumors and the other for malignant tumors in the brain. Interestingly, their suggested DL model performed better than the rest, obtaining 100% precision during training and 98% test precision [34]. Many ML models, such as random forests, decision trees, neural networks, KNNs, and naive Bayes, have been recommended by Garg et al. as instruments for identifying tumors in the brain. They used a dataset of 2556 images of brain tumors, with 85% of the collected information designated for use in training and 15% in testing, to evaluate these ML models' performance. Using Stationary Wavelet transform (SWT), principal component analysis (PCA), and gray-level co-occurrence matrix (GLCM), they discovered 13 unique features. When the recommended approach of detecting and categorizing brain tumors was tested, the results revealed more than 97% reliability, sensitivity, specificity, accuracy, and precision [35].

Stadlbauer et al. have developed an MRI-based approach that employs nine ML models for the early detection of brain tumors. ML-based radiophysiomics could be useful in identifying brain tumors in a clinical setup [36]. Additionally, Aamir et al. have presented an automated methodology for detecting brain tumors using MR images. They evaluated the proposed ML model and observed that it outperformed the existing methods with an impressive 98.95% accuracy [37].

Kumar and colleagues have used particle swarm optimization (PSO) to identify the most relevant features and SVM for tumor classification. Their proposed system introduced a hybrid model, Particle Optimization Stratergy (POS)-SVM, which effectively selects essential features crucial for tumor detection. Consequently, this approach reduced the number of features required, leading to heightened accuracy, enhanced efficiency in classification, and reduced processing time. The POS-SVM model attained an accuracy of 95.23%, notably surpassing the simpler SVM model, which attained 86.82%. Moreover, the POS-SVM model exhibited 94% specificity and 100% sensitivity [38].

16.5 Deep Learning in Brain Tumor Detection

DL, a segment of ML methods, employs a multi-layered approach for extracting a structured hierarchy of more complex patterns from the raw inputs (Fig. 16.1). In ML, the primary goal is to uncover suitable data representations for their input [39].

16.5.1 Neural Networks

In neural network learning, three key components are involved: (a) an error function that assesses output quality, (b) a search function determining how to modify the network to minimize errors, and (c) an update function that adjusts network weights based on the search function. Nodes for inputs, hidden layers, and nodes for outputs comprise a typical neural network. During training, examples are presented, errors are computed, and weights are updated iteratively until minimal improvement is achieved. Real-world networks are often more intricate, with multiple hidden layers and complex node functions [30].

16.5.1.1 Convolutional Neural Networks

Convolutional neural network (CNN) is the most popular and widely used algorithm in the DL domain. A significant advantage it offers over previous approaches is its capacity to discern relevant features without human intervention

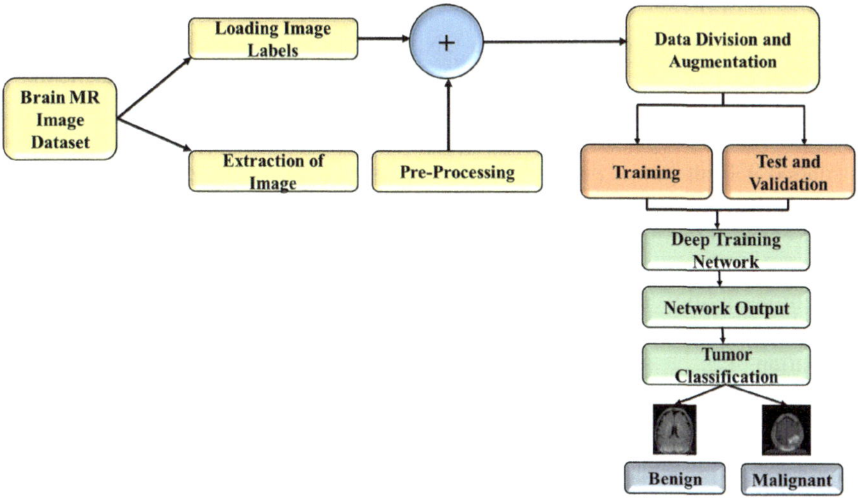

Fig. 16.1 Framework for the DL system entails the following structure. (Figure adapted from [40], Creative Commons License 4.0)

autonomously. CNNs have found extensive use in diverse domains, including computer vision, speech processing, and facial recognition. CNNs' architectural layout closely resembles a conventional neural network modeled after the neural structure in human brains [41].

Neural networks possess a crucial characteristic—they can precisely approximate mathematical expressions, making them versatile function approximators. This feature has fueled interest in neural networks, allowing the modeling of various processes. However, it does not provide a clear way to determine the ideal characteristics for a specific dataset or guarantee precise predictions for new data. Neural networks are synthetic neurons that process input numbers using specific mathematical expressions to generate outcomes [22].

The study utilized DL with MR and CT scans, combining DeepMedic and 3D U-Net models tailored for small lesion detection and tumor segmentation, enhancing overall performance. It addressed concerns in brain tumor stereotactic radiosurgery (SRS) quality assurance, improving target delineation agreement among readers. Sensitivity for lesion detection increased from 82.6% to 91.3%, particularly for smaller tumors, representing a significant advancement in brain SRS. This study marks progress in using DL for radiotherapy planning and optimizing clinician–tool interactions for better patient care [42].

Soheila et al. used CNNs to categorize these tumors. Five distinctive CNN models were tested for classifying brain tumors in their study, and one architecture in particular showed the highest accuracy. This architecture, called Architecture 2, used maximum pooling with 64 encoded neurons and a rectified linear unit (ReLU) layer and achieved validation and training accuracies of 84.19% and 98.51%, respectively. Although the study's other networks performed better in testing than Architecture 2, it is important to remember that this architecture can overfit when using a low learning rate or insufficient training and information validation (Fig. 16.2) [43].

In 2018, Abiwinanda employed a CNN to diagnose the most prevalent brain tumor types. During training, they utilized the "Adam" optimizer, a stochastic optimization method derived from stochastic gradient descent. The dataset included T-1 weighted MRI images of tumors in the brain obtained from Cheng. These images included 930 images of pituitary gland tumors, 708 images of gliomas, and 1426 images of meningiomas. A total of 700 images were used for each class, of which 500 were used to train and 200 to validate the data. The CNN architecture involved 32 filters in all convolutional layers, ReLU activation, 2×2 maxpool kernels, and 64 neurons in fully connected layers. The output layer comprised three neurons with softmax activation, and the highest reported 98% accuracy for training and 84% for validation [44].

Amin ul Haq et al. conducted several investigations to determine whether the suggested model was feasible within an Internet-of-things healthcare system. They used a dataset of photographs of brain tumors to evaluate their model. They used previously trained models, such as Xception, DenseNet201, VGG-16, Inception V3, ResNet-50, and MobileNet, to obtain well-trained weight parameters to improve the precision of predictions of the CNN model. Subsequently, they individually

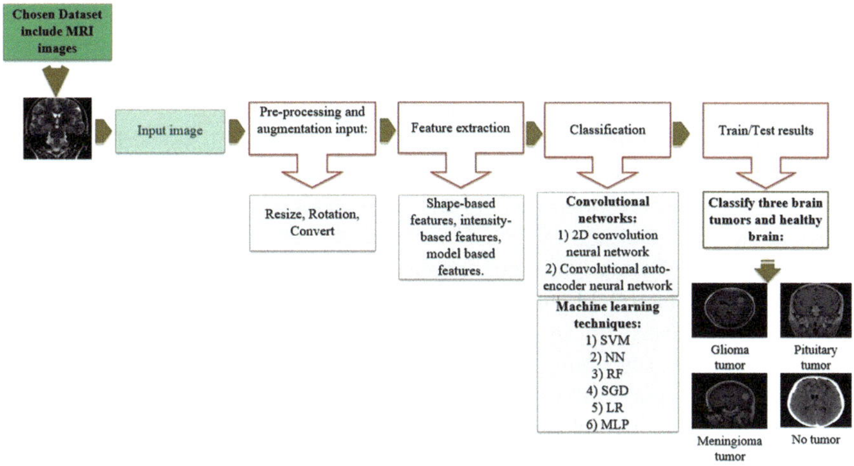

Fig. 16.2 Illustrative representations of the ML model employed in automating brain tumor segmentation. (Figure adapted from [43], Creative Commons License 4.0)

transferred these previously trained model weights to their CNN model to facilitate effective training. To further fine-tune the CNN model for classification, they utilized a small brain tumor dataset consisting of 233 areas and 3064 scans, classified into three groups: meningioma, glioma, and pituitary tumors. Given the limited size of the dataset, they employed data augmentation methods to augment the original data. The experimental outcomes showcased that their combined diagnostic framework, ResNet-CNN, attained an outstanding accuracy rate of 99.90% when contrasted with the baseline methods [45].

Glioblastoma was analyzed using a radiomics model, which uses deep features and transfer learning to retrieve many attributes from a CNN's hidden layers. It obtained a C-index of 0.71 when radiomic features were taken into account alone, which rose to 0.74 when clinical predictors were added. A DL architecture with a 3D CNN was also used to evaluate high-grade gliomas. This architecture included diffusion-tensor imaging, functional MRI, and conventional MRI, among other imaging modalities. An analysis of 69 individuals with high-grade gliomas showed that CNN 89.9% accurately predicted the patients' overall survival. However, it is important to note that this accuracy figure has no independent validation [46].

Pashaei and colleagues proposed multiple approaches for identifying "meningioma, glioma, and pituitary tumors." They employed a CNN framework consisting of different layers to extract significant features from images. The model was trained on Cheng's dataset and employed a tenfold cross-validation strategy, dividing the data into 70% training and 30% testing. Through comparative analysis with other methods such as MLP, Stacking, XGBoost, SVM, and RBF, their proposed approach demonstrated superior accuracy, achieving a rate of 93.68% [47].

A group of researchers led by P.M. Siva Raja has developed a novel image segmentation approach that uses statistical modeling with DL for distinguishing benign/malignant brain tumors. To begin with, the pictures are processed by the researchers by implementing a non-local mean filter for cleaning and a Bayesian FCM algorithm for segmentation. To extract features from the images, various information-theoretic measures are utilized, such as wavelet packet Tsallis entropy and scattering transforms. Finally, a deep autoencoder that utilizes the Jaya optimization algorithm and SoftMax regression is used to classify the images. This technique has demonstrated remarkable precision, with a success rate of 98.5% [48].

Using an MRI dataset, Talo et al. introduced a previously trained ResNet34 model for distinguishing tumor/non-tumor conditions. This approach yielded precise imaging of abnormal brain tumors. A radiologist routinely examined patients' abnormal brain scans, yet analyzing numerous MRI images within a constrained timeframe proved exceedingly challenging. A computer-aided diagnostic tool facilitated swift and accurate analysis of MRI images. This model was tested on >600 MRI images using a fivefold cross-validation technique and attained a 100% accuracy rate [49].

Nyoman Abiwinanda et al. investigated CNNs, which were adapted to identify three prevalent types of brain tumors. Among the five distinct CNN architectures examined, architecture 2 demonstrated the highest accuracy. It is worth noting that the architecture achieved high accuracy levels—98.51% during training and 84.19% during validation [44].

16.5.1.2 Recurrent Neural Networks

A subset of "ANNs" are called "RNNs" that excel in modeling temporal sequences and display dynamic behavior over time as they process data. RNNs are a direct evolution of traditional neural networks, leveraging an internal memory mechanism known as the hidden state. The hidden state plays a central role in RNNs, enabling them to preserve crucial information about sequences and rendering them highly adaptable for various applications, including but not restricted to segmentation, image recognition, and speech recognition. RNNs' capacity to manage inputs of varying lengths and capture temporal dependencies establishes them as a potent asset in the DL [50].

Begum et al. developed an algorithm based on statistical characteristics to categorize efficiently and segment brain tumors. The method uses run-length attributes of textures and GLCM matrix analysis to retrieve characteristics of statistical significance after first preprocessing MRI images to remove background noise. To classify these features as abnormal growth or not, they are refined further using an oppositional gravitational search algorithm (OGSA), after which they are fed into an RNN. The second phase of the process involves identifying and isolating the region of interest in tumor scans. When tested on the dataset, the results show that the suggested algorithm attains an impressive 96% accuracy in classification [51].

In their study, K. Kalaiselvi et al. addressed the classification of brain tumors, treating it as a constraint classification problem aimed at distinguishing among pituitary tumors, gliomas, and meningiomas. The classification process involved several stages, including preprocessing, segmentation, and classification, to accurately identify and segregate these different tumor types. The methodology involved training the RNN classifier on the extracted features to discern the various brain tumors effectively. The experiments used a threefold cross-validation over an MRI brain image dataset. The outcomes of the study demonstrated the efficacy of the proposed RNN classifier, reaching a remarkable mean classification accuracy of 98% [52].

Damandeep et al. combined independent component analysis (ICA), RNN, and a hybrid optimization approach and produced the highest values of peak signal-to-noise ratio (PSNR) and accuracy, reaching 64 dB and 98%, respectively. The study used various optimization methods, such as cuckoo search, firefly, lion, and bat algorithms, followed by Naive Bayes classifier (NBC) and RNN for image recognition tasks. The preprocessing steps included contrast-limited adaptive histogram equalization and thresholding techniques while featuring extraction utilizing PCA and ICA. The study observed that when ICA, hybrid optimization, and RNN are combined, the achieved PSNR is 64, whereas ICA, cuckoo search, and NBC resulted in a mean square error of 1. Based on the results, the study concluded that an accuracy of 98 can be achieved by using ICA, hybrid optimization, and RNN together [53].

The proposed computational approach by Jalluri Gnana Siva Sai et al. utilizes the fuzzy recurrent neural network (FRNN) for segmenting and identifying brain tumors. The process involves accessing MR images through file paths followed by their transformation into grayscale images. A technique called adaptive bilateral filtering is utilized during the preprocessing stage to handle the noise present in the initial images. After reducing the noise in the image, it is processed through binary thresholding and segmentation. The FRNN algorithm is then used to detect and pinpoint areas in MRI images containing tumors. The proposed model has achieved an 87% accuracy rate, which is very promising due to the minimal number of errors observed and the significant reduction in computational time required for the analysis [54].

16.6 Computer Vision in Brain Tumor Detection

"Computer vision, a subset of AI, instructs machines to decipher and comprehend visual data from the physical world" [8]. In medical imaging, computer vision algorithms can scrutinize intricate patterns within radiological images, including CT and MRI, to recognize irregularities that suggest the presence of brain tumors. Numerous research endeavors have underscored computer vision's effectiveness in analyzing medical images [8]. Computer vision algorithms demonstrated exceptional proficiency in spotting subtle alterations in image data that might be challenging for humans to discern. These algorithms are adept at identifying anomalies in tissue density, shape, and texture, which greatly assist in pinpointing and describing tumors [18]. The DS-Detect dataset includes MRI scans of 99 individuals, including 62 patients diagnosed with glioma and 37 individuals in good health. The preprocessing of images involved the conversion of DICOM scans into NifTI format. The computer-aided diagnosis (CAD) system's primary objective is detecting and classifying gliomas. Glioma segmentation is essential as it enables a quantitative assessment of clinical factors concerning the morphological features of these tumors. Tumor ontology is constructed using tumor-related data to understand brain tumors comprehensively. Each voxel in the MRI images is classified into distinct categories through segmentation. A pair of models were created and trained using the DS-Detect and DS-Grading records. This training resulted in an impressive 86.3% accuracy rate for glioma detection and a 76.3% accuracy rate for glioma grading. These achievements greatly aid in identifying and classifying gliomas within the brain [55].

Ulku et al. developed a CAD framework with four key stages: preprocessing, segmentation, feature analysis, and classification. The preprocessing phase used histogram equalization, while segmentation utilized dilation and erosion techniques. They evaluated six algorithms: SVM, KNN, decision trees, naive Bayes, and PSO. The study involved a dataset of more than 120 brain MRI scans from eleven subjects, with three normal scans and eight abnormal ones. After segmentation, 350 ROIs were identified, of which 86 were positive and 264 were negative. The "SVM-PSO and KNN algorithms" achieved 100% accuracy [56].

Ghahfarrokhi et al. have presented a system for diagnosing benign and malignant brain tumors using computer-assisted techniques. Their procedure involves segmenting images using FCM and extracting features using the GLCM. They then used three classifiers—SVM, KNN, and Pattern Net—to diagnose the malignancies accurately. Notably, the Pattern Net classifier demonstrated the most accurate results among the three classifiers, achieving a maximum accuracy of 98.9% [57].

16.7 Datasets Used for Brain Tumor Detection

Effectively employing AI, particularly DL, for brain tumor detection necessitates extensive and diverse datasets. These large datasets enable AI models to grasp intricate patterns, enhancing their accuracy and adaptability in medical imaging tasks.

However, a significant hurdle in AI-based brain tumor detection lies in the scarcity of well-annotated medical imaging datasets, especially those with detailed segmentation and ground-truth annotations. Given the varying sizes, shapes, and locations of brain tumors, it is imperative to have comprehensive datasets encompassing this diversity. While researchers and institutions acknowledge the necessity for curated, openly accessible datasets, the creation process is laborious and resource-intensive due to the meticulous annotation and segmentation requirements [58, 59].

16.7.1 Types of Datasets

Internet Brain Segmentation Respository (IBSR) provides a brain MRI dataset (Fig. 16.3) essential for automated brain segmentation research, while Open Access Series of Imaging Studies (OASIS) offers brain-imaging data to support neuroimaging and neuroscience studies. The Whole Brain Atlas is an online resource offering comprehensive information on detailed brain anatomy and pathology. Brain Tumor Segmentation (BRATS) supplies a specialized brain tumor dataset used to develop segmentation algorithms for brain tumors, and Reference iamges Database to Evaluate Therapy Responce (RIDER) provides valuable imaging data for assessing cancer therapy response. Radiopedia is a collaborative platform rich in radiological information, aiding medical imaging professionals and learners. The Cancer Imaging Archive (TCIA) is a repository housing a diverse collection of cancer-related imaging datasets, whereas FigshareCJData offers a variety of research datasets across various domains.ischemic Stroke Lesion Segmentation (ISLES) features a dataset intended to develop algorithms to detect ischemic stroke lesions. The Cancer Genome Atlas (TCGA) is a cornerstone resource, providing comprehensive genomic and clinical data across different cancer types. Kaggle, a versatile data science platform, hosts many datasets and competitions, offering opportunities for data scientists to tackle real-world challenges and refine their modeling skills [60].

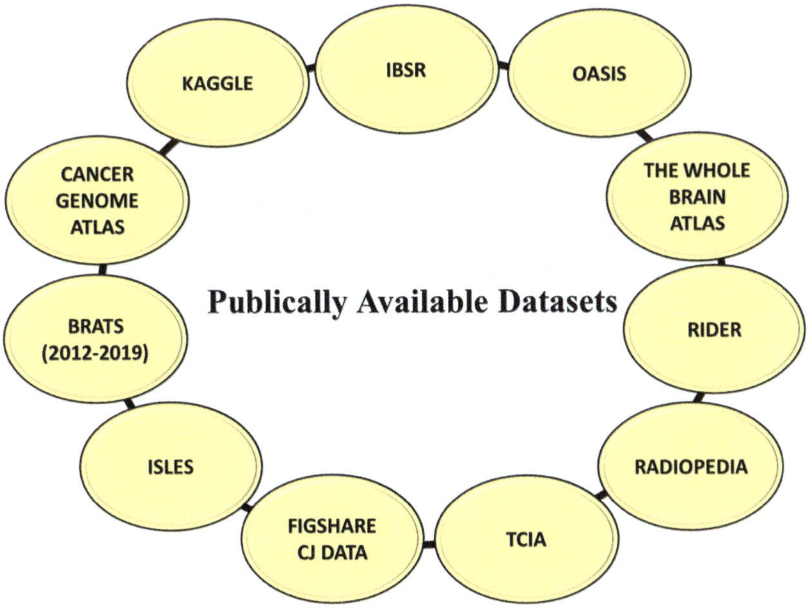

Fig. 16.3 Datasets used for brain tumor detection. (Figure adapted [60] with permission from Nazir et al. [60])

Cheng et al. experimented using a brain tumor dataset accessible on Figshare. Their methodology included choosing the augmented tumor area as the focus region, subsequently segmenting it into smaller sub-regions through an adaptive spatial division approach. Subsequently, they derived various characteristics, including the GLCM, intensity histogram, and attributes rooted in a bag-of-words model of these sub-regions. Their results demonstrated significant accuracy rates of 87.54%, 89.72%, and 91.28% when employing a ring-form partition method on these extracted features [61]. Megha H C et al. utilized the Kaggle MRI brain tumor dataset and applied CNN architecture to extract image features. Their CNN model uses MRI scans as input, conducts preprocessing, carries out segmentation, and retrieves relevant characteristics. Following these steps, the image is categorized as depicting a tumor. Furthermore, the developed graphical user interface demonstrated more than 90%accuracy level when evaluated on the Kaggle dataset [62].

Abhishek Anil et al. compiled a dataset comprising thousands of images sourced from existing datasets such as OASIS and the BRATS2018 challenge. These datasets are classified as tumors/not tumors. Out of the total images, more than 20,000 were selected. Of these, 20% were set aside for evaluation, and the remaining 80% were used for learning. The DL toolbox in MATLAB 2018b was utilized to facilitate transfer learning, focusing on existing networks like AlexNet and VGGnet. These three networks demonstrated the highest efficiency, achieving 95% accuracy [63].

Amin et al. methodology involved a Weiner filter for MRI image noise reduction, image segmentation, and feature extraction using local binary pattern and GLCM wavelet transform. Different classification algorithms were utilized to identify MRI images containing tumors. The devised framework underwent testing across three distinct datasets: local, BRATS 2013, and BRATS 2015. Remarkably, the framework attained accuracy rates of 90% (local), 93% (BRATS 2013), and 97% (BRATS 2015) [35].

Ramdas Vankdothu et al. thoroughly analyzed and evaluated their planned approach using the Kaggle dataset. The data was partitioned into training (2870 MR images) and testing sets (394 MR images) sourced from more than 200 patients. Additionally, the dataset included synthetic data characterized by minimal variance in energy values within similar classes, which facilitated the classification process, particularly in accurately discerning patient data across various classifications. Comparative analysis revealed that the proposed region-based CNN (RCNN) demonstrated a 98% sensitivity rate than previous methods such as backpropagation, U-Net, and RCNN [64].

Mesut Togacar et al. classified malignant/non-malignant brain MRI images by using an enhanced version of RESnet50. This modified iteration integrates eight supplementary layers into the original RESnet50 framework and undergoes training using a dataset sourced from Kaggle. The comparative assessment was conducted against well-known CNN architectures such as DENSEnet, Alexnet, and Googlenet. The proposed model was superior in efficacy, with 97% improved accuracy compared to other DL models [65].

16.8 Challenges and Opportunities

16.8.1 Improved Data Quality and Annotation

Ongoing efforts are underway to advance the quality and diversity of datasets for training AI models in brain tumor detection. These initiatives, exemplified by endeavors like the Cancer Imaging Archive or collaborations between medical institutions and AI developers, have a common objective of furnishing more extensive and diverse datasets with well-annotated information. The aim here is to mitigate bias and enhance the precision of AI algorithms, ultimately advancing their capabilities in detecting brain tumors [66].

16.8.2 Enhancing Interpretability and Explainability

Scientists are actively addressing the issue of AI model interpretability by focusing on methods to improve the clarity of AI decision-making. A key approach in this effort is advancing explainable AI (XAI). The primary aim of XAI techniques is to make AI algorithms more understandable, particularly for healthcare professionals. XAI facilitates better understanding and fosters trust in AI-driven diagnoses by providing insights into the reasoning behind a specific decision [67].

16.8.3 Real-World Generalization

Researchers are actively working to enhance AI models for practical medical applications. However, adapting AI algorithms to handle the unpredictability of real-world data remains a challenge. With ongoing refinement and extensive evaluation across various clinical scenarios, AI models are expected to become more dependable in real healthcare settings. AI's ability to continuously analyze data makes it ideal for real-time monitoring of brain tumor progression. In the future, AI systems may alert healthcare providers to significant changes in patient data, potentially leading to early interventions and improved patient outcomes [46, 68].

16.8.4 Collaborative Research and Interdisciplinary Research

Interdisciplinary collaboration among AI researchers, radiologists, oncologists, and healthcare professionals is vital for advancing AI in healthcare. Education and training programs are essential to ensure healthcare providers effectively use AI tools. Such collaborations promote a shared understanding of AI's potential and limitations in clinical settings, driving innovative solutions for brain tumor detection. These efforts will make AI an indispensable tool in neuro-oncology, benefiting diagnostic precision and patient outcomes. "The collaboration of these teams with diverse expertise helps to connect the AI research with real-world clinical applications. This is evident from initiatives such as the RSNA AI Challenge" [69].

16.8.5 Ethical Considerations

Utilizing AI in healthcare raises ethical and regulatory issues, particularly safeguarding patient privacy and data security. As emphasized, handling medical data strictly with ethical principles is imperative to safeguard patient confidentiality. The continuous challenge lies in adhering to regulatory frameworks and ensuring the responsible deployment of AI in healthcare [70].

16.9 Future Prospects and Conclusion—Unveiling the Horizon of AI-Powered Neuro-Oncology

The marriage of AI with neuro-oncology, specifically in brain tumor detection, marks a momentous stride forward, encapsulating staggering potential and formidable challenges. The repercussions of AI's impact on brain tumor diagnosis and patient care are far-reaching, promising elevated outcomes, heightened diagnostic precision, and streamlined tumor identification. These advancements are orchestrated by the intricate dance of AI-driven algorithms rooted in computer vision, DL, and ML methodologies. Nurtured by expansive datasets, these algorithms unlock the gates to nuanced tumor identification, pinpointing anomalies with finesse and revealing prognostic intricacies.

Yet, the path to realizing AI's complete potential within the clinical landscape is paved with obstacles. The demand for high-quality annotated datasets, the need to mitigate biases in training data, interpretability dilemmas, and the seamless integration into established clinical workflows form formidable barricades. The solution lies in the harmonious collaboration among AI visionaries, seasoned radiologists, adept neuro-oncologists, and astute data scientists. This collaborative synergy aligns cutting-edge AI algorithms with the delicate nuances of clinical demands, birthing ethical, validated, and profoundly patient-centric solutions. Upholding the sanctity of ethical standards, patient confidentiality, and data fortification is paramount, meticulously overseen by vigilant institutional review boards and stringent adherence to ethical guidelines.

The panorama of AI in brain tumor detection unfurls with tantalizing promise, beckoning us to refine algorithms with meticulous precision, embark on expansive clinical voyages, delve deep into the enigmatic realms of radiogenomics, pioneer

AI-driven histopathological analyses, and democratize AI through pioneering population-scale screening initiatives. In this intricate tapestry of potential and challenges, interdisciplinary collaboration and the conscientious evolution of AI emerge as guiding stars. Together, they empower us to harness AI's unparalleled potential, bestowing upon brain tumor patients the invaluable gift of pinpoint precision in diagnoses and personalized treatment regimens. This legacy is not merely technological but profoundly ethical, upholding the highest standards and fortifying the bastions of data privacy.

As technology strides forward and partnerships strengthen, we find ourselves on the brink of an era where the complexities of brain tumors yield to the relentless advance of scientific prowess. In this era, AI emerges not merely as a tool but also as a beacon of hope, illuminating the path toward a future where every individual grappling with the labyrinth of brain tumors finds solace in the brilliance of precision medicine and the compassionate touch of humane AI. It is a future where AI, working tirelessly, guides us toward a world free from the clutches of this formidable adversary, transcending the barriers of disease and heralding an age of unparalleled medical excellence and compassion.

References

1. Kantelhardt S. New strategies in diagnosis and treatments for brain tumors. Cancers (Basel). 2023;15:1–5.
2. Ayadi W, Elhamzi W, Charfi I, Atri M. Deep CNN For brain tumor classification. Neural Process Lett [Internet]. 2021;53:671–700. Available from: https://doi.org/10.1007/s11063-020-10398-2
3. Bangare SL, Pradeepini G, Patil ST. Brain tumor classification using mixed method approach. 2017 Int Conf Inf Commun Embed Syst ICICES. 2017;2017:0–3.
4. Alqudah AM, Alquraan H, Qasmieh IA, Alqudah A, Al-Sharu W. Brain tumor classification using deep learning technique - A comparison between cropped, uncropped, and segmented lesion images with different sizes. Int J Adv Trends Comput Sci Eng [Internet]. 2019;8:3684–91. Available from: https://arxiv.org/abs/2001.08844
5. Cè M, Irmici G, Foschini C, Danesini GM, Falsitta LV, Serio ML, et al. Artificial intelligence in brain tumor imaging: A step toward personalized medicine. Curr Oncol. 2023;30:2673–701.
6. Aldape K, Brindle KM, Chesler L, Chopra R, Gajjar A, Gilbert MR, et al. Challenges to curing primary brain tumours. Nat Rev Clin Oncol [Internet] 2019;16:509–20. Available from: https://doi.org/10.1038/s41571-019-0177-5.
7. Ardila D, Kiraly AP, Bharadwaj S, Choi B, Reicher JJ, Peng L, et al. End-to-end lung cancer screening with three-dimensional deep learning on low-dose chest computed tomography. Nat Med [Internet] 2019;25:954–61. Available from: https://doi.org/10.1038/s41591-019-0447-x.
8. Litjens G, Kooi T, Bejnordi BE, Setio AAA, Ciompi F, Ghafoorian M, et al. A survey on deep learning in medical image analysis. Med Image Anal. 2017;42:60–88.
9. Romaisa A, Safa A, Mustafa AM. Brain tumor detection using deep learning. 2023 14th Int Conf Inf Commun Syst ICICS. 2023;2023
10. ZainEldin H, Gamel SA, El-Kenawy ESM, Alharbi AH, Khafaga DS, Ibrahim A, et al. Brain tumor detection and classification using deep learning and sine-cosine fitness Grey wolf optimization. Bioengineering. 2023;10:1–19.

11. Özkaraca O, Bağrıaçık Oİ, Gürüler H, Khan F, Hussain J, Khan J, et al. Multiple brain tumor classification with dense CNN architecture using brain MRI images. Life. 2023;13
12. Gordillo N, Montseny E, Sobrevilla P. A new fuzzy approach to brain tumor segmentation. 2010 IEEE World Congr Comput Intell WCCI. 2010;2010
13. Ari A, Hanbay D. Deep learning based brain tumor classification and detection system. Turkish J Electr Eng Comput Sci. 2018;26:2275–86.
14. Özyurt F, Sert E, Avci E, Dogantekin E. Brain tumor detection based on convolutional neural network with neutrosophic expert maximum fuzzy sure entropy. Meas J Int Meas Confed [internet]. 2019:147. Available from: https://doi.org/10.1016/j.measurement.2019.07.058
15. Singh P, A. Proceedings of the international conference on recent cognizance in Wireless Communication & Image Processing; 2016. p. 123–31.
16. Raza A, Ayub H, Khan JA, Ahmad I, Salama AS, Daradkeh YI, et al. A hybrid deep learning-based approach for brain tumor classification. Electron. 2022;11:11.
17. Nikam RM, Yue X, Kaur G, Kandula V, Khair A, Kecskemethy HH, et al. Advanced neuroimaging approaches to pediatric brain tumors. Cancers (Basel). 2022;14:14.
18. Bakas S, Akbari H, Sotiras A, Bilello M, Rozycki M, Kirby JS, et al. Advancing the cancer genome atlas glioma MRI collections with expert segmentation labels and radiomic features. Sci Data. 2017;4:1–13.
19. Mabray MC, Barajas RF, Cha S. Modern brain tumor imaging. Brain Tumor Res Treat. 2015;3:8.
20. Monsour R. Neuroimaging in the era of artificial intelligence: current applications. Fed Pract. 2022:14–20.
21. Al-Khawari H, Athyal RP, Al-Saeed O, Sada PN, Al-Muthairi S, Al-Awadhi A. Inter- and intraobserver variation between radiologists in the detection of abnormal parenchymal lung changes on high-resolution computed tomography. Ann Saudi Med. 2010;30:129–33.
22. Iqbal S, Qureshi AN, Li J, Mahmood T. On the Analyses of Medical Images Using Traditional Machine Learning Techniques and Convolutional Neural Networks. In: Archives of computational methods in engineering. Netherlands: Springer; 2023.
23. Helskyaho H, Yu J, Yu K. Machine learning for Oracle database professionals. 2021.
24. Sheikh H, Prins C, Schrijvers E. Mission AI: the new system technology. In: J. E. J. (Corien) Prins F. W. A. (Frans) Brom (ed), editor. Mission AI. Spring; 2023.
25. Iuga AI, Carolus H, Höink AJ, Brosch T, Klinder T, Maintz D, et al. Automated detection and segmentation of thoracic lymph nodes from CT using 3D foveal fully convolutional neural networks. BMC Med Imaging. 2021;21:1–12.
26. Battineni G, Sagaro GG, Chinatalapudi N, Amenta F. Applications of machine learning predictive models in the chronic disease diagnosis. J Pers Med. 2020;10:1–11.
27. Lao J, Chen Y, Li ZC, Li Q, Zhang J, Liu J, et al. A deep learning-based Radiomics Model for prediction of survival in glioblastoma Multiforme. Sci Rep [Internet] 2017;7:1–8. Available from: https://doi.org/10.1038/s41598-017-10649-8.
28. Williams S, Horsfall HL, Funnell JP, Hanrahan JG, Khan DZ, Muirhead W, et al. Artificial intelligence in brain tumour surgery—an emerging paradigm. Cancers (Basel). 2021;13:1–25.
29. Kapitanova K, Son SH. Machine learning basics. Intell Sens Networks Integr Sens Networks: Signal Process Mach Learn; 2012. p. 3–29.
30. Erickson BJ. Panagiotis KorfiatisZeynettin Akkus TLK. Machine learning for medical imaging. J Healthc Eng. 2017:1–11.
31. Mahmud MI, Mamun M, Abdelgawad A. A deep analysis of brain tumor detection from MR images using deep learning networks. Algorithms. 2023;16:1–19.
32. Liu Z, Yang C, Huang J, Liu S, Zhuo Y, Lu X. Deep learning framework based on integration of S-mask R-CNN and inception-v3 for ultrasound image-aided diagnosis of prostate cancer. Futur Gener Comput Syst. 2021;114:358–67.
33. Simonyan K, Zisserman A. Very deep convolutional networks for large-scale image recognition. In: 3rd Int Conf Learn Represent ICLR 2015 - Conf Track Proc; 2015. p. 1–14.

34. Almadhoun HR, Samy S, Naser A. Detection of brain tumor types using deep learning. Proc 2nd Int Conf Artif Intell Smart Energy, ICAIS. 2022;6:459–65.
35. Amin J, Sharif M, Haldorai A, Yasmin M, Nayak RS. Brain tumor detection and classification using machine learning: a comprehensive survey. Complex Intell Syst. 2021;8:3161–83.
36. Stadlbauer A, Marhold F, Oberndorfer S, Heinz G, Buchfelder M, Kinfe TM, et al. Radiophysiomics: brain tumors classification by machine learning and physiological MRI data. Cancers (Basel). 2022;14:1–23.
37. Aamir M, Rahman Z, Dayo ZA, Abro WA, Uddin MI, Khan I, et al. A deep learning approach for brain tumor classification using MRI images. Comput Electr Eng [Internet]. 2022;101:108105. Available from: https://doi.org/10.1016/j.compeleceng.2022.108105
38. Kumar A, Ashok A, Ansari MA. Brain tumor classification using hybrid model of pso and svm classifier. Proc - IEEE 2018 Int Conf Adv Comput Commun control networking, ICACCCN. 2018;2018:1022–6.
39. Magadza TSV. Deep learning for brain tumor segmentation: A survey of state-of-the-art. J Imaging. 2021;908:189–201.
40. Mehrotra R, Ansari MA, Agrawal R, Anand RS. A transfer learning approach for AI-based classification of brain tumors. Mach Learn with Appl. 2020;2:100003.
41. Alzubaidi L, Zhang J, Humaidi AJ, Al-Dujaili A, Duan Y, Al-Shamma O, et al. Review of deep learning: concepts, CNN architectures, challenges, applications, future directions [Internet]. J. Big Data. 2021; Springer International Publishing. Available from: https://doi.org/10.1186/s40537-021-00444-8.
42. Lu SL, Xiao FR, Cheng JCH, Yang WC, Cheng YH, Chang YC, et al. Randomized multi-reader evaluation of automated detection and segmentation of brain tumors in stereotactic radiosurgery with deep neural networks. Neuro-Oncology. 2021;23:1560–8.
43. Saeedi S, Rezayi S, Keshavarz H, Niakan SR, Kalhori. MRI-based brain tumor detection using convolutional deep learning methods and chosen machine learning techniques. BMC Med Inform Decis Mak. 2023;23:1–17.
44. Abiwinanda N, Muhammad Hanif S, Hesaputra T, Astri Handayani and TRM. Brain tumor classification using convolution neural network [internet]. Lect Notes networks Syst. Singapore: Springer ; 2022. Available from: https://doi.org/10.1007/978-981-10-9035-6_33.
45. Haq A u, Li JP, Khan S, Alshara MA, Alotaibi RM, Mawuli CB. DACBT: deep learning approach for classification of brain tumors using MRI data in IoT healthcare environment. Sci Rep. 2022;12:1–14.
46. Park JE, Kickingereder P, Kim HS. Radiomics and deep learning from research to clinical workflow: neuro-oncologic imaging. Korean J Radiol. 2020;21:1126–37.
47. Pashaei A, Sajedi H, Jazayeri N. Brain tumor classification via convolutional neural network and extreme learning machines. 2018 8th Int Conf Comput Knowl Eng ICCKE. 2018;2018:314–9.
48. Siva Raja PM, rani AV. Brain tumor classification using a hybrid deep autoencoder with Bayesian fuzzy clustering-based segmentation approach. Biocybern Biomed Eng [Internet]. 2020;40:440–53. Available from: https://doi.org/10.1016/j.bbe.2020.01.006
49. Talo M, Baloglu UB, Yıldırım Ö, Rajendra Acharya U. Application of deep transfer learning for automated brain abnormality classification using MR images. Cogn Syst Res [Internet] 2019;54:176–88. Available from: https://doi.org/10.1016/j.cogsys.2018.12.007.
50. Li L, Wang JZ. DDIT - A tool for dicom brain images de-identification. 5th Int Conf Bioinforma Biomed Eng iCBBE. 2011;2011:1–5.
51. Begum SS, Lakshmi DR. Combining optimal wavelet statistical texture and recurrent neural network for tumour detection and classification over MRI. Multimed Tools Appl. 2020;79:14009–30.
52. Kalaiselvi K, Karthikeyan C, Devi MS, Kalpana C. Improved classification of brain tumor in MR images using RNN classification framework. Int J Innov Technol Explor Eng. 2020;9:1098–101.

53. Kaur D, Singh S, Mansoor W, Kumar Y, Verma S, Dash S, et al. Computational intelligence and metaheuristic techniques for brain tumor detection through IoMT-enabled MRI devices. Wirel Commun Mob Comput. 2022;2022:1.
54. SivaSai JG, Srinivasu PN, Sindhuri MN, Rohitha K, Deepika S. An automated segmentation of brain MR image through fuzzy recurrent neural network [internet]. Stud Comput Intell Springer Singapore. 2021; Available from: https://doi.org/10.1007/978-981-15-5495-7_9
55. Chen T, Hu L, Lu Q, Xiao F, Xu H, Li H, et al. A computer-aided diagnosis system for brain tumors based on artificial intelligence algorithms. Front Neurosci. 2023;17:1–12.
56. Ulku EE, Camurcu AY. Computer aided brain tumor detection with histogram equalization and morphological image processing techniques. 2013 Int Conf Electron Comput Comput ICECCO. 2013;2013:48–51.
57. Salem Ghahfarrokhi S, Khodadadi H. Human brain tumor diagnosis using the combination of the complexity measures and texture features through magnetic resonance image. Biomed Signal Process Control [Internet]. 2020;61:102025. Available from: https://doi.org/10.1016/j.bspc.2020.102025
58. Arabahmadi M, Farahbakhsh R. Deep learning for smart healthcare—A survey on brain tumor detection from medical imaging. Sensors. 2022;22:1–27.
59. Menze BH, Jakab A, Bauer S, Kalpathy-cramer J, Farahani K, Kirby J, et al. The Multimodal Brain Tumor Image Segmentation Benchmark (BRATS). 2015;34:1993–2024.
60. Nazir M, Shakil S, Khurshid K. Role of deep learning in brain tumor detection and classification (2015 to 2020): A review. Comput Med Imaging Graph. 2021;91:101940.
61. Cheng J, Huang W, Cao S, Yang R, Yang W, Yun Z, et al. Enhanced performance of brain tumor classification via tumor region augmentation and partition. PLoS One. 2015;10:1–13.
62. Megha HC. Evaluation of brain tumor MRI imaging test detection and classification. Int J Res Appl Sci Eng Technol. 2020;8:124–31.
63. Anil A, Raj A, Aravind Sarma H, R NC, P L D. Brain tumor detection from brain MRI using deep learning. Int J Innov Res Appl Sci Eng. 2019;3:458.
64. Vankdothu R, Hameed MA. Brain tumor MRI images identification and classification based on the recurrent convolutional neural network. Meas Sensors [Internet]. 2022;24:100412. Available from: https://doi.org/10.1016/j.measen.2022.100412
65. Toğaçar M, Ergen B, Cömert Z. BrainMRNet: brain tumor detection using magnetic resonance images with a novel convolutional neural network model. Med Hypotheses [Internet]. 2020;134:109531. Available from: https://doi.org/10.1016/j.mehy.2019.109531
66. Koh D-M, Papanikolaou N, Bick U, Illing R, Kahn CE, Kalpathi-Cramer J, et al. Artificial intelligence and machine learning in cancer imaging. Commun Med. 2022;2:1–14.
67. Saranya A, Subhashini R. A systematic review of explainable artificial intelligence models and applications: recent developments and future trends. Decis Anal J [Internet]. 2023;7:1–14. Available from: https://doi.org/10.1016/j.dajour.2023.100230
68. Philip AK, Samuel BA, Bhatia S, Khalifa SAM, El-Seedi HR. Artificial intelligence and precision medicine: A new frontier for the treatment of brain tumors. Life. 2023;13:1–16.
69. Chokshi FH, Flanders AE, Prevedello LM, Langlotz CP. Fostering a healthy ai ecosystem for radiology: conclusions of the 2018 rsna summit on ai in radiology. Radiol Artif Intell. 2019;1:1–6.
70. Shelatkar T, Urvashi D, Shorfuzzaman M, Alsufyani A, Lakshmanna K. Diagnosis of brain tumor using light weight deep learning Model with fine-tuning approach. Comput Math Methods Med. 2022;2022:1–9.
71. Patil S, Kirange D. Ensemble of Deep Learning Models for brain tumor detection. Procedia Comput Sci. 2022;218:2468–79.
72. Lohmann P, Kocher M, Ceccon G, Bauer EK, Stoffels G, Viswanathan S, et al. Combined FET PET/MRI radiomics differentiates radiation injury from recurrent brain metastasis. NeuroImage Clin. 2018;20:537–42.

Chapter 17
Disease Detection and Diagnosis Using Artificial Intelligence Techniques for Sustainable Economic Growth

Vijayetta Sharma

17.1 Introduction

The timely and proactive detection and diagnosis of diseases are crucial for the sustainable development of the country and for steadfast progress toward sustainable development goals (SDGs). The healthcare care indicators of the country need to be continuously monitored and evaluated to achieve the desired targets as set in the action plans of different policy documents aimed at strengthening healthcare access, delivery, and equity. The invention and augmentation of artificial intelligence (AI) and allied technologies play a stupendous role in ameliorating the action plans and provide a valid justification for the agreed social indicators of health. The burgeoning realization of the benefits posed by AI technologies and their potential uses has high value in terms of the realization of gains in the health sector and patient outcomes. The usage of AI-led algorithms in disease detection poses varied benefits and drawbacks in the application and use of AI devices in the health sector. Many studies noted increasing interest among the information science and healthcare communities in using AI for sustainability and economic growth [3, 5, 12, 13, 16, 19, 23, 30, 34].

The concept of AI-led social sustainability in healthcare is perpetually augmenting with the pace of increasing utilization in disease diagnosis and treatment facilitation both in terms of growth in the process as well as techno-medical analysis of health problems. Furthermore, outcomes can be measured in terms of patient satisfaction, the rate of change in diseased state, absorption of AI tools in disease diagnosis, and sustainable AI practices in organizations. The augmentation of sustainable

V. Sharma (✉)
Public Policy, Department of Social and Political Studies, School of Behavioural and Social Sciences, Manav Rachna International Institute of Research and Studies, Faridabad, Haryana, India
e-mail: vijayettasharma.sbss@mriu.edu.in

T. Choudhury, A. Katal (eds.), *Machine Learning for Disease Detection, Prediction, and Diagnosis*, https://doi.org/10.1007/978-981-96-4241-0_17

AI practices in organizations would depend upon the potential use of algorithms in the detection of diseases and their diagnosis, the methodology used for the detection of diseases using different AI techniques, the use of AI in measuring the time span of a disease, and advanced AI interventions in the detection of rare diseases (RDs) which are crucial to achieving a sustained growth of the AI-led health sector. It further pushes for the growth of healthcare enterprises adding up to the economic growth of the nation. The contribution of AI in the medical diagnostic market is finding a new meaning and has gone through a paradigm shift with the fast-paced invention of new AI-led technologies. However, it has been noted that around one-fifth of healthcare organizations worldwide have been in an early stage of adoption of AI models since 2021. The AI techniques when used copiously in disease prediction, diagnosis, and risk management would aid in better decision-making and selection of the right course of treatment and ultimately help in sustainable economic growth by providing a healthier population.

17.2 Benefits of AI-Based Algorithms in Disease Detection and Diagnosis

AI-based algorithms have come into vogue for the detection of rare as well as conventional diseases. Integration of various algorithms into the routine and regular functions of the medical departments can bring an enormous amount of efficiency and effectiveness in the medical service operations. Algorithms can analyze the lifestyle, environmental, and genetic data to assess the risk of developing certain diseases which provides tremendous scope for medical innovation and treatment successes. It can be noted that, in the future, the evolving techniques of AI will be increasingly used to find, assess, and predict patterns in large volumes of medical data and patient histories. This technology-informed social innovation will aid in disease prediction and prevention before symptoms appear. The following broader categories of benefits can be perceived from the AI-based algorithms used in disease detection and diagnosis:

- **Improved veracity in the diagnosis of unrelated and unpredicted patterns**: AI has led to technological breakthroughs through the use of supervised and unsupervised learning techniques to predict unusual patterns in diseases. It helps in understating the emergence and formation of different mutations leading to the disease symptoms.
- **Personalized treatment plans:** Highly personalized treatment plans can be generated for patients using AI technologies that dwell upon the patient's medical history, genetics, and other biological aspects. It has aided the process of knowledge acquisition and providing tailor-made solutions depending upon the specific medical needs of the people.
- **Development of expert systems:** Expert systems developed using AI technologies can speed up prototyping, facilitate maintenance, and produce more impor-

tant information to be used for the detection as well as diagnosis of RDs. The added advantage of expert systems for generating forecasts under close monitoring, collection of accumulated experiences, and development of knowledge systems has inherent advantages in the development of repositories for long-term data retention, usage, and analysis of patterns.

- **Disease monitoring through wearable devices:** Wearable devices loaded with specific algorithms aid in monitoring and treatment of regular and routine ailments, as well as detecting patterns and sending signals in case of RDs. Wearable devices are getting into vogue due to their usability and easy operability requiring lesser expert supervision. The sensors used in wearable devices work for transmitting patient data to a control device wirelessly in a healthcare setup. The wearable technology functions with the help of microprocessors and batteries and shares data via internet connectivity. The advantage of wearable devices is that the collected data can be synced with other electronic devices or transferred to mobile devices or laptops.

Social benefits of the use of AI in disease detection and diagnosis observed at the macro level:

AI technology leads to innovations in social governance and improvement in demographic disease chart indicators count of the nations. It aids in providing viable ground by developing, retaining, and propagating codes based on algorithmic decisional analysis which can be effectually used for the promotion of preventive medicines to patients. The society benefits on a broader whole in terms of clarity, access, transparency, and explainability offered by the algorithmic decision-making approach used by AI which is utilized in finding appropriate solutions for the prevalent disease burden. Furthermore, AI can be vouched for providing a reduction in the global disease burden via a better understanding of demographic and disease patterns prevalent in different regions, available human resources, infrastructure, and other resources based on the adoption and absorption of AI techniques. With the use of AI, an augmented evidence-oriented improvement in health outcomes can be observed in the healthcare industry, both public and private, which is crucial for the achievement of SDGs.

17.3 The Methods for Disease Detection Using AI

AI poses several unique value propositions for disease detection using specialized techniques such as artificial neural networks, rule-based intelligent systems, deep learning, computer vision, and so on. The methodology for disease detection, along with its potential uses, determines the success of the interventions in the treatment utilization of the different ailments. The following are the prominent methods used for disease detection using AI:

- **Federated learning:** Federated learning builds accurate models for healthcare applications, including health data management, remote health monitoring, and medical imaging which has a huge potential for medical innovation and obtaining the initial signals of metabolic changes leading to a diseased state in a timely manner. Federated learning seeks to address the problems of data governance and training accuracy which is highly desirable in disease detection and creating a database of various features leading to disease or non-disease variables.
- **Deep learning:** The novel techniques of deep learning improve perception accuracy and aid in computer-aided diagnosis and detection of various diseases. Deep learning involves the use of neural networks which work by connecting the neurons to each other, updating new data using hyper-parameters, and creating tailor-made solutions. A lot of AI algorithms have been proposed and developed by computer scientists for the detection and diagnosis of diseases such as cancer, heart disease, Alzheimer's disease, Parkinson's disease (PD), tuberculosis, kidney stone detection, dermatological diseases, glaucoma, triage patients, diabetic retinopathy, diseases of the lung, dengue, rheumatoid arthritis, hepatitis, liver and skin disease, and so on. These algorithms perform more precisely compared to specialists without human error [14].

The specific AI techniques in the detection of various diseases are discussed as follows:

- **Neurological diseases:** Neurological disorders use neuroimaging which utilizes mass-univariate statistical methods and has revolutionized the treatment interventions for Alzheimer's disease [33]. Deep learning techniques and support vector machines give higher accuracies in the identification of Alzheimer's disease [2]. It has been noted that AI is a potential bridge between real-world clinical treatment options by practitioners across the globe using neuroimaging findings. The use of supervised algorithms using hyperplanes to classify linear and non-linear datasets is beneficial in the detection of PD patients from vocal signals [18]. Magnetic resonance imaging is utilized for brain tumor mapping with the help of the mean gray-level method [1].
- **Heart diseases:** Algorithmic machine learning (ML) tools are integrated into heart monitoring applications which can predict whether a person can develop heart disease [11]. It can provide a pretest probability of disease which is used to screen patients for angiography [8]. The transformative effect of AI on cardiovascular medicine can be gained via a standard test, that is, the electrocardiogram, a cloud database, and an AI platform for cardiac disease detection which ought to provide inputs for the standardization of medical procedures and better patient outcomes [25, 27, 32]. It may be noted that, with the advent of new guided therapy for the treatment of cardiovascular diseases, the physician can provide prompt, individualized, and customized treatment to the patients which will have a significant psychological effect as the patients feel safe while continuously monitored [38].
- **Eye diseases:** It can be further noted that, in AI, deep learning has been applied to bring stupendous results in ophthalmology to achieve robust classification per-

formance in the detection of diabetic retinopathy, glaucoma-like disc, age-related macular degeneration, and macular edema [35]. The commonly used diagnostic modalities incorporating fundus photography, gonioscopy, and standard automated perimetry are the standard AI practices used in providing treatment interventions to glaucoma patients [6, 8].

- **Kidney, liver, lungs, and skin diseases:** The detection of kidney stones is done using artificial neural networks and segmentation [26]. The use of dual-stage approach comprising computer vision and ML on clinically evaluated histopathological attributes is used to accurately identify dermatological diseases [24]. The efficacy of various ML algorithms has posed results in liver disease diagnosis through prediction [15, 21]. For the detection and diagnosis of lung diseases, deep learning methodologies are used by chest radiographers [29]. Computer vision has a role in the detection of skin diseases and works on the inputs of a color image, and extraction of features using a pretrained convolutional neural network [7].

AI methods and their integration with fuzzy logic and wavelength analysis provide an enhanced ability for data interpretation and modeling, along with interpretations of the crucial information and converting it into organized information with optimized features for diagnostics [20, 21, 37].

17.4 The Role of AI in Early Disease Detection

AI plays a commendable role in the early and timely detection of diseases by diagnosing and showing any hidden signs of risk that individuals may possess. Some studies have shown that AI has accurately predicted heart disease with 86.7% accuracy in diagnosis and prognosis and early stages of lung cancer with 94% predictability [10, 22]. It can be noted that AI via its specialized techniques can help in the accurate diagnosis of colon cancer more than a trained pathologist [40]. Furthermore, AI technologies are able to detect tumors in scans of different body parts of patients with greater veracity than professional radiologists [9]. The specialized techniques of AI in the form of algorithmic integration foster the early detection of disease posing and add a number of benefits for the patients, providers, and governance actors. The detection of early disease has several advantages in terms of operationalization of the medical process in time to save the lives of patients, decrease the mortality counts as per the SDG indicators, provide better access to services in terms of early reaching the hospital or intrahospital movement of patients for service utilization, reduce the delay in treatment, early recovery, lower burden on the health ecosystem, and ensure healthier economies with sustainable health advantages. Sustainable healthcare outcomes attained through AI interventions can entrust better elasticity and long-term maintenance of health outcomes bringing stability to the healthcare governance action plans envisaged under health policies.

17.5 AI Interventions in the Detection of Rare Diseases

AI technologies can be relied upon by the governance actors when it comes to the detection and diagnosis of RDs. The RD has an element of obscureness, hidden mutation of unknown biological cells, or an unpredictable tension between the cells which could be the cause of future known or unknown patterns. RDs, on the other hand, are also not diagnosed at an early stage by physicians due to the involvement of common symptoms which also occur in many of the known ailments. AI by way of its stupendous supervised and unsupervised algorithmic integrations such as K-nearest neighbor, neural networks, k-means clustering, anomaly detection, and so on tries to reach the patterns that are earlier not known, or known in the context of some other disease with another abnormality dimension differentiating it from the earlier known patterns with a different accuracy score, or posing a new untracked dimension to the earlier known patterns differentiating it drastically from the earlier known disease patterns creating a high chances of interposed pattern which could be classified as a new disease. RD identification has become much easier with the help of AI interventions as the patterns can become more discrete, their relationship can be studied in more detail, the classification can be made in a greater elaborated fashion, and social prevalence can be observed with a greater clarity. Furthermore, both the practitioners and the patients can leverage the AI innovations for the facilitation of treatment in case of identification of RDs in terms of operationalization of hospital care, deciding the care interventions on time, and facilitating the movement of patients within the hospital or between the hospitals for having the required care options.

In the healthcare clinical research associations and management groups, RDs are also referred to as "zebras" due to their unusual and unexpected occurrence along with or without the symptoms of common ailments. A set of algorithms powered with AI called PANDAs are used by researchers from the University of Florida and Penn Medicine to find rare "zebras." These PANDAs are investigated using deep learning techniques and help patients obtain diagnostic and treatment options more effectively and in a timely manner (Diana [36]). A number of times, an isolated symptom of an RD can be misdiagnosed and classified into a category of a common disease, which many times leads to "premature closure" of the treatment intervention leading to aggravation of the health condition. On the contrary, the prolonged diagnostic journey encompassing various steps of consultations with different physicians accompanied by a plethora of examinations and treatment formalities impairs diagnostic judgment [28]. The recent advancements in AI have empowered researchers to train models on large datasets and thereafter fine-tune these models on smaller datasets containing evidence of RDs [39]. Studies also showed that the interaction between AI technology and RDs is an impactful event [17]. At the same time, when there is suboptimal performance of AI technologies in patients having uncommon conditions, it can retard the clinical utility of AI medical devices for routine clinical practices [31].

17.6 Global AI in Medical Diagnostic Market

Healthcare sectors in India and China, which are emerging economies, are spurring at a fast pace with the introduction of new medical technologies. It can also be noted that the untapped markets in countries such as Brazil, India, and China are poised to provide significant expansion opportunities in the medical diagnostic market as a result of their increasing patient populations. The growth of AI in medical diagnostics is due to the profuse influx of big data in different segments of health which subsequently impacts the economy, government, and private sector-led initiatives to gear up AI-based technologies, funding options for startups endorsing AI in the healthcare domain, and cross-industry collaboration and partnerships for development of novel AI technologies. As per the recent market trends, the contribution of AI in the medical diagnostic market is estimated to be around $1.3 billion in 2023 and is expected to reach $3.7 billion in 2028. It is expanding at the transcending speed of CAGR of 23.2% from 2023 to 2028 (see Fig. 17.1).

There is enough scope for expanding opportunities for the use of AI and catching attractive investments by entrepreneurs in the medical diagnostic market. There is immense potential in the medical image diagnosis offered by AI in the radiology sector in terms of improving image triage, better analysis of image data, faster analysis, and efficient interpretation. The software segment accounted for the largest share of AI in the medical diagnostic industry in 2022 from three major components of AI in the medical diagnostic market, that is, software, hardware, and services. North America grasps for the largest share of global AI in the medical diagnostic market in 2022, when compared regionally. Some of the significant contributors in the diagnostic market are Microsoft (United States), GE Healthcare (United States), Digital Diagnostics Inc. (United States), InformAI (United States), Koninklijke

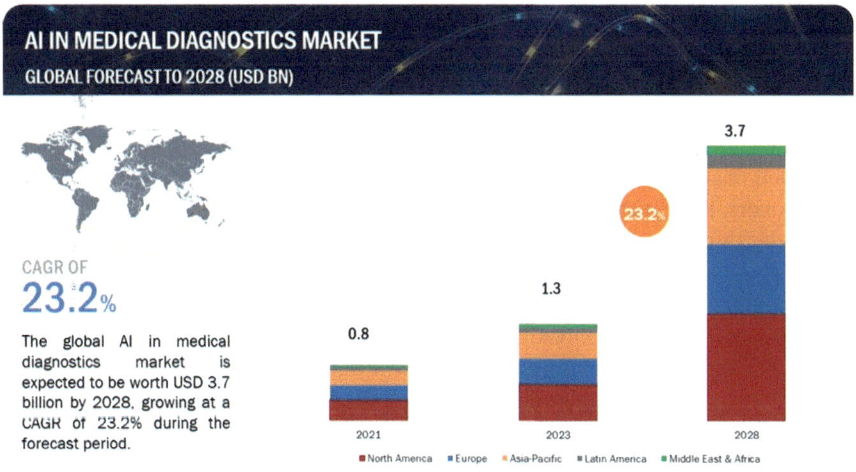

Fig. 17.1 AI in medical diagnostics market (Source: Market and Markets, Global Forecast [4])

Philips NV (Netherlands), Brainomix (UK), icometrix (Belgium), Aidence (Netherland), Lunit Inc. (South Korea), Nano X imaging (Israel), Quibim (Spain), Qure.ia (India), Therapixel (France), and so on.

A few recent developments of AI in the medical diagnostic market are worth noting. In January 2022, Siemens Healthineers (Germany) and Ohio State Wexner Medical Centre (United States) partnered to provide cutting-edge radiation oncology and advanced imaging modalities in outpatient care. In October 2022, Google Cloud (United States) launched an AI-enabled medical imaging suite that provides new industry solutions rendering precision in image diagnostics, operational efficiency, improved care access, and better organizational and patient outcomes. In January 2023, Intel Corporation (United States), launched Fourth-Generation Scalable Processors providing augmented results in the data center performance, defining new capabilities of AI rendering efficiency, and addressing security concerns. Furthermore, in September 2023, Mayo Clinic (United States) and GE Healthcare (United States) developed a collaboration in product development to better equip clinicians to effectually diagnose and provide viable solutions to medical conditions through the use of AI.

17.7 Conclusion

The sustainable economic growth of different nations depends upon the integration and the pace of adoption of AI technologies in the medical sector whose primal function lies in the detection and diagnosis before giving valuable solutions to the treatment of various diseases. If the well-researched technologies of AI are utilized to fulfill the developmental objectives of the nations aimed at reducing the disease burden under the SDGs and the other national goals and priorities, half of the task of attaining inclusive healthcare is achieved. The AI technologies veritably provide for strengthening the health systems and providing inputs for process improvements leading to efficient detection and diagnosis of diseases before they occur and also after the emergence of symptoms. AI, with the help of farsighted technologies and the ability to provide clarity, explainability, and transparency, is bound to bring farsighted results in the reduction of disease burden in the years to come. Medical practitioners, policymakers, and health technology enthusiasts should be cognizant of the development of AI technologies, processes, strategies, operational procedures, governmental interventions, entrepreneurial initiatives, and national and international regulations to utilize the transcending wave of automation and artificial techniques learning to maximum potential.

The healthcare space in the coming age will redefine its processes and priorities to an advanced level of elaboration and interpretation earmarked with digitization and expansive use of AI providing impetus to sustainability, governance, and economic growth marked by the healthcare industry. AI-driven decision-making offers care outcomes by lending specialized algorithmic combinations for disease detection and diagnosis and is bound to achieve national priorities for the attainment of

SDGs in an augmented and pervasive manner. AI technologies are revolutionizing the healthcare sector and providing an added advantage of superior techno-medical diagnostic abilities to the healthcare systems lending excellence, equity, and healthcare access in service provisions.

References

1. Abdalla HEM, Esmail MY. Brain tumor detection by using artificial neural network. In: International conference on computer, control, electrical, and electronics engineering (ICCCEEE). IEEE; 2018. p. 1–6.
2. Afzal S, Maqsood M, Khan U, Mehmood I, Nawaz H, Aadil F, ... Yunyoung N. Alzheimer disease detection techniques and methods: a review. 2021 .
3. Ahmed M, Zubair S. Explainable artificial intelligence in sustainable smart healthcare. In: Explainable artificial intelligence for cyber security: next generation artificial intelligence. Cham: Springer International Publishing; 2022. p. 265–80.
4. AI in Medical Diagnostics Market Size, Share, Trends and Revenue Forecast. 2023. marketsandmarkets.com
5. Alami H, Rivard L, Lehoux P, Hoffman SJ, Cadeddu SBM, Savoldelli M, Fortin JP. Artificial intelligence in health care: laying the Foundation for Responsible, sustainable, and inclusive innovation in low-and middle-income countries. Glob Health. 2020;16:1–6.
6. Al-Aswad LA, Ramachandran R, Schuman JS, Medeiros F, Eydelman MB, Abramoff MD, et al. Artificial intelligence for glaucoma: creating and implementing artificial intelligence for disease detection and progression. Ophthalmology Glaucoma. 2022;5(5):e16–25.
7. ALEnezi NSA. A method of skin disease detection using image processing and machine learning. Procedia Computer Science. 2019;163:85–92.
8. Alizadehsani R, Khosravi A, Roshanzamir M, Abdar M, Sarrafzadegan N, Shafie D, et al. Coronary artery disease detection using artificial intelligence techniques: A survey of trends, geographical differences and diagnostic features 1991–2020. Comput Biol Med. 2021;128:104095.
9. Ardila D, Kiraly AP, Bharadwaj S, Choi B, Reicher JJ, Peng L, Tse D, Etemadi M, Ye W, Corrado G, Naidich DP. End-to-end lung cancer screening with three-dimensional deep learning on low-dose chest computed tomography. Nat Med. 2019;25(6):954–61.
10. Cancer Research UK. Lung Cancer survival. 2020;. Available at: https://www.cancerresearchuk.org/about-cancer/lung-cancer/survival
11. Chang V, Bhavani VR, Xu AQ, Hossain MA. An artificial intelligence model for heart disease detection using machine learning algorithms. Healthcare Analytics. 2022;2:100016.
12. Damoah IS, Ayakwah A, Tingbani I. Artificial intelligence (AI)-enhanced medical drones in the healthcare supply chain (HSC) for sustainability development: A case study. J Clean Prod. 2021;328:129598.
13. Dicuonzo G, Donofrio F, Fusco A, Dell'Atti V. Big data and artificial intelligence for health system sustainability: the case of Veneto region. Management. Control. 2021, 2021; Suppl. 1
14. Ghaffar Nia N, Kaplanoglu E, Nasab A. Evaluation of artificial intelligence techniques in disease diagnosis and prediction. Discov Artif Intell. 2023;3(1):5.
15. Ghazal TM, Rehman AU, Saleem M, Ahmad M, Ahmad S, Mehmood F. February. In: Intelligent model to predict early liver disease using machine learning technique. International conference on business analytics for technology and security (ICBATS). IEEE; 2022. p. 1–5.
16. Hadley TD, Pettit RW, Malik T, Khoei AA, Salihu HM. Artificial intelligence in global health—a framework and strategy for adoption and sustainability. Int J Maternal Child Health AIDS. 2020;9(1):121.

17. Hasani N, Farhadi F, Morris MA, Nikpanah M, Rhamim A, Xu Y, Pariser A, Collins MT, Summers RM, Jones E, Siegel E, Saboury B. Artificial intelligence in medical imaging and its impact on the rare disease community: threats, challenges and opportunities. PET Clin. 2022;17(1):13–29. https://doi.org/10.1016/j.cpet.2021.09.009. PMID: 34809862; PMCID: PMC8764708
18. Hoq M, Uddin MN, Park SB. Vocal feature extraction-based artificial intelligent model for Parkinson's disease detection. Diagnostics. 2021;11(6):1076.
19. Hsu LP, Huang YK, Chang YC. The implementation of artificial intelligence in dentistry could enhance environmental sustainability. J Dental Sci. 2022;17(2):1081.
20. Huang Y. Advances in artificial neural networks–methodological development and application. Algorithms. 2009;2(3):973–1007.
21. İlhan HO, Çelik E. The mesothelioma disease diagnosis with artificial intelligence methods. In: 2016 IEEE 10th international conference on application of information and communication technologies (AICT). IEEE; 2016, October. p. 1–5.
22. Jiang F, Jiang Y, Zhi H, Dong Y, Li H, Ma S, Wang Y, Dong Q, Shen H, Wang Y. Artificial intelligence in healthcare: past, present, and future. Stroke Vascular Neurol. 2017;2(4):230.
23. Kar AK, Choudhary SK, Singh VK. How can artificial intelligence impact sustainability: a systematic literature review. J Clean Prod. 2022:134120.
24. Kumar VB, Kumar SS, Saboo V. Dermatological disease detection using image processing and machine learning. In: 2016 third international conference on artificial intelligence and pattern recognition (AIPR). IEEE; 2016. p. 1–6.
25. Lin YJ, Chuang CW, Yen CY, Huang SH, Huang PW, Chen JY, Lee SY. Artificial intelligence of things wearable system for cardiac disease detection. In: 2019 IEEE international conference on artificial intelligence circuits and systems (AICAS). IEEE; 2019. p. 67–70.
26. Nithya A, Appathurai A, Venkatadri N, Ramji DR, Palagan CA. Kidney disease detection and segmentation using artificial neural network and multi-kernel k-means clustering for ultrasound images. Measurement. 2020;149:106952.
27. Omarov B, Saparkhojayev N, Shekerbekova S, Akhmetova O, Sakypbekova M, Kamalova G, et al. Artificial intelligence in medicine: real time electronic stethoscope for heart diseases detection. Computers, Materials & Continua. 2022;70(2):2815.
28. Priyom B. Pioneering AI-based approach identifies rare disease patients using electronic health records. News-Medical. 2023; Retrieved on October 19, 2023 from https://www.news-medical.net/news/20230511/Pioneering-AI-based-approach-identifies-rare-disease-patients-using-electronic-health-records.aspx
29. Qin C, Yao D, Shi Y, Song Z. Computer-aided detection in chest radiography based on artificial intelligence: a survey. Biomed Eng Online. 2018;17(1):1–23.
30. Sciarretta E, Mancini R, Greco E. Artificial intelligence for healthcare and social services: optimizing resources and promoting sustainability. Sustainability. 2022;14(24):16464.
31. Sharma N, Aggarwal LM. Automated medical image segmentation techniques. J Med Phys. 2010;35:3–14.
32. Siontis KC, Noseworthy PA, Attia ZI, Friedman PA. Artificial intelligence-enhanced electrocardiography in cardiovascular disease management. Nature Rev Cardiol. 2021;18(7):465–78.
33. Subasi A. Use of artificial intelligence in Alzheimer's disease detection. In: Artificial Intelligence Precision Health; 2020. p. 257–78.
34. Suha SA, Sanam TF. Exploring dominant factors for ensuring the sustainability of utilizing artificial intelligence in healthcare decision making: an emerging country context. Int J Info Manag Data Insights. 2023;3(1):100170.
35. Ting DSW, Pasquale LR, Peng L, Campbell JP, Lee AY, Raman R, et al. Artificial intelligence and deep learning in ophthalmology. Br J Ophthalmol. 2019;103(2):167–75.
36. Tonnessen D. Researchers will use AI to predict who may develop certain rare diseases. College of Medicine, University of Florida; 2022. October 25, 2022; https://ufhealth.org/news/2022/researchers-will-use-ai-predict-who-may-develop-certain-rare-diseases

37. Vashistha R, Chhabra D, Shukla P. Integrated artificial intelligence approaches for disease diagnostics. Indian J Microbiol. 2018;58:252–5.
38. Visco V, Ferruzzi GJ, Nicastro F, Virtuoso N, Carrizzo A, Galasso G, et al. Artificial intelligence as a business partner in cardiovascular precision medicine: an emerging approach for disease detection and treatment optimization. Curr Med Chem. 2021;28(32):6569–90.
39. Wojtara M, Rana E, Rahman T, Khanna P, Singh H. Artificial intelligence in rare disease diagnosis and treatment. Clin Transl Sci. 2023;16:2106.
40. Yu G, Sun K, Xu C, Shi XH, Wu C, Xie T, Meng RQ, Meng XH, Wang KS, Xiao HM, Deng HW. Accurate recognition of colorectal cancer with semi-supervised deep learning on pathological images. Nat Commun. 2021;12(1):1–13.

Chapter 18
Developing a COVID-19 Prediction Kit Using Machine Learning

Tanupriya Choudhury, Sumit Aich, Avita Katal, Subhangi Sati, Purvika Joshi, and Ayan Sar

18.1 Introduction

The coronavirus disease revealed the truth about what is real and what matters—namely, to have the best tools that will precisely quantize and fasten the task of forecasting the spread of infectious diseases. With virus mutations and the emergence of new variants, the need for an accurate method of prediction becomes a momentous concern to define strategies to fight infections including the allocation of healthcare resources. The classical epidemiological models might be satisfactory, but most of them are based on obvious assumptions and fail to involve the true dynamics of the virus spread at any given time. Amid this situation lies the answer to real-time forecasting and prediction using machine learning (ML). This is where tackling this challenge becomes interesting and easy. As a subset of artificial intelligence (AI), ML endows computational systems with the capacity to learn from data and Lakota through these processes, data is collected every day and accurately informs future decisions. Utilizing ML algorithms, researchers, along with healthcare professionals, can overcome the challenge of analyzing huge volumes of heterogeneous data, which may include demographic details, clinical reports, mobile phone tracking data, and expression of genes simultaneously to provide relative risks of COVID-19 acquisition and severity. ML capability in making predictions enables models to adjust and refine their estimations over time when new data is available, thus facilitating more

T. Choudhury (✉) · S. Aich · A. Katal · S. Sati · P. Joshi · A. Sar
School of Computer Science, University of Petroleum and Energy Studies (UPES), Dehradun, Uttarakhand, India
e-mail: tanupriya@ddn.upes.ac.in; akatal@ddn.upes.ac.in

T. Choudhury, A. Katal (eds.), *Machine Learning for Disease Detection, Prediction, and Diagnosis*, https://doi.org/10.1007/978-981-96-4241-0_18

dynamic and effective decision-making as the course of the future coronavirus changes. An alternative approach to predicting the COVID-19 outbreak lies in developing an AI kit for prevention, which may present a breakthrough in how we prepare for and contain pandemics. Such a toolbox gives the authorities, as well as healthcare professionals and ordinary people chance to process immediate details of the impending outbreak, affected areas, and the control methods. Besides, ML-driven detection kits may help ascertain whether emerging strains are capable of igniting the next pandemic, wherein immediate counterattacks may prove helpful in preventing their spread. Indeed, accompanying these potential ML benefits are several problems, including data quality ditches, explainable ML, ethical inspections, and concerns of equity. Solving these challenges requires an interdisciplinary team member of data scientists, epidemiologists, healthcare workers, policymakers, and leaders within the communities. In the process of dealing with these obstacles, we have a chance to evolve ML into the most advanced tool capable of forecasting and reducing the effects of COVID-19 as well as the possibility of any pandemic.

18.2 Literature Review

The study of ML in predicting a COVID-19 illness kit longer reports a flourishing area of research that intends to use computational models for diagnosing, prognosing, and managing the disease. Deep learning algorithms have been demonstrated to be effective in various molecular tasks related to COVID-19 DLanes, such as antigen and drug development. The initial investigation was concerned with ML models analyzing medical imaging data, which included the cases of chest X-rays and computed tomography (CT) scans, to identify COVID-19 infection and to classify the severity of the condition. Likewise, Wang et al. [1] took the deep learning approach to the detection of COVID-19 from chest X-ray images, and (that) led to a remarkably high sensitivity and accuracy. On the other side, machine-learning-based technologies are another field researchers have considered for predicting COVID-19 outcomes such as disease progression and mortality risk. Liang and his colleagues [2] proposed a model combining ML and persuasive factors, such as the age of a patient, any existing health problems, or laboratory test results, in order to forecast the risk of critical illness of patients with COVID-19. Apart from that, we see that MS for seeing where and how COVID-19 outbreaks will appear and where we might be defenseless becomes an important part of these projections. Lippi and the co-authors [3] designed an ML model enabling to forecast of the development of the COVID-19 outbreak's path based on the available epidemiological data and providing more accurate diagnosis and treatment (prevention and protection especially) for the general public. Additionally, the researchers examined combining multiple data sources, consisting of clinical, demographic, and environmental factors, which were made use of to increase the predictive precision of ML models that were designed for COVID-19 prediction. As an example, the research of Yan et al. [4] utilized clinical and laboratory data to develop ML tools that anticipate the intensity of COVID-19 patients and, in due course, they will need hospital care. Another problem arising in the creation of ML-based COVID-19 diagnosis kits is that, first of all, they need large raw data samples, which are supposed to be diverse. Second, there should be an explanation of the diagnosis the model has achieved. Third, there have to be no distinctions between the population groups. Jain et al. suggested a method that was based on gradients and statistics from chest X-ray image grades [5], whereas Chen et al. developed a reward system using CT images for diagnosis and prediction purposes [6]. Alneyadi et al. constructed a model that predicted the regression of daily COVID-19 cases, which were based on the population that had been vaccinated [7], while Nor Iskandar et al. applied the ML techniques for risk factor prediction using the data from South Sudan and Eastern Democratic Republic of the Congo [8]. Moreover, Mathur and Datta constructed a learning machine model for COVID-19 in India [9], and Jaleel et al. suggested a K-nearest neighbor classifier model that can be used for the prediction of COVID-19 in Iraq [10]. While Deva et al. focused on mathematical models for the analysis and prediction of COVID-19, Tahir et al. compared the models on the basis of the performance of the patients' registration slips as well as the performance of each model [11, 12]. Nastiti et al. performed a systemic literature review about the ML models for the prediction of the spread of COVID-19, while Gode et al. analyzed the risk prediction via the use of an ensemble boosting algorithm [13, 14]. Also, Haritha et al. used convolutional neural networks (CNNs) to diagnose COVID-19 cases from X-rays [15]. As to that, Bi et al. achieved a competitive performance in volume prediction of pulmonary damages with an embedded uncertainty estimation in the Trans-UNet [16]. Bai used regression and deep learning models to create a forecasting model for epidemic cases of COVID-19 [17], and Chen and Zhou brought forward a multitone learning one for pandemic prediction [18]. Bhardwaj et al. introduced a global prediction model by means of ML which predicts a number of cases and deaths [19]. On the other hand, Rana et al. highlighted secure communication systems for outbreak prediction in Internet of things [20]. In Indonesia, the prediction model was built by Yaqin et al. involving support vector regression (SVR) and sampling importance resampling (SIR) algorithms [21]. The SIR model was also used in mortality prediction by Chen et al. in their analysis [22]. Kumar and Kaur presented a hybrid self organizing maps (SOM)-fuzzy time series technique as well to make COVID-19 forecasts [23], and Shi et al. used autoregressive integrated moving average (ARIMA) models for predicting the trend of the pandemic in America [24]. On the other side, Kasturi et al. proposed a study related to the usage of the Facebook Prophet method for the time series data forecasting in the context of Bontang City, whereas Ayan et al. made a study on the cellular network traffic based on the broadcasted data [25]. Pramanik et al. created a deep learning model to study lung scans and assess the risk of COVID-19 [26], in comparison to Dai and Zhu, who applied ML methods to predict mortality age during the pandemic in India [27]. The last study by Devi et al. also adopted ML for the diagnosis of COVID-19 using symptoms [28].

However, the collection of all this data raises concerns. However, the literature is developing faster and highlights the value of ML methods in aiding the management and control of COVID-19, making the process more clarified for healthcare providers and policymakers. For the resolution of numbered issues and confirmation of the usefulness of ML models in clinical sites and public health monitoring, a future study should be conducted. The bibliometric study is also shown in Fig. 18.1.

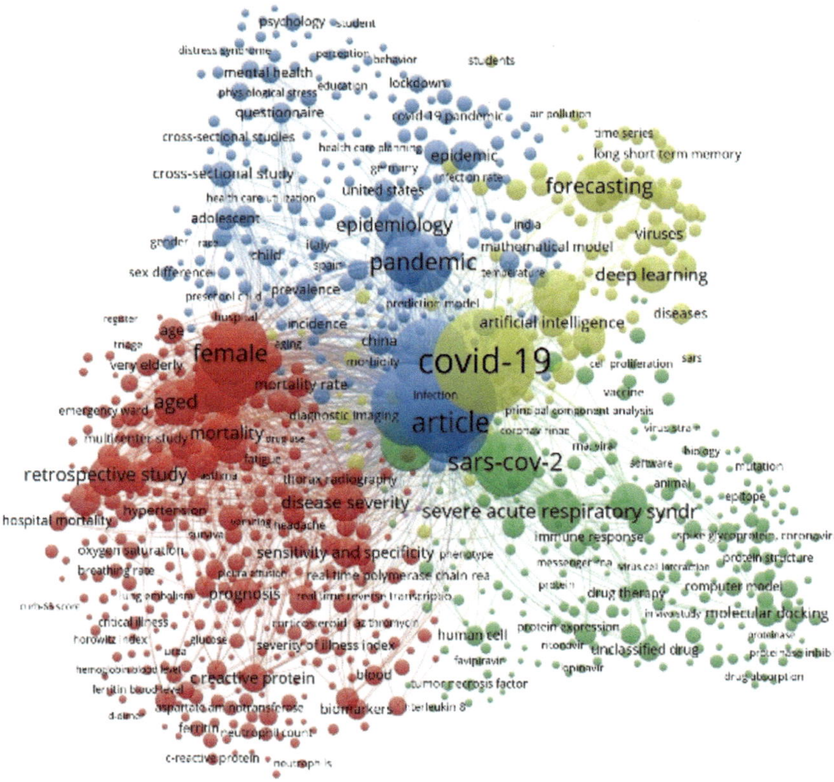

Fig. 18.1 Bibliometric study

18.3 Methodology

The sawtooth of coronavirus prediction methods has been covered after its eruption since the 11th month of 2019. At first, modeling epidemiologic models based on the mathematical equations' compartmental models (such as SIR and SEIR) were popular tools for getting predictions of virus spread across the globe. While they gave the theoretical view of the transmission dynamics, there was a lot of assumption that the models could not mirror the aspects of the pandemic which were complex. ML approaches that have come forward as luckier models as compared to traditional epidemiological models tend to add the power of data-driven techniques which, as a result, produce an accurate health forecast. AI ML models can run on big varied data such as demographic data, patient records, movement records, and genetic sequences for identifying patterns and predictions. There are several ML-based COVID-19 prediction models developed that use different algorithms such as random forest, support vector machines (SVMs), gradient boosting machines, CNN, and long short-term memory deep learning architectures. These models have also been used to predict the patterns of the coronavirus pandemic. Mathematical models with AI are the most efficient among the old epidemiological models in terms of speed and adaptability. However, a droplet of honey forbidding, ML as a method still encounters some challenges. One major problem that has to be solved is the quality and quantity of data which is accessible for analysis. COVID-19 data, most times, are very noisy, inadequate, and biased especially in reporting them, resulting in a challenge when training and validating models. However, the fast-developing COVID-19 pandemic does not allow for a long time for researching and refining models by using continuous updates to models and adaptive learning techniques. Another challenge is the nature of ML models for generalization. Although it may be the case that these models provide high predictive accuracy, it is essential not just to understand the underlying mechanisms responsible for such predictions but also to ensure stakeholders (e.g., healthcare professionals and policymakers) trust and accept that models can be completely reliable. Another important point is the validation (quality assurance) of data and ML systems that should be made before this system is deployed in real-world situations and integrated into the existing healthcare systems. Scalability, reliability, and compliance with regulatory standards need to be the main targets that have to be achieved for legally mandated adoption. The diagrammatic representation of the proposed methodology is shown in Fig. 18.2.

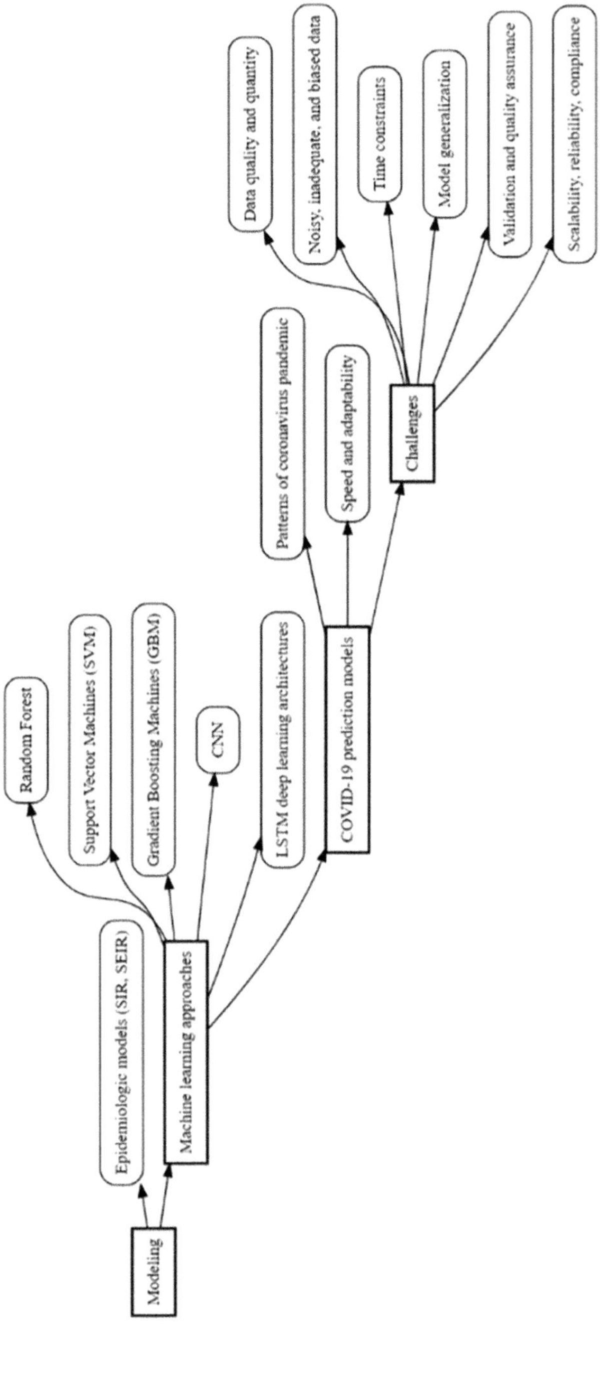

Fig. 18.2 Methodological diagram of the proposed system

18.4 Data Acquisition and Preprocessing

Accurate forecasting of COVID-19 necessitates the use of different data sources with reliable data that include demographic information, clinical data, and epidemiological statistics, and there may be included also data from wearable devices or environmental sensors. Despite this, the quality and the accessibility of this information are inconsistently valid, creating obstacles to the construction of advanced prediction models. Data acquisition encompasses a range of ways through which data is drawn from multiple sources; examples of these sources include public health databases, hospital records, research studies, and even government reports. These datasets do not only hold information that is inaccurate because of different data collection protocols and data recording methods but also reflect variations in the individual hospitals' protocols. Concerningly, data privacy and sharing protocols might prevent the owning of full airline and insurance datasets, thus rendering complex forecasting models impossible. Through this data-receiving stage, ML needs to start preprocessing to avoid problems that could arise during the next step. This involves several steps as shown in Fig. 18.3.

18.4.1 Cleaning

Removal of redundancies, understanding of errors and their removal, and perception of dealing with missing values are essential steps of data cleaning. For example, imputation methods such as mean imputation or predictive modeling can be employed in filling missing values but with the data integrity shield intact.

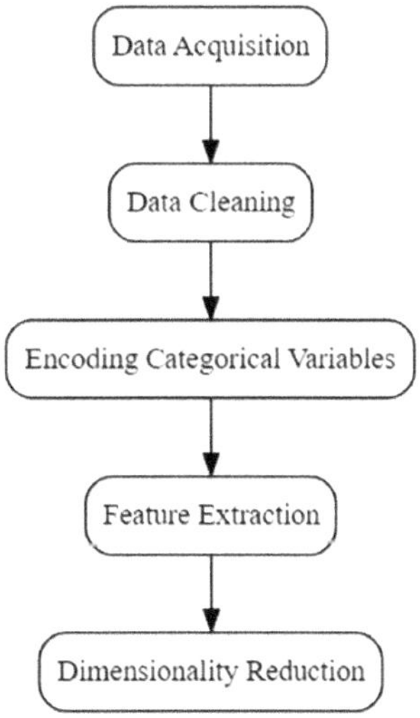

Fig. 18.3 Data acquisition diagrammatic representation

18.4.2 *Encoding Categorical Variables*

Most algorithms used in ML demand categorical variables to be encoded into numerical representations, because if they were input as categories, the machine would be unable to process them. One-hot representation, one-hot encoding, or ordinal encoding are preferred for this encoding stage.

18.4.3 *Feature Extraction*

In the feature extraction process, data are condensed and made more informative by the process of extracting features. Such may include eliminating supremely valuable features from text data, time-series data, and images, by way of word embeddings, Fourier transforms, and CNNs, respectively.

18.4.4 *Dimensionality Reduction*

High dimensionalities may impede the analysis and approach too computationally demanding, providing the researchers with an increased number of combinations of independent variables, which results in the overfitting possibility. With multidimensionality reduction techniques such as principal component analysis or t-distributed stochastic neighbor embedding, one can have the right amount of features to extract their essence with minimum loss.

18.5 Machine Learning Algorithms

A range of ML algorithms is what makes up COVID-19 predicting systems, helping to forecast behavioral patterns in the data. This section delves into various algorithms suitable for this task, their functionalities, and considerations for selection: This section delves into various algorithms suitable for this task, their functionalities, and considerations for selection.

18.5.1 *Algorithm Overview*

Random Forest It is a team learning method that is adaptable enough to create each decision tree and combine their predictions. The capability to tackle large datasets, as well as the non-linearity of relations, makes the ML method appropriate for the predetermination of COVID-19. A sample random forest structure is shown in Fig. 18.4.

SVM SVM does well by finding a plane that not only separates data from one class to the other but also maximizes the distances of data points between the hyperplane and classes. Through its kernel trick, it can perform nonlinear classification, one of the many features that make it compatible with complex interactions between environmental as well as biological variables during this pandemic. This system of SVMs is shown in Fig. 18.5.

Neural Networks Through their deep learning methodologies, such as CNNs and recurrent neural networks, which are capable of learning implicit deep feature representations hierarchically, these architectures have shown a high ability to capture complex patterns in COVID-19 data.

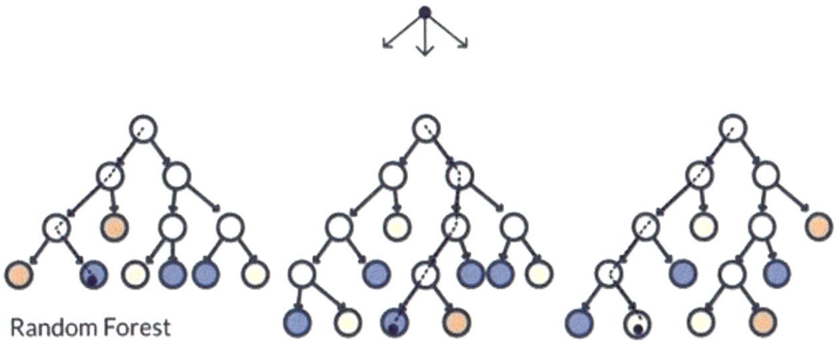

Fig. 18.4 Sample structure of random forest

Fig. 18.5 Sample structure of SVMs

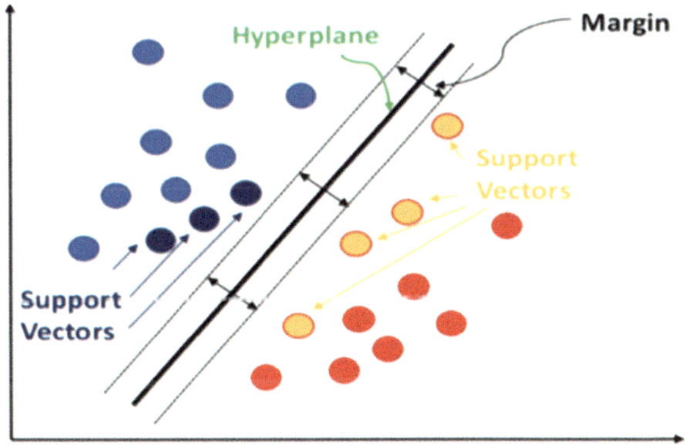

18.5.2 Comparative Analysis

Performance Metrics The evaluation metrics including, accuracy, precision, recall, and F1-score are used to determine the quality of the algorithms between them.

Scalability Implication of computational stipulations for the algorithms and scaling to large deployment can be taken into consideration.

Interpretability It is a balance between model complexity and interpretability; therefore, it should lead to insights about the predictive factors of COVID-19.

18.5.3 Selection Criteria

Data Complexity Whether an ML algorithm uses supervised learning, unsupervised learning, or some combination of these is determined by the nature and complexity of COVID-19 data. An example of this is that neural networks cope very well with handling vector data.

Resource Constraints Accessing of computational resources, time pressure, and scalability concerns are some of the factors that will determine the selection of the algorithms.

Accuracy Requirements The appropriate rate of prediction accuracy and the compromise areas between bias and variance determine the model type. Ensemble methods such as random forest may be performed by incorporating the essence that ensures resilience in case of uncertainties.

18.6 Feature Selection and Engineering

Attributes selection and engineering are very important stages in building a multi-task ML model for COVID-19 predictive kits. This phase is dedicated to selecting and specifying the most suitable data attributes (features) pertinent to the result of precise prediction. Here is a breakdown of the sub-points as in Fig. 18.6.

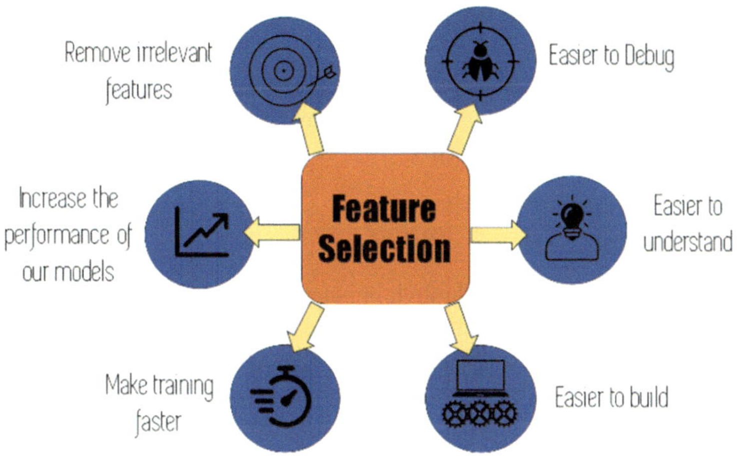

Fig. 18.6 Six pillars of feature selection and engineering

18.6.1 Identification of Relevant Features

- Figuring out which data attributes regarding COVID-19 are ideally to be used to determine the stage and progression of the disease.
- Tele scripts may encompass subsets of data such as patient demographics (age, gender), symptoms, comorbidities, laboratory test results (e.g., blood markers), imaging findings (e.g., chest X-rays), and epidemiologic data.
- Select the features that offer relatively high accuracy and relevance to the disease, as well as the features that are present in the dataset.

18.6.2 Techniques for Feature Selection

- Various methods can be employed to select the most informative features while reducing dimensionality: Various methods can be employed to select the most informative features while reducing dimensionality.
- Univariate feature selection: Measuring the relative importance of each feature that has the most amount of influence on the prediction.
- Feature importance ranking: Through algorithms such as random forest and gradient boosting, to assess the acceptable level of each feature for the model, its contribution to the model performance is taken into account.
- Wrapper methods: Gradually pick features from the subsets, by contemplating the model's performance.
- Embedded methods: We incorporate a feature selection method during the training stage of the model (e.g., Lasso regression).

18.6.3 Feature Engineering

- Corresponds to modifications of the model architecture for identifying areas that can be improved and bringing model performance to a new level.
- **Techniques include the following**:

    1. Encoding categorical variables.
    2. Transforming numerical features' values to bounds within the same size range.
    3. Interrupting the lost data through imputation or fitting missingness markers. Use our AI to write for you about any topic! Enjoy our sample by starting a conversation. The process of aging is a natural and inevitable part of our lives, but as the years pass, we may begin to notice a range of changes occurring in our bodies. From wrinkles to slowed reflexes, aging can bring about a list.
    4. Besides making interaction terms out of polynomials to cover the complex connections.
    5. To have the model learn domain-specific feature transformations based on medical knowledge (e.g., creating a depiction of the severity of an illness by adding individual symptoms).

18.7 Model Training and Validation

Establishing model training and validation is a very important part of the entire process of the development of an ML-based COVID-19 screening. In these stages doing so, the forecasts remain true, reliable, and powerful enough to make clear preferences. Here is a breakdown of sub-points within this stage.

18.7.1 Methodologies for Training

Learning methodologies include supervised, unsupervised, and semi-supervised, each for a different job. The predictive activity of COVID-19 occurs with the help of supervised learning that involves learning from labeled information that contains data points such as cases, demography, symptoms, and relevant factors. The model learns to optimize its parameters when training to minimize the prediction errors, thus maximizing the accuracy.

18.7.2 Cross-Validation Techniques

This is a fundamental task for bias analysis of ML algorithm models. Techniques of k-fold cross-validation or leave-one-out cross-validation are popular tools in the champion's toolkit. In k-fold cross-validation, the dataset is broken down into k subsets. The model is trained on each subset, k–1 times, with each time using k–1 subsets for training and the remaining subset for validation. Through this process, we can estimate the model's generalization capability and detect any problem, for instance, the model's ability to learn and forget.

18.7.3 Strategies for Mitigating Overfitting and Underfitting

The overfitting case takes place when the model remembers the training data pieces rather than learning general rules from them and yields a poor performance based on unseen data. While overfitting occurs when the model is too detailed to capture how data is structured, underfitting seems to be the opposite scenario in which the model is too simple. One of the strategies undertaken to address these weaknesses is carrying out procedures such as regularization, dropout, and ensembles. Adding an extra penalty to the loss function creates a regularization, which prevents the model from becoming too complex, while dropout randomly shuts off some neurons that prevent the model from being over-reliant on certain features. Such methods include ensembles of multiple models designed to produce a more accurate and robust result than each model can do on its own.

18.8 Deployment and Integration

Integrating ML-powered COVID-19 prediction kits coming in use with the real-world healthcare provision system involves facing many complex considerations to make them effective and acceptable to users. Here is a breakdown of the sub-points.

18.8.1 Technical Challenges

The prediction model implementation into the existing healthcare infrastructure must bear the technical considerations such as addressing data formats' compatibility, electronic health records in-build capabilities, and collaborating with other healthcare systems with the scope of interoperability in mind. If the solution is to make interoperability possible, then perhaps standardized APIs and data exchange protocols should be developed.

18.8.2 Real-Time Prediction

To realize timely actions and decision-making, kits need to provide real-time outputs or outputs provided in a near real-time mode. Therefore, we will construct the data pipelines that are optimized, the inference models and we might have to consider the distributed computing architecture if the volume of data is too large.

18.8.3 User Interface Design

An easy-to-use and user-friendly interface of a prediction kit, available to health workers, is a must. Factors of interest, such as providing meaningful predictions, logical steps leading to a specific action, and visual screens to simplify comprehension and expand decision-making procedures are part of the design. User feedback and an iterative design process are the ultimate basis for the optimization of the interface's usability.

18.8.4 Regulatory Compliance

Respect for regulatory needs and data privacy should be ensured at any cost to rule out the ethical and legal assimilation of prediction kits. Meeting those requirement regulations including the Health Insurance Portability and Accountability Act in the US and the General Data Protection Regulation in Europe necessitates having a data security system in place and ensuring that your data policies are clear to customers.

18.8.5 Scalability and Resource Optimization

Scalability becomes a requirement and an issue for such kits as they address different numbers of elements and loads of data. Methods such as model compression, distributed computing, and cloud-based solutions can help to distribute computation power rather than managing all resources on a single conventional hardware that ensures scaling and cost-saving simultaneously.

18.9 Ethical and Societal Implications

While implementing the ML-based COVID-19 predicting kits, their further use should include ways and means of tackling ethical and societal implications that accompany them. Navigating and treading the safety lines of ethics through the enforcement of predictive tools on the other side sparks the issue of privacy and data security. Existing patient health data utilized in this fashion must be protected against a breach in secure storage systems or misuse by promoting confidentiality. In addition, the possibility of perpetuating the biases in already present data cannot be ruled out, so disparities in healthcare outcomes among traditionally disadvantaged communities cannot be ruled out. How one can Fairness and equality in the referral are one of the things we should do to avoid increasing the existing healthcare inequalities. From a social perspective, a significant number of prediction kits used by the population would be noticed and highly likely to influence public behavior and policy measures. To serve this purpose, wisdom would be required to make people fully aware of what such a forecast entails, and the limitations of the truth behind it, to prevent situations like that of needless panic and carelessness. Furthermore, providing equal access to these resources is essential to avoid that nobody is left behind or vulnerable populations. On the one hand, precise prediction accuracy should be achieved, and, on the other hand, privacy, fairness, and transparency are vital aspects that should be considered mindfully along with the governance principle. The collective aspiration of healthcare professionals along with bioethicists, policymakers, and technologists is the necessity via proper navigation and remain on the right track in terms of the above-mentioned complicated ethical and societal problems. In retrospect, we can respond to this issue in a well-thought-out manner and at the same time pursue ethics in ML prediction for COVID-19 while protecting societal welfare.

18.10 Real-Time Case Studies

Due to the recently surged COVID-19 outbreak worldwide, a research team started the implementation of a forecasting kit with ML algorithms during the last announcement. In concert with healthcare establishments and data experts, a team began to consolidate precise datasets which consisted of inpatient and outpatient records, demographic information as well as diagnostic test results of confirmed cases. Utilizing this wealth of data, they trained ML models that picked up endpoints representing patterns and correlations linked to the disease. Utilizing a thoughtful and clever feature selection approach and the careful refinement of algorithms, they were able to make their models more predictable. This permits them to forecast an individual's risk of contracting the virus based on many factors, including age, coexisting conditions, and geographical area. To prove the validity of the prediction kit, the team carried out stringent testing on both retrospective and prospective datasets and managed to acquire considerable results with commendable sensitivity and specificity. The kit was validated on-site and afterward was deployed in different health settings where it provided clinicians with the capability for early detection of those with a high risk of COVID-19 and allowed for better allocation of resources. This kit also provided us with hyper-targeted intervention and prevention measures. As a result, the virus is continuously controlled within communities. Beyond its direct importance, the emergence of such a kit has been a monumental move at the intersection of healthcare and ML, which represents how data-driven techniques can defeat infectious diseases. Adjustment and improvement of the prediction kit will also undergo as the pandemic constantly changes. This is required to keep on par with the development of the prediction kit to ensure the continued safety of the general public.

18.11 Future Directions and Challenges

Surpassing the current situation, COVID-19 prediction with the help of the ML process, the researchers are met with future problems and challenges, which signify the pathway for further study and methodology. The key issue is to model the robustness and generalizability of predictions properly. In the future, research could be oriented toward developing models that can easily change the missing data quality, demographic nature, and viral strain variations. Furthermore, investigating ensemble learning as a way to integrate many models into one model will be of great value, especially in the improvement of the models' accuracy and reliability. Besides that, new data sources added to the structure of the model will contribute to the enhancement of the forecasting competencies. Integrating data from wearable gadgets, social networks, and environmental circumstances remains critical in understanding the mechanism of infection and transmission patterns of diseases. Nevertheless, data interoperation becomes the mainstay with great challenges of data privacy, data interoperability, and data heterogeneity that must be tackled. The next aspect to be addressed is the creation of interpretable and defensive ML algorithms. As the prediction kits are introduced to the clinical setting, it becomes vital to investigate the decision-making condition in this process. Further work should be oriented toward making the AI models more intelligible for offline validation so that healthcare providers can have accurate and reliable predictions. Additionally, fixing the problem of data scarcity and receiving flawless information is a major difficulty. The issue of arriving at adequate data representative enough for training these models lies in the health lacunas and the inadequately populated regions. Collaborated and interdisciplinary teams that include scientists from several disciplines as well as governmental organizations and partnerships both beyond and within the national borders are imperative to address the inequalities in terms of predictive tools availability and fair access.

18.12 Conclusion

Finally, the construction of ML-based COVID-19 prediction kits marks a pronounced achievement in our fight against the pandemic by triumphing over it. From the beginning of this article, we have highlighted core principles, approaches, and challenges impacting the ML algorithm practice of COVID-19 prognosis. Such prediction kits seem to be a perfect additional tool for public health such as to anticipate the dynamics of epidemic shed light on vulnerable groups and prioritize resource allocations. By using the capabilities of large-scale data processing, and advanced algorithms, to transform these raw numbers into meaningful information, that can be used to make informed decisions by policymakers, healthcare professionals, and public health authorities, is made possible. In addition, the utilization of a prediction kit calls for wide public participation and highlights the interdisciplinary nature of mitigation of public health disasters. Teamwork and cooperation among data scientists, epidemiologists, healthcare providers, and policymakers are necessary to make the right choice and to make these tools relevant, accurate, and ethically acceptable. Moreover, such progress should be looked at through the perspective of constant research and development which is crucial for the development of high-performance and reliable models that can address changes in viral variants and the progression of epidemiological dynamics. Nevertheless, along with the wide use of MLP-devised prediction kits, clear ethical and social challenges arise. Trust building and the fair dispersion of these technologies call for concrete measures like privacy safety, transparency, and bias reduction. Also considering that the mission will be to close the gap between the digital divide of the guarded communities and give access to the merits of predictive analytics.

Acknowledgments We want to acknowledge VosViewer for providing us with the visualization of the literature survey and bibliometric study.

References

1. Wang L, Lin ZQ, Wong A. COVID-Net: a tailored deep convolutional neural network design for detection of COVID-19 cases from chest X-Ray images. Sci Rep. 2020;10(1):1–12.
2. Liang W, Liang H, Ou L, et al. Development and validation of a clinical risk score to predict the occurrence of critical illness in hospitalized patients with COVID-19. JAMA Intern Med. 2020;180(8):1081–9.
3. Lippi G, Henry BM, Bovo C, et al. Predictive values of established risk factors for COVID-19 severity and outcomes: a systematic review and meta-analysis. Int J Med Sci. 2020;17(15):2265–73.
4. Yan L, Zhang HT, Goncalves J, et al. A machine learning-based model for survival prediction in patients with severe COVID-19 infection. medRxiv. 2020.
5. Jain A, Tiwari S, Choudhury T, Dewangan BK. Gradient and statistical features-based prediction system for COVID-19 using chest X-ray images. Int J Comput Appl Technol. 2021;66(3/4):362. https://doi.org/10.1504/ijcat.2021.120464.
6. Chen S, et al. Reinforcement learning based diagnosis and prediction for COVID-19 by optimizing a mixed cost function from CT images. IEEE J Biomed Health Inform. 2022;26(11):5344–54. https://doi.org/10.1109/JBHI.2022.3197666.
7. AlNeyadi A, AlMarri A, AlKuwaiti S, Ahmad A. A Study of the effect of vaccinated populations on the prediction accuracy of the regression models for daily Covid-19 cases prediction. In: 2023 15th international conference on innovations in information technology (IIT), Al Ain, United Arab Emirates, 2023, pp. 257–61. https://doi.org/10.1109/IIT59782.2023.10366482.

8. Nurhakim Nor Iskandar MF, Usman S, Ahmad N, Amran ME. Covid-19 risk factors prediction using machine learning: south sudan and eastern democratic republic of the congo dataset. In: 2023 IEEE 2nd National Biomedical Engineering Conference (NBEC), Melaka, Malaysia, 2023, pp. 116–20, https://doi.org/10.1109/NBEC58134.2023.10352636.
9. Mathur S, Datta K. Prediction of Covid-19 cases in India through machine learning using Python. In: 2021 9th International Conference on Reliability, Infocom Technologies and Optimization (Trends and Future Directions) (ICRITO), Noida, India, 2021, pp. 1–6. https://doi.org/10.1109/ICRITO51393.2021.9596116.
10. Jaleel RA, Burhan IM, Jalookh AM. A proposed model for prediction of COVID-19 depend on K-Nearest neighbors classifier:Iraq case study. In: 2021 International conference on electrical, communication, and computer engineering (ICECCE), Kuala Lumpur, Malaysia, 2021, pp. 1–6. https://doi.org/10.1109/ICECCE52056.2021.9514171.
11. Deva R, Chauhan SPS. Machine learning-based prediction & analysis of COVID-19 using clinical data. In: 2022 2nd International conference on technological advancements in computational sciences (ICTACS), Tashkent, Uzbekistan, 2022, pp. 457–61, https://doi.org/10.1109/ICTACS56270.2022.9988594.
12. Tahir H, Iftikhar A, Mumraiz M. Forecasting COVID-19 via registration slips of patients using ResNet-101 and performance analysis and comparison of prediction for COVID-19 using Faster R-CNN, Mask R-CNN, and ResNet-50. In: 2021 International conference on advances in electrical, computing, communication and sustainable technologies (ICAECT), Bhilai, India, 2021, pp. 1–6. https://doi.org/10.1109/ICAECT49130.2021.9392487.
13. Nastiti FE, Musa S, Yafi E, Chauhan R. Systematic literature review: machine learning prediction model for Covid-19 spreading. In: 2022 4th International conference on cybernetics and intelligent system (ICORIS), Prapat, Indonesia, 2022, pp. 1–5. https://doi.org/10.1109/ICORIS56080.2022.10031414.
14. Gode A, Hancer E, Cesmeli MS, Covid-19 risk prediction with ensemble boosting algorithms. In: 2023 Innovations in intelligent systems and applications conference (ASYU), Sivas, Turkiye, 2023, pp. 1–4. https://doi.org/10.1109/ASYU58738.2023.10296689.
15. Haritha D, Swaroop N, Mounika M. Prediction of COVID-19 cases using CNN with X-rays. In: 2020 5th International conference on computing, communication and security (ICCCS), Patna, India, 2020, pp. 1–6. https://doi.org/10.1109/ICCCS49678.2020.9276753.
16. Bi Y, Yin S, Wang P, Hong Y, Yang Z, Liu Y. Improving COVID-19 pulmonary damage volume prediction with uncertainty estimation and prediction refinement embedded in Trans-UNet. In: 2023 International conference on applied physics and computing (ICAPC), Ottawa, ON, Canada, 2023, pp. 103–6. https://doi.org/10.1109/ICAPC61546.2023.00026.
17. Bai Y. Epidemic case prediction of COVID-19: using regression and deep based models. In: 2020 2nd International conference on machine learning, big data and business intelligence (MLBDBI), Taiyuan, China, 2020, pp. 40–5. https://doi.org/10.1109/MLBDBI51377.2020.00015.
18. Chen D, Zhou L. A rolling-based multipeak learning model for COVID-19 pandemic predicting. In: 2022 IEEE International conference on industrial engineering and engineering management (IEEM), Kuala Lumpur, Malaysia, 2022, pp. 732–6. https://doi.org/10.1109/IEEM55944.2022.9989642.
19. Bhardwaj S, Bhardwaj H, Bhardwaj J, Gupta P. Global prediction of COVID-19 cases and deaths using machine learning. In: 2021 Sixth international conference on image information processing (ICIIP), Shimla, India, 2021, pp. 422–6. https://doi.org/10.1109/ICIIP53038.2021.9702560.
20. Rana MM, Abdelhadi A, Ahmed MRU, Ali A. Secure IoT communication systems for prediction of COVID-19 outbreak: an optimal signal processing algorithm. In: 2020 Third international conference on smart systems and inventive technology (ICSSIT), Tirunelveli, India, 2020, pp. 135–9. https://doi.org/10.1109/ICSSIT48917.2020.9214124.
21. Yaqin A, Rahardi M, Abdulloh FF, Kusnawi, Budiprayitno S, Fatonah S. The prediction of COVID-19 pandemic situation in Indonesia using SVR and SIR algorithm. In: 2022 6th

International conference on information technology, information systems and electrical engineering (ICITISEE), Yogyakarta, Indonesia, 2022, pp. 570–3. https://doi.org/10.1109/ICITISEE57756.2022.10057813.
22. Chen Z, Zhang C, Chun N. Visualization analysis and XGBoost-based prediction of Covid-19 mortality. In: 2023 IEEE International conference on image processing and computer applications (ICIPCA), Changchun, China, 2023, pp. 70–4. https://doi.org/10.1109/ICIPCA59209.2023.10257874.
23. Kumar A, Kaur K. A hybrid SOM-fuzzy time series (SOMFTS) technique for future forecasting of COVID-19 cases and MCDM based evaluation of COVID-19 forecasting models. 2021 International conference on computing, communication, and intelligent systems (ICCCIS), Greater Noida, India, 2021, pp. 612–7. https://doi.org/10.1109/ICCCIS51004.2021.9397216.
24. Shi Y, Wu K, Zhang M. COVID-19 pandemic trend prediction in America using ARIMA model. In: 2022 International conference on big data, information and computer network (BDICN), Sanya, China, 2022, pp. 72–9. https://doi.org/10.1109/BDICN55575.2022.00022.
25. Kasturi K, Putera MIA, Natasia SR. Prediction of Bontang City COVID-19 data time series using the facebook prophet method. In: 2021 4th International seminar on research of information technology and intelligent systems (ISRITI), Yogyakarta, Indonesia, 2021, pp. 247–251. https://doi.org/10.1109/ISRITI54043.2021.9702874.
26. Pramanik A, Sinha A, Aluvala S. Prediction of COVID-19 from lung scans using deep learning. In: 2023 6th International conference on contemporary computing and informatics (IC3I), Gautam Buddha Nagar, India, 2023, pp. 833–7. https://doi.org/10.1109/IC3I59117.2023.10398007.
27. Dai Q, Zhu X. Age prediction of death in India' Covid-19 pandemic using machine learning methods. In: 2021 IEEE 3rd International conference on frontiers technology of information and computer (ICFTIC), Greenville, SC, USA, 2021, pp. 242–5. https://doi.org/10.1109/ICFTIC54370.2021.9647228.
28. Devi EA, et al., A diagnostic study on prediction of Covid-19 by symptoms using machine learning. In: 2022 International conference on electronics and renewable systems (ICEARS), Tuticorin, India, 2022, pp. 1416–21. https://doi.org/10.1109/ICEARS53579.2022.9752301.

Chapter 19
Plant Disease Detection: Comprehensive Review of Methods and Techniques

Madhavi, Thipendra P. Singh, and Shailendra Pratap Singh

19.1 Introduction

In agricultural development, crop health security and food safety are highly dependent on the identification and classification of plant and leaf diseases. As agriculture is essential for economic stability, the early and precise identification of plant diseases can greatly lower crop yields while also increasing productivity and profitability. Modern advances in machine learning (ML) and deep learning (DL) have created new avenues for the automatic identification and categorization of plant diseases, providing encouraging answers to this long-standing problem.

The use of ML and DL models in agriculture goes beyond conventional techniques by offering quick, precise, and affordable ways to identify problems with plant health. With the help of this revolutionary method, which processes and analyzes enormous collections of plant photos, disease patterns and symptoms can be identified very accurately. Among them, the application of convolutional neural networks (CNNs), recurrent neural networks (RNNs), and other specialized algorithms has shown good efficacy in differentiating between damaged and healthy plant tissues in a variety of crop kinds, such as melons, olives, and more.

For example, by analyzing image data of olive [17] leaves, the novel use of decision tree and Naive Bayes models has been employed to predict olive diseases, attaining remarkable accuracy. The stacked RNN [14] model is a DL model that is used to categorize melons according to disease and distinguishes the disease as a

Madhavi (✉) · T. P. Singh
School of Computer Science Engineering & Technology, Bennett University, Greater Noida, India
e-mail: e22soep0077@bennett.edu.in; shailendra.singh@bennett.edu.in

S. P. Singh
Madan Mohan Malaviya University of Technology, Gorakhpur, India

T. Choudhury, A. Katal (eds.), *Machine Learning for Disease Detection, Prediction, and Diagnosis*, https://doi.org/10.1007/978-981-96-4241-0_19

healthy or unhealthy leaf. The developments in ML and DL models [31] transform the agricultural sector completely. The models automate the process of detecting diseases, and farmers can monitor crop health more effectively and control possible attacks by having access to important resources. This reduces the chances of huge crop failures and, by suggesting the use of fewer chemical treatments, supports sustainable farming methods. Plant disease identification and classification can be done with earlier unheard-of levels of accuracy and efficiency but due to the ongoing evolution of ML and DL models. These models can strengthen our agricultural systems against the continuously changing challenges posed by plant diseases and open the chances to a future in which nature and technology work together to promote strength and growth.

With the merging of ML and DL, plant disease detection is advancing toward automated, accurate, and user-friendly diagnostic techniques. The author [21] uses a federated learning CNN model for the privacy-sensitive and decentralized identification of maize diseases with an accuracy of 89.4%. The author [29] identifies grape leaf diseases by processing images with complex backgrounds through thresholding, noise removal, and K-means clustering, with classification performed by a trained feed-forward backpropagation neural network with an accuracy of 100%. The author [13] identifies disease using grab-cut segmentation and various classification techniques, achieving a high accuracy of 93%. The author [19] uses InceptionV3, transfer learning, and traditional approaches like support vector machine (SVM) to classify betel leaf diseases with an accuracy of 94.83%. The author [6] identifies banana plant diseases using image processing and SVM classification, extracting color, shape, and texture features and achieving an average accuracy of 85%. A table summary is provided in Table 19.1.

Table 19.1 Summary of ML & DL models in agriculture

References	Model	Value of metric	Contribution
[15] 2023	Assessment-based CNNs (A-CNNs)	Accuracy 92%	A-CNN, which significantly improves the detection of plants
[8] 2020	KNN	Accuracy 89.9%	Feature extraction and classification using the KNN algorithm for the efficient detection and treatment recommendation of tomato plant diseases
[28] 2021	SVM	–	Novel machine-learning approach for accurately predicting and classifying plant diseases.
[4] 2022	SVM and KNN	Accuracy 96.5%	SVM for the effective identification and precise localization of leaf diseases, leveraging K-means clustering for segmentation and SVM for classification, proving highly successful in detecting leaf illnesses.
[34] 2023	VGGNet19 and CNN	Accuracy 85.4% and 83%	VGG net 19 and CNN, achieving accuracies of 85.4% and 83%, respectively, thereby facilitating early disease identification and suggesting remedies to mitigate crop damage

(continued)

Table 19.1 (continued)

References	Model	Value of metric	Contribution
[22] 2019	DL models	Accuracy 96.5%	This study leverages DL and computer vision to create a system capable of detecting multiple diseases across various plant types
[3] 2013	ResNet-9	Accuracy 99.25%	ResNet-9 model that accurately detects early and late blight diseases in potato and tomato crops with high accuracy
[26] 2021	Fast R-CNN and YOLOv3	Accuracy 96.7%	Introduces an automated system using a portable video camera and DL models to detect abnormal potato plants
[25] 2021	VGG16	Accuracy 99.98%	ICRMBO algorithm, to automatically classify four types of tomato leaf diseases with high accuracy
[23] 2015	SVM	Accuracy 99.83%	Utilizing GLCM for feature extraction and SVM with linear kernel for classification, achieving a high accuracy of 99.83%
[32] 2023	InceptionResNetV2	Accuracy 87%	CNN-based InceptionResNet V2 model for detecting cassava leaf diseases
[33] 2019	Single shot multibox	Accuracy 85.3%	(CNN) with the TensorFlow library to classify healthy and diseased cassava leaves in Indonesia.
[12] 2021	Improved CNN	Accuracy 99.70%	CNN with a gated pooling function andbatch normalization to classify ginger powder images, detecting adulteration with up to 99.70% accuracy
[2] 2022	AlexNet, Resnet, and DensNet	Accuracy 90.12%	Olive leaf diseases, enhanced by an adaptive genetic algorithm for parameter selection, achieving superior accuracy
[20] 2021	ANN	Accuracy 95.9%	DOST for image processing, achieving a high classification accuracy of 95.9% with an ANN classifier when using ELBP features

19.2 Datasets

A few well-known datasets for the identification of plant diseases in this field are presented in Tables 19.2, 19.3, and 19.4. There are several benefits to using computational models to manage large datasets for disease identification:

- Their ability to process enormous amounts of data quickly is especially handy in emergencies.
- Because there is less space for human mistakes, these models are more accurate at identifying patterns and anomalies.
- They can predict the spread of diseases before they get out of control.
- They contribute to maximizing the use of resources for disease control and treatment by correctly diagnosing conditions.

Table 19.2 PlantDoc dataset [30]

Crop name	Total images
Apple	273
Bell pepper	132
Blueberry	174
Cherry	207
Corn	134
Grape	133
Peach	229
Potato	170
Raspberry	130
Soybean	65
Squash	93
Strawberry	63
Tomato	613

Table 19.3 PlantVillage dataset [10]

Crop name	Total images
Apple	1741
Blueberry	225
Cherry	1522
Corn	1162
Grape	423
Orange	5507
Peach	3607
Bell pepper	2478
Potato	2150
Raspberry	1000
Soybean	5090
Squash	1835
Strawberry	456
Tomato	14,508

- Computational models make it simple to handle both local and global data.
- As ML models are exposed to more data, they can get better over time, which could lead to better results for disease prediction and diagnosis.

Table 19.4 Rice disease dataset [27]

Class	Images
Healthy	791
Brown spot	167
Hispa	344
Leaf blast	166

19.3 Machine Learning Algorithm

Plant disease classification and prediction depend on ML algorithm methods such as SVMs and random forests. These algorithms efficiently notice agricultural diseases through the analysis of visual data, thereby enabling prompt intervention and disease control [11]. By using data analysis techniques, they improve results and make plant disease identification more straightforward [11]. Furthermore, ML models might be able to fast identify diseases even from smartphone photos by utilizing techniques like CNN, which have been trained on massive datasets of photos of plant ailments [24].

This section will manage the widely used ML models and algorithms that are appropriate for the classification, identification, and detection of plant and leaf disease categories. The most often used algorithms across different categories are as follows.

19.3.1 Linear Regression

In supervised learning, linear regression is a technique that is used to indicate a continuous dependent variable from one or more independent variables. When it comes to the identification and detection of plant and leaf disease types, the linear regression predicts continuous outcomes such as the level of a disease under specific circumstances and classifies input data into discrete labels. Plant diseases may not be directly classified by linear regression, but it can be useful in related tasks that can assist in predicting the disease under different environmental conditions or evaluating the impact of a disease.

Limitations

- Classification versus Regression: Classification methods such as logistic regression, decision trees, and CNNs are more suitable for the identification and detection of plant and leaf disease categories.
- As linear regression works with continuous data, it is less suitable for classifying diseases but may be useful for associated continuous outcomes.
- Homoscedasticity is necessary for the success of linear regression but may not apply to other sides of disease impact analysis.

19.3.2 Logistic Regression

Logistic regression is more directly suited to the identification and detection of plant and leaf disease groups in the agriculture area since logistic regression, in contrast to linear regression, is a statistical method used for binary classification tasks. It is capable of classifying several kinds of plant diseases by expanding to multiclass classification using methods like multinomial logistic regression. The challenge could be phrased as either distinguishing the specific type of disease affecting a plant or leaf means multiclass classification or determining whether the plant or leaf is healthy or diseased known as binary classification.

A variety of data sources, such as visual traits from leaf photos, and environmental factors, could be used to generate each input feature for the model. A logistic regression model calculates the likelihood that an input falls into a specific category. It outputs a value between 0 and 1, which indicates the likelihood of the positive class (e.g., the presence of a certain disease), using a logistic function.

Advantages

- The significance of each characteristic for disease prediction is indicated by the weights given to it.
- Logistic regression can be applied to large-scale or real-time agricultural applications due to its high scalability and efficiency when working with huge datasets.

Limitations

- Logistic regression may have trouble with complicated or nonlinear correlations between features and classes, despite its strength for linearly separable classes. The models, such as decision trees or neural networks, may function better in some situations.

19.3.3 Decision Trees

Decision trees are supervised learning ML algorithms used for classification and regression tasks. In the field of agriculture, decision trees provide a simple method of detecting disease by using a sequence of binary or categorical choices for the identification and detection of plant and leaf disease categories. A decision tree models decisions and their potential outcomes using a structure resembling a tree. Each leaf node symbolizes a disease category, each branch indicates the test's outcome, and each internal node represents a "test" on a feature (such as a leaf's color, texture, or size). Plant-specific characteristics such as the plant's age, growth stage, temperature, humidity, spots, discoloration, and leaf shape deformation could all be utilized to identify plant illnesses.

Advantages

- Decision trees are particularly useful in agricultural applications where farmers need to be able to interpret and understand predictions like the type of disease detected to make decisions. This is because decision trees are simple to understand and analyze.
- They can capture non-linear patterns without requiring data alteration, and decision trees are ideal for disease patterns.
- Decision trees can manage both numerical and categorical data, and they are flexible enough to accommodate a wide variety of data types collected in agriculture.

Limitations

- With decision trees, overfitting to the training set is a prevalent issue. By employing techniques such as pruning and establishing a maximum depth for the tree, this can be reduced.
- This problem can be resolved by using ensemble techniques like random forests, which construct several trees and integrate their predictions.

19.3.4 Random Forest

Multiple decision trees' outputs are combined into a single output using the common ML technique known as random forest. The random forest is used in regression and classification algorithms. A random forest blends multiple trees to enhance forecasts. Large datasets can be handled by it, which helps reduce overfitting. Random forests can be used for both category and constant variables. It selects a small number of features and data points to create a decision tree. A small number of samples are used to form the decision tree, and the majority vote of the decision tree is used to determine whether to classify or regression the output.

There are some important hyperparameters in random forest:

- Number of trees.
- Maximum number of features.
- Minimum number of sample leaves.
- Maximum leaf nodes.

Advantages

- Nodes in random forests rectify each other's errors.
- Random forests can handle noisy data.
- Random forests are less inclined to overfit.
- It can determine which features are most important for disease detection.
- Random forests are capable of handling classification and regression tasks.

Challenges

- Random forest ensemble approach makes it more difficult.
- Training multiple trees on extensive datasets is computationally tough.

19.3.5 Support Vector Machines

SVM [9] is a supervised ML algorithm that works on complex and smaller datasets. It can be used for both classification and regression, but it works efficiently on classification tasks. In SVM, we try to find the hyperplane which can separate two classes. SVMs can have multiple hyperplanes, but it is mandatory to select the best hyperplane that can have the maximum margin between the two classes. It is basically of two types: linear SVM and non-linear SVM. Since the margin is the distance between the hyperplane and the support vector is the points that are closest to the hyperplane, the support vector is crucial in SVM [5].

SVMs can have several components:

- A kernel function.
- Decision boundary.
- Slack variable.
- Input features.
- Support vectors.

Advantages

- SVMs have high accuracy in classification tasks.
- SVMs are flexible to a broad range of data sources and classification problems.
- SVMs can be used for in-depth examination of plant and leaf images to detect diseases.

Limitations

- The choice of kernel and regularization parameters can significantly affect the model's performance; therefore, careful adjustment is required.
- As dataset sizes grow, SVMs might become more computationally demanding.

19.3.6 Naive Bayes

The Naive Bayes ML model is a straightforward and efficient probabilistic AI algorithm. The name "naive" comes from the strong assumption that is rarely true in practice that the features used to forecast the output class are independent of one another. This assumption makes the computations easier. When it comes to identifying and detecting plant and leaf disease categories in the agricultural field, Naive Bayes can perform well. The Bayes theorem determines the likelihood of a hypothesis. It combines past understanding of circumstances with the probability of seeing the evidence if the hypothesis is correct.

Advantages

- Naive Bayes is a good choice for real-time or on-device illness detection applications since it is easy to build and uses less processing power than more sophisticated models.

- It can get good results even with a small quantity of training data, which is useful in situations where obtaining large-scale labeled datasets is difficult.
- By ignoring the missing values during probability calculations, Naive Bayes is still able to produce predictions in situations when some feature data is absent.

Limitations
- The assumption of feature independence is rarely satisfied in real-world situations, particularly when dealing with image data where pixel values are frequently associated. Naive Bayes can still function effectively generally; however, in particular situations, this can result in inaccurate results.
- Naive Bayes' accuracy for some illness detection types may be limited since it may not be as good at capturing complicated relationships between variables as some other models, including neural networks.

19.3.7 K-Nearest Neighbors

K-nearest neighbor (KNN) is a non-parametric ML technique that can be applied to regression and classification tasks. To classify and recognize plant and leaf disease categories in the agricultural domain, KNN determines which of the "k" training samples is nearest to a new sample and predicts its class based on which of these neighbors has the majority class. KNN identifies new, unlabeled closest neighbors by calculating the distance between points in the feature space. Manhattan, Hamming, and Euclidean distance are examples of common distance measures. The algorithm's performance can be greatly affected by the distance metric selected, so it is important to match it with the features being employed.

The "k" in KNN refers to the number of closest neighbors that need to be considered. While a large value of "k" boosts the model's resistance to outliers but may also lower its accuracy, a low amount of "k" shows that the model is sensitive to noise in the data. Typically, cross-validation is employed to determine the optimal value of "k."

Advantages
- It is simple to comprehend and apply.
- It is appropriate for a variety of disease detection scenarios because of its ability to handle binary, multiclass, and even regression jobs.
- KNN does not assume anything about the underlying data distribution, and it can be extremely effective in situations when the decision border is irregular.

Limitations
- KNN calculates the distance to each point in the dataset for each prediction, and it can become computationally expensive as the dataset expands.
- The curse of dimensionality can cause performance to decline in extremely high-dimensional settings.
- The KNN algorithm is directly affected by the size of the features since distance computations determine its scale.

19.3.8 *Gradient Boosting Machine*

Gradient boosting machine (GBM) is an effective ensemble method applied to regression and classification problems. Similar to other boosting techniques, they construct their models in phases and then allow for the optimization of any differentiable loss function. Because GBMs are trained to handle a variety of data formats and complex, non-linear feature correlations, they can be extremely useful in identifying and detecting plant and leaf disease categories in agriculture. GBMs generate a single tree at a time, in contrast to bagging techniques (like random forests) that build models in parallel. Every new tree fixes errors achieved by trees that were built earlier. By adding weak learners one after the other, the ensemble model's ability to predict the target variable steadily improves. GBMs use gradient descent to minimize the loss when adding new models. Since the loss function in GBMs may be changed, the model can be customized to fulfill the unique requirements of plant disease detection. A multiclass logarithmic loss function can be used for multi-class classification (identifying certain diseases), while a binary cross-entropy loss function can be used.

for binary classification (healthy versus diseased).

Advantages

- GBMs frequently have a high predictive accuracy, which allows them for the difficult task of identifying plant diseases.
- They can deal with a variety of data formats and can adjust to varied regression and classification issues.

Disadvantages

- GBM model training can be costly and time-consuming.
- GBMs overfit when the feature space data is noisy.

19.4 Deep Learning Algorithm

DL significantly improves the efficiency and accuracy of plant disease detection, particularly when applied to CNN architectures. These models perform exceptionally well when processing and evaluating visual data. They can identify disease signs with remarkable accuracy rates, sometimes reaching as high as 99.35% on test sets, by analyzing leaf attributes including color and damage [35]. Research has demonstrated that DL techniques work better than typical ML models, providing enhanced diagnostic precision through superior data processing capabilities [18]. Moreover, several DL models, like ResNet, have demonstrated remarkable precision in classifying plant diseases, indicating the potential of these technologies to improve farming methods [7].

DL's architecture draws inspiration from the neural networks that make up the brain, in terms of both structure and functionality. The well-liked DL technique and models—which are suitable for the identification, detection, and categorization of plant and leaf disease categories—will be covered in this section. Some of the well-known and frequently applied DL algorithms are provided below.

19.4.1 Convolutional Neural Networks

CNNs are a subclass of deep neural networks that excel in processing data with a topology resembling a grid, like pictures. Convolutional layers are the fundamental elements of a CNN [16]. They process the images through several learnable filters. These filters are made to identify particular characteristics, such as edges, colors, or textures, that point to the health or disease of a plant. The filters can identify complex patterns as the network gets deeper.

After convolutional layers, pooling layers are utilized to reduce the dimensionality of the feature maps. Overfitting is addressed and the processing load is decreased by encoding the features present in the convolutional layers. One or more fully linked layers, where each neuron is coupled to every other neuron from the layer before, are used at the end of the network. These layers do the classification based on the high-level characteristics retrieved by the convolutional and pooling layers.

Advantages

- CNNs frequently achieve high accuracy in image classification tasks, such as the identification and detection of plant and leaf diseases, because of their capacity to learn complex patterns and features from images.
- CNNs extract the most relevant features for disease classification straight from the images, saving time and effort by doing away with the requirement for manual feature extraction.
- CNNs are appropriate for large-scale agricultural monitoring systems because they can be trained on massive datasets.

Disadvantages

- CNN training can be computationally demanding, using a huge amount of memory and computing power, particularly for big networks and datasets.
- Risk of overfitting on a small dataset.

19.4.2 Recurrent Neural Networks

RNNs are a subclass of neural networks that are specifically designed to handle sequences. These can be text sequences, time series data sequences, or any other type of data where context and order are important. RNNs are superior to CNNs at capturing temporal or sequential dependencies, which makes them less direct when applied to the identification and detection of plant and leaf disease categories based on static images. RNNs process input sequences by iterating over each sequence element one at a time, unlike typical neural networks that process inputs in isolation.

This enables each output to have a "memory" of the inputs. The hidden state functions as a type of memory in RNN. Information regarding the processing completed up to this point is captured. At every stage of the process, the hidden state is updated using the most recent hidden state as well as any new input. RNNs memorize weight matrices, during training. Generally, there are three primary sets of weights: one for the hidden state's input, one for its output, and one for the hidden state's input from the previous time step to the present time step.

Limitations
- Difficulties such as vanishing and exploding gradients make training RNNs difficult. To address these issues, variations such as gated recurrent units and long short-term memory (LSTM) networks have been developed.
- Adequate sequential data is necessary for the efficient application of RNNs; however, not all agricultural scenarios may allow for the collection of this data.

19.4.3 Long Short-Term Memory Networks

RNNs with the ability to learn long-term dependencies are known as LSTM networks. LSTMs were introduced to address vanishing and exploding gradients that restrict RNN's ability to process long sequence data efficiently.

A complex computational unit called a gate that controls the flow of information is used in the design of LSTM devices. The LSTM network can handle a longer context and retain information for a longer time. It determines what data should be retained and deleted at each stage of the process. In agricultural areas, LSTMs are helpful in situations where the environment affects plant health while diagnosing and detecting different types of plant and leaf diseases.

Limitations
- A substantial amount of sequential data is necessary for the effective application of LSTMs for disease diagnosis and detection, and this data may be difficult to get in some agricultural situations.
- LSTM training can be computationally demanding and needs a significant amount of hardware, particularly for models that hybrid LSTMs with CNNs for sequence and picture processing.
- Compared to simpler models, the complexity of LSTM structures can make tuning and optimization more difficult.

19.4.4 Autoencoders

Neural networks called autoencoders are used to learn effective codings of input data unsupervised, usually for feature learning or dimensionality reduction. An autoencoder compresses the input into a lower-dimensional code, from which it then learns to reconstruct the input as closely as feasible. When an autoencoder is trained to minimize reconstruction error, it can pick up important properties from the data. Autoencoders are very helpful in the agriculture domain for several activities related to classifying and diagnosing plant and leaf disease categories. These tasks include data denoising, anomaly detection, and feature extraction. Autoencoders can learn to represent plant and leaf images or other relevant agricultural data in a more compact form while preserving the key information necessary for disease identification.

Limitations

- The quality and diversity of the data autoencoders that are trained on can have a big impact on their performance; even though they are unsupervised, they do not require labeled data to learn.
- Depending on the requirements of the task and the data, the autoencoder's design must be carefully evaluated.

19.4.5 Generative Adversarial Networks

Generative adversarial networks (GANs) are implemented by a system of two neural networks that compete against each other for use in unsupervised ML. GANs have a discriminator and a generator that can produce new data that is similar to your training data. Creating similar images of unhealthy leaves is one of the main uses of GANs [1] in agriculture. It can be helpful when there are not many images for plant diseases. GANs can produce real images to increase the size of the dataset for training purposes. In terms of orientation and appearance, the artificial images generated are different from the real images in the dataset, but, by adding these images to the dataset, the models can generalize and categorize the disease in the images.

Limitations

- GANs are difficult to train.
- Using synthetic images to train models for disease diagnosis raises concerns regarding the validity of the models produced as well as the possibility of overfitting to artificial features that may not exist in actual situations.

19.4.6 Transformer Models

Transformer models have completely changed how computers understand and produce human language. They were first developed for use in natural language processing (NLP) applications. They are particularly good at sequence-to-sequence tasks because of their fundamental process, the self-attention mechanism, which enables them to independently assess the significance of various input data elements. Transformer models have been developed for usage in a variety of other fields, including computer vision and agriculture, particularly in tasks like identification and detection of plant and leaf disease categories, despite their most notable accomplishments being in natural language processing.

The self-attention mechanism is the core of Transformer models, which enables the model to assess the significance of various input data components differently. This means that, when processing a sentence, a Transformer can focus more on words that are relevant and less on terms that are not. Transformers employ many sets, or "heads," of attention processes in place of a single set, enabling the model to focus on various input components at once. As a result, the model can capture a deeper comprehension of the data. Layer normalization is the next step after residual connection for every sub-layer in a Transformer, including completely connected and attention layers. Because it makes training smoother, this helps to mitigate the vanishing gradient issue and permits deeper models. The data passes through feed-forward neural networks separately for each point after the attention layers. Although the properties of these networks vary from layer to layer, they remain the same in various situations.

Advantages

- Transformers are well-suited for difficult pattern recognition tasks in disease detection because they can handle long-range dependencies in data.
- Transformer models are powerful tools in the agriculture domain because of their intrinsic flexibility and ability to be successfully adapted to a wide number of tasks beyond their initial NLP uses.
- With the rising availability of high-resolution photos and vast environmental monitoring data, transformer models are highly scalable and perform well with huge datasets, a common scenario in agriculture.

Limitations

- Transformer models are resource-intensive and require a lot of memory and processing power.
- Transformer models need labeled data.

19.5 Drawbacks

Several issues can make it difficult for ML and DL models to generalize across various crop species or environmental circumstances. **Data Discrimination:** Models often display features found in the training set. The model's performance across data sources might be harmed by homogenous training data.

Overfitting Models that fit the training data too closely may not be able to generalize well to new data when the training environment is not representative of the variety of real-world settings.

Transferability A model trained on one crop species might not be able to successfully transfer its knowledge to another due to biological and morphological variances.

Independence Under Conditions Environmental variables such as lighting, soil composition, temperature, and camera angles can significantly affect the model's performance under different circumstances when gathering data.

19.6 Conclusion

An important first step in leveraging technology to improve agriculture and ensure food security is the investigation of ML and DL algorithms in the identification of plant diseases. Since these computer models can quickly, reliably, and non-invasively diagnose issues connected to plant health, they have the potential to be extremely powerful. This allows for prompt reactions and lowers crop losses. Notwithstanding these challenges, which include the need for huge and diverse datasets for training and specific limitations in cross-species or cross-environment applications, the efficiency, scalability, and precision of ML and DL models are clear. To fully reap the benefits of ML and DL in agriculture, we need to keep pushing the boundaries of algorithmic design, data collection methods, and mobile and Internet-of-things device integration.

The importance of interdisciplinary research and collaboration between computer scientists, agronomists, and farmers is emphasized in this chapter to advance these technologies and increase their applicability and accessibility for global agricultural practices. The clever application of ML and DL algorithms in plant disease identification and management promises a new era of agricultural productivity and sustainability.

References

1. Abbas A, et al. Tomato plant disease detection using transfer learning with C-GAN synthetic images. Comput Electron Agric. 2021;187:106279.
2. Alshammari H, et al. Optimal deep learning model for olive disease diagnosis based on an adaptive genetic algorithm. Wirel Commun Mobile Comput. 2022;2022:1–13.
3. Anim-Ayeko AO, Schillaci C, Lipani A. Automatic blight disease detection in potato (Solanum tuberosum L.) and tomato (Solanum lycopersicum, L. 1753) plants using deep learning. Smart Agric Technol. 2023;4:100178.
4. Ashwini J, Anushalini R, Divya G, et al. Image segmentation enhanced with machine learning techniques for automatic detection of plant leaf diseases. In: 2022 8th International conference on smart structures and systems (ICSSS). IEEE. 2022, pp. 1–6.
5. Calderon R, Navas-Cortes JA, Zarco-Tejada PJ. Early detection and quantification of verticillium wilt in olive using hyperspectral and thermal imagery over large areas. Remote Sens. 2015;7(5):5584–610.
6. Chaudhari V, Patil M. Banana leaf disease detection using K-means clustering and feature extraction techniques. In: 2020 International conference on advances in computing, communication materials (ICACCM). 2020, pp. 126–130. https://doi.org/10.1109/ICACCM50413.2020.9212816.
7. Chopra N, et al. Plant disease detection using deep learning and machine learning algorithms. In: 2023 3rd International conference on technological advancements in computational sciences (ICTACS). 2023, pp. 162–5. https://doi.org/10.1109/ICTACS59847.2023.10390154.
8. Geetha G, et al. Plant leaf disease classification and detection system using machine learning. J Phys Conf Ser. 2020;1712(1):012012.
9. Hou C, et al. Recognition of early blight and late blight diseases on potato leaves based on graph cut segmentation. J Agric Food Res. 2021;5:100154.
10. Hughes D, Salathe M, et al. An open access repository of images on plant health to enable the development of mobile disease diagnostics. arXiv preprint arXiv:1511.08060. 2015.
11. Jackulin C, Murugavalli S. A comprehensive review on detection of plant disease using machine learning and deep learning approaches. Meas Sens. 2022;24:100441.
12. Jahanbakhshi A, et al. Detection of fraud in ginger powder using an automatic sorting system based on image processing technique and deep learning. Comput Biol Med. 2021;136:104764.
13. Jaisakthi SM, et al. Grape leaf disease identification using machine learning techniques. In: 2019 International conference on computational intelligence in data science (ICCIDS). 2019, pp. 1–6. https://doi.org/10.1109/ICCIDS.2019.8862084.
14. Jayakumar D et al. Automatic prediction and classification of diseases in melons using stacked RNN based deep learning model. In: 2020 international conference on system, computation, automation and networking (ICSCAN). IEEE. 2020, pp. 1–5.
15. Jerome NJ, et al. An effective approach for plant disease detection using assessment-based convolutional neural networks (A-CNN). J Adv Res Appl Sci Eng Technol. 2023;31(3):155–72.
16. Karthik R, et al. Attention embedded residual CNN for disease detection in tomato leaves. Appl Soft Comput. 2020;86:105933.
17. Decision Tree Learning Models. A prediction olive diseases using machine learning models, decision tree and naive bayes models. J Theor Appl Inf Technol. 2021;99(18)
18. Li F. Effectiveness of deep learning model for plant disease detection. Highlights Sci Eng Technol. 2023;39:356–61.
19. Abdul Malek MA, Basak S, Reya SS. An approach to identify diseases in betel leaf using deep learning techniques. In: 2022 4th International conference on sustainable technologies for industry 4.0 (STI). 2022, pp. 1–5. https://doi.org/10.1109/STI56238.2022.10103348.
20. Mathew D, Kumar CS, Cherian KA. Application of discrete orthonormal Stockwell transform and local neighborhood patterns for leaf disease classification in banana. In: 2021 International Conference on Communication, Control and Information Sciences (ICCISc). Vol. 1, 2021, pp. 1–5. https://doi.org/10.1109/ICCISc52257.2021.9484860.

21. Mehta S, Kukreja V, Gupta A. Revolutionizing maize disease management with federated learning CNNs: a decentralized and privacy-sensitive approach. In: 2023 4th International conference for emerging technology (INCET). 2023, pp. 1–6. https://doi.org/10.1109/INCET57972.2023.10170499.
22. Militante SV, Gerardo BD, Dionisio NV. Plant leaf detection and disease recognition using deep learning. In: 2019 IEEE Eurasia conference on IOT, communication and engineering (ECICE). 2019, pp. 579–582. https://doi.org/10.1109/ECICE47484.2019.8942686.
23. Mokhtar U, et al. SVM-based detection of tomato leaves diseases. In: Intelligent systems' 2014: proceedings of the 7th IEEE international conference intelligent systems IS'2014, September 24–26, 2014, Warsaw, Poland, Volume 2: tools, architectures, systems, applications. Springer. 2015, pp. 641–652.
24. Sajitha N, et al. The plant disease detection using CNN and deep learning techniques merged with the concepts of machine learning". In: 2022 2nd International conference on advance computing and innovative technologies in engineering (ICACITE). 2022, pp. 1547–1551. https://doi.org/10.1109/ICACITE53722.2022.9823921.
25. Nandhini S, Ashokkumar K. Improved crossover based monarch butterfly optimization for tomato leaf disease classification using convolutional neural network. Multimed Tools Appl. 2021;80(12):18583–610.
26. Oishi Y, et al. Automated abnormal potato plant detection system using deep learning models and portable video cameras. Int J Appl Earth Observ Geoinf. 2021;104:102509.
27. Prajapati HB, Shah JP, Dabhi VK. Detection and classification of rice plant diseases. Intell Decis Technol. 2017;11(3):357–73.
28. Rashmi N, Shetty C, et al. A machine learning technique for identification of plant diseases in leaves. In: 2021 6th international conference on inventive computation technologies (ICICT). IEEE. 2021, pp. 481–4.
29. Sannakki SS, et al. Diagnosis and classification of grape leaf diseases using neural networks. In: 2013 Fourth international conference on computing, communications and networking technologies (ICCCNT). 2013, pp. 1–5. https://doi.org/10.1109/ICCCNT.2013.6726616.
30. Singh D, et al. PlantDoc: a dataset for visual plant disease detection. In: Proceedings of the 7th ACM IKDD CoDS and 25th COMAD. 2020, pp. 249–53.
31. Singh G, Yogi KK. Performance evaluation of plant leaf disease detection using deep learning models. Arch Phytopathol Plant Protect. 2023;56(3):209–33.
32. Singh R, et al. A deep learning-based InceptionResNet V2 model for cassava leaf disease detection. In: International conference on emerging trends in expert applications & security. Springer; 2023. p. 423–32.
33. Surya R, Gautama E. Cassava leaf disease detection using convolutional neural networks. In: 2020 6th International conference on science in information technology (ICSITech). 2020, pp. 97–102. https://doi.org/10.1109/ICSITech49800.2020.9392051.
34. Tandekar D, Dongre S. Identification of various diseases in plant leaves using image processing and CNN approach. In: 2023 14th International conference on computing communication and networking technologies (ICCCNT). 2023, pp. 1–6. https://doi.org/10.1109/ICCCNT56998.2023.10306979.
35. Thotad PN, Kallur S, Nandeppanavar A. An efficient model for plant disease detection in agriculture using deep learning approaches. In: 2023 4th IEEE global conference for advancement in technology (GCAT). 2023, pp. 1–6. https://doi.org/10.1109/GCAT59970.2023.10353343.

GPSR Compliance

The European Union's (EU) General Product Safety Regulation (GPSR) is a set of rules that requires consumer products to be safe and our obligations to ensure this.

If you have any concerns about our products, you can contact us on ProductSafety@springernature.com

In case Publisher is established outside the EU, the EU authorized representative is:

Springer Nature Customer Service Center GmbH
Europaplatz 3
69115 Heidelberg, Germany

Batch number: 10726058

Printed by Printforce, the Netherlands